TRANSFER RNA

Cell Monograph Series

1. *Concepts in Mammalian Embryogenesis,* edited by Michael I. Sherman
2. *Transfer RNA,* edited by Sidney Altman

TRANSFER RNA

Edited by Sidney Altman

The MIT Press
Cambridge, Massachusetts, and London, England

This book was set in Compugraphic English Times by Wilkins & Associates. It was printed and bound by Murray Printing Company in the United States of America.

Library of Congress Cataloging in Publication Data
Main entry under title:
Transfer RNA.
  (Cell monograph series ; 2)
  Includes bibliographies and index.
  1. Ribonucleic acid, Transfer. I. Altman,
Sidney. II. Series.
QP623.T7   574.8'732   78-15688
ISBN 0-262-01056-9

CONTENTS

PREFACE

The study of transfer RNA involves all aspects of molecular biology. The function of tRNA is critical in the process of translation; and tRNA engages, as Crick envisioned twenty years ago, in highly specific interactions with both nucleic acids and proteins. This book is intended as an introduction to tRNA and recent research regarding it. Thus the audience addressed includes graduate students and others who wish to familiarize themselves with tRNA in a more than superficial manner. Scientists engaged in tRNA research should also find this volume a comprehensive, though not encyclopedic, review to early 1977 when the writing of most of the articles was completed.

In any book of this kind it is difficult to achieve uniformity of exposition and style. If there is a small measure of success here, it is largely attributable to Dr. Ann Körner who gave unstinting help with editorial matters. My colleagues at Yale and elsewhere also provided generous assistance in many ways.

Sidney Altman

CONTRIBUTORS

S. Altman
Department of Biology
Yale University
New Haven, CT 06520

B. F. C. Clark
Department of Chemistry
Aarhus University
Langelansgade 140
DK-8000 Aarhus C
Denmark

P. E. Cole
Department of Chemistry
Columbia University
New York, NY

F. Cramer
Max-Planck-Institut für Experimentelle
Medizin
Hermann-Rein-Strasse 3
D-3400 Göttingen
Federal Republic of Germany

D. M. Crothers
Department of Chemistry
Yale University
New Haven, CT 06520

S. I. Feinstein
Radiobiological Laboratories
Yale University School of Medicine
New Haven, CT

G. L. Igloi
Max-Planck-Institut für Experimentelle
Medizin
Hermann-Rein-Strasse 3
D-3400 Göttingen
Federal Republic of Germany

S.-H. Kim
Department of Biochemistry
Duke University Medical Center
Durham, NC 27710

A. Körner
Department of Biology
Yale University
New Haven, CT 06520

R. LaRossa
Department of Molecular Biophysics
and Biochemistry
Yale University
New Haven, CT 06520

S. Nishimura
Biology Division
National Cancer Center Research
Institute
Tsukiji 5-chome
Chuo-ku, Tokyo
Japan

O. Pongs
Ruhr-Universität Bochum
Abteilung für Chemie
463 Bochum, Postfach 2148
Federal Republic of Germany

D. Söll
Department of Molecular Biophysics
and Biochemistry
Yale University
New Haven, CT 06520

H. G. Zachau
Institut für Physiologische Chemie,
Physikalische Biochemie und Zellbio-
logie der Universität München
Goethestrasse 33
8000 München 2
Federal Republic of Germany

TRANSFER RNA

1 INTRODUCTION: TRANSFER RNA COMING OF AGE

H. G. Zachau

In the early 1950s molecular biology was a small field populated mostly with X-ray crystallographers, bacteriophage workers, and a few biochemists who studied the biosynthesis of nucleic acids and proteins. There may have been about as many people working in all areas of molecular biology as are now involved in tRNA research.

Transfer RNA was discovered about 20 years ago. I remember a seminar which M. B. Hoagland gave at MIT during the winter of 1956–1957. Hoagland reported on the results of P. C. Zamecnik's group at the Massachusetts General Hospital, Boston. A "pH 5 enzyme" incorporated radioactive leucine into an RNA component of this enzyme fraction. In a protein-synthesizing system the RNA-bound leucine could be transferred to protein. The RNA was called soluble RNA or S-RNA (1). The "pH 5 enzyme" had been prepared from rat liver cell sap, that is from a high-speed supernatant of a homogenate, by the addition of acid to pH 5.2, centrifugation, and redissolving of the pellet in neutral buffer. As we know today, the precipitate contained practically all the tRNA of the cell and most aminoacyl-tRNA synthetases; pH 5.2 happens to be close to the isoelectric point of the ribonucleoprotein complexes. At the time I was working at MIT as a "straight chemist" and certainly did not understand the details of Hoagland's lecture. I was impressed, however, by the prospects of a newly emerging field and, luckily enough, I was accepted by F. Lipmann to join his group and to work on the soluble RNA fraction.

The discovery of tRNA was not the outcome of a one-shot experiment. The foundations had been laid during several years of systematic work in Zamecnik's group and in other groups. It was not a unique experiment either; independently K. Ogata and H. Nohara in Japan had made similar observations (2) but, not being tuned in on the wavelength of international communication, they had not expressed them as clearly.

A remarkable feature in the discovery of tRNA was its "prediction" by F. H. C. Crick. In the early fifties, G. Gamov had founded an "RNA tie club" with 20 members and 4 honorary members. Manuscripts on the coding of the 20 amino acids by the 4 nucleobases and on related topics were circulated among the members of the club. Early in 1955 Crick circulated a note for the RNA tie club entitled "On Degenerate Templates and the Adaptor Hypothesis." Since he could not conceive of any structure acting as a direct template for amino acids, he postulated in this note that "each amino acid would combine chemically, at a special enzyme, with a small molecule which, having a specific hydrogen-bonding surface, would combine specifically with the nucleic acid template. This combination would also supply the energy necessary for polymerisation. In its simplest form there would be 20 different kinds of adaptor molecule, one for each amino acid, and 20 different enzymes to join the amino acid to their adaptors." A year or two later the soluble RNA and the enzymes were found and had essentially the properties postulated by Crick. Crick's adaptor hypothesis and his wobble hypothesis are prominent examples of the interplay between theoretical deduction and experimental results that was characteristic of the first years of the development of the tRNA field.

The note for the RNA tie club has not been published in full, but the most pertinent paragraph was reprinted in an article by Hoagland (3). This article is the most comprehensive review of the initial phases of research on protein synthesis and tRNA. It was written in the days before the discovery of messenger RNA, so ribosomal RNA was depicted as the partner of tRNA in the adapting process.

In the decade 1956–1966 the first individual tRNAs were isolated and sequenced, and the genetic code and anticode were elucidated. When, in 1957 in F. Lipmann's laboratory at the Rockefeller University, I was assigned to work on the pH 5 fraction, its RNA component was the center of interest. Soon, in this lab and in a number of others, the first few milligrams of what is now called unfractionated tRNA were isolated. We proved by chemical means that the activated amino acid is bound by an ester linkage to the ribose moiety of the terminal adenosine of tRNA (4). R. W. Holley and S. H. Merrill achieved the first fractionation of tRNA by counter-current distribution, which resulted in the partial purification of alanine-specific tRNA (5). After my return to Germany we continued to work on the amino acid esters of RNA (6) but were mainly involved in the development of systems for the purification of the serine-specific tRNAs (7). Much work, in several laboratories including our own, was required to refine the purification procedures and to develop nucleotide sequencing methods. In 1965 Holley and his co-workers announced the elucidation of the nucleotide sequence of an alanine-specific tRNA from yeast (8). A year later we completed work on the nucleotide sequence of two serine-specific tRNAs from yeast (9). The sequence of a tyrosine-specific tRNA soon followed (10), and by the time of the 1966 Cold Spring Harbor Symposium, which was devoted to the genetic code and anticode, the general principles of primary and secondary (cloverleaf) structure of tRNA were clear (11). The location of the anticodon was established. Several modified nucleotides had been found in tRNA, and the function of one of them, inosinic acid, had been established: it serves as a "wobble base" in the anticodon (12).

By 1966 the role of tRNA in protein synthesis was basically understood—that is, with regard to the aminoacylation reaction and also in a first approximation to the functioning of aminoacyl-tRNA on the ribosome. Binding of aminoacyl-tRNAs to oligonucleotide- or RNA-ribosome complexes and amino acid incorporation in RNA-dependent ribosomal systems had been the major experimental approaches in the unraveling of the genetic code. But in lectures on protein synthesis the ribosome, with its attached aminoacyl-tRNAs, mRNA, and protein factors, was usually still referred to as a black box.

Physical measurements on tRNA in solution had produced much data, as had studies on the modification of tRNA by chemical and physical means. Much of this work, including some of our own on the UV irradiation of tRNA (13), was carried out with unfractionated tRNA and was therefore of rather limited value. In our case, the work served also an additional purpose: the grant that financed most of our sequence work was for the irradiation of tRNA. In retrospect, however, the main achievement of the early chemical and physical work was the development of methods later applied to the study of pure tRNA species. By 1966 the genetic work on nonsense and missense suppression by tRNA had also borne first fruit, and there were already some hints as to functions of tRNA outside protein synthesis. On the whole it was a time just before a veritable boom in tRNA research. The work of the first decade and the state of knowledge in the early phases of the boom period have been described in a fairly comprehensive review (14).

The second decade of tRNA research brought an abundance of nucleotide sequence data. The fingerprint method for sequencing radioactive tRNAs (15) superseded the column chromatographic procedures used in the elucidation of the first tRNA structures. Transfer RNA of sufficiently high specific radioactivity was obtained by uniform labeling with [^{32}P] phosphate in vivo. Whenever this was not easily feasible, as in the case of tRNA from higher organisms, nonradioactive tRNAs were degraded and the degradation products made radioactive by postlabeling procedures (see, for example, refs. 16 and 17). As of summer 1976, 96 tRNA sequences had been determined (18). The cloverleaf model has been confirmed as the apparently universal representation of the secondary structure, and many interesting aspects have emerged.

A race toward the elucidation of the three-dimensional structure of tRNA started in 1968, when six independent groups managed almost simultaneously to obtain tRNA crystals (for an account see ref. 19). It still took several years of difficult X-ray crystallographic analysis, with disappointment in some quarters and triumph in others, until the three-dimensional structure of yeast tRNA^{Phe} was established by groups from Cambridge, Massachusetts (20), and Cambridge, England (21). Some of the elements of the three-dimensional structure of tRNA^{Phe} may be common to many or even all tRNA species. But some of the extensions and generalizations of the tRNA^{Phe} structure remind me of the hypothetical tRNA sequences in the early 1960s and the hypothetical models of the three-dimensional tRNA structure in the late 1960s and early 1970s. They stimulate thought and sometimes experimentation but are forgotten as soon as hard facts become known.

Is the structure of tRNA in solution identical to its structure in the crystal? The analogous question had existed in the protein field for a long time, and it had to be asked anew for the first crystallized nucleic acid. The results of eight to ten years of chemical and biochemical studies on "exposed" and "protected" regions of the tRNA molecule agreed fairly well with the expectations from the crystal structure; discrepancies could rather easily be explained by invoking special solution conditions or secondary changes by the modifying agent. The study of tRNA in solution by absorption spectroscopy including hyperchromicity measurements, by circular dichroism, and by nuclear magnetic resonance techniques provided values for the numbers of hydrogen-bonded and stacked bases. Because of the underlying assumptions of the methods, these results were not unequivocal enough to disagree with X-ray crystallographic data. But in recent years the interpretations have become safer. So the results of the physical, chemical, and biochemical methods together now allow the conclusion that a tRNA structure similar or identical to the one in the crystal predominates in Mg^{++}-containing solutions of sufficient ionic strength.

The physical, chemical, and biochemical methods of investigating tRNA in solution usually do not provide absolute structural information. But they are very well suited to detecting and monitoring changes in tRNA conformation. And such changes do occur (22). Already in 1966 denatured conformations of tRNA were discovered (23, 24), and it turned out that at least in some tRNAs the conformational changes during denaturation involve the dihydrouridine and the miniloop regions of the molecules (25). The changes could serve as a model for more interesting, naturally occurring changes of the tRNA conformation, as they may take place, for instance, upon aminoacylation, peptide bond formation, translocation, and release from the ribosome. There are now good indications that at least some processes in protein synthesis are accompanied by conformational changes in tRNA, but clearly more work is required along these lines.

The interactions between tRNA and the other components of the protein-synthesizing system touch on important biological questions and have therefore been widely studied. In the beginning the problem of aminoacyl synthetase-tRNA recognition was the center of interest. How does an aminoacyl-tRNA synthetase recognize and aminoacylate its cognate tRNA, and how does it discriminate against other tRNAs? What are the reasons for the high degree of specificity? The question has been investigated not only for its own sake, but also as a model for specific protein–nucleic acid interactions since the synthetase-tRNA system was for a long time the only one in which both components were available in pure and well-characterized form. When the first tRNA sequences became known, people searched within the molecules for small, defined synthetase recognition sites analogous to the anticodon, which can be considered the recognition site for the codon in the messenger RNA. There was talk of a "second genetic code." Chemical modification of tRNA (14, 22, 26, 27) as well as partial nuclease digestion of tRNA (for example, 28, 29) and of synthetase-tRNA complexes (for example, 30), heterologous aminoacylation experiments (summary: 31), studies on mutant tRNAs (32), and a number of other methods were applied to the problems of synthetase-tRNA interaction and recognition. The proposal, in a particular case, of the stem of the dihydrouridine region as synthetase recognition site (33) was stimulating. It soon became apparent, however, that the synthetase-tRNA recognition is a complex process. On the basis of misacylation studies, the importance of kinetic factors for the specificity of recognition was stressed (for example, 31, 34, 35), and experiments such as those in which C-C-A halves (3′-fragments) of tRNAs are aminoacylated either alone or in combination with pG halves (5′-fragments) from other tRNAs (36) indicated the limits of a formalistic recognition site concept. It may be that the recognition site problem will not be solved without X-ray analysis of synthetase-tRNA cocrystals. Despite many attempts, such crystals have not yet been obtained. A number of synthetases have been crystallized in the absence of tRNA, and recently the X-ray crystallographic analysis of a tyrosyl-tRNA synthetase at 2.7 Å resolution was reported (37). Attempts at fitting three-dimensional models of tRNA to those of synthetases may soon give some clues about sites of interaction.

The full description of an interacting system like the synthetase-tRNA system requires, in addition to the structural data, knowledge of all pertinent kinetic parameters. Ideally the kinetics of all individual molecular steps in the interaction (and reaction) between synthetase, tRNA, ATP, and amino acids should be known. Present studies on the interactions between synthetases and tRNAs (for example, 38–40) and on the aminoacylation mechanism (41–43) are still far from this stage, but they have revealed several interesting aspects of general biological importance.

Some concepts from the work on synthetase-tRNA interactions are being extended, and some of the techniques are being applied now to the study of interactions between tRNAs and other components of the protein-synthesizing system, notably to tRNA-ribosome interactions (for example, 44). One function of this book is to give an up-to-date account of the role of tRNA in protein synthesis, since so much information has been gathered recently on all facets of the chain initiation, elongation, and termination processes.

In the second decade of tRNA research the genetic analysis of tRNA genes has reached a high degree of sophistication (32). *Escherichia coli* and its phages and, more recently, yeast have been the organisms of choice. Mutational changes have been found to affect codon or synthetase recognition or to influence the processing of tRNA precursors. Functional studies of tRNA as well as the investigation of its biosynthesis have benefited greatly from the advances in tRNA genetics.

A new stage was reached when tRNA genes were isolated (45, 46) and when the momentous task of manufacturing a tRNA gene by way of total synthesis was complete (47). Insertion of the genes into plasmids or phages and application of the in vitro–recombinant DNA technology should allow not only the production of these genes in sufficient quantity but also the investigation of possible regulatory DNA sequences linked to the tRNA genes. This will be genetics largely by biochemical instead of genetic methods, genetics by gene-isolation and gene-handling. There now exists an extensive knowledge of the primary transcription products of tRNA genes and of the processing pathways (48, 49). The availability of the proper DNA templates for transcription studies will greatly facilitate further work on the biosynthesis of tRNA.

Transfer RNA genes and their primary transcription products may also help in solving one of the most puzzling problems in tRNA research: the function of the modified nucleotides. In the early days nucleotides with methyl (50), acetyl (51), and isopentenyl substituents (52, 53) were found; more and more exotic modifications have since become known (54). The function of some modified nucleotides located within the anticodon has been clarified. Others might stabilize certain tRNA conformations or facilitate conformational changes which, in turn, might be important for the interactions between the tRNA and synthetases, mRNA, ribosomes, enzymes, or regulatory proteins. Modified nucleotides may act through their effect on tRNA conformation, and they may stabilize or prevent, through their side groups, the interactions between tRNAs and other components of the cell. Experiments showing striking effects in defined systems are still lacking, however, for many modified nucleotides. Several such comparisons should now be possible—between tRNAs with and without a certain modification, between completely unmodified tRNAs and tRNAs with one, two, or three of the modifications. Several modifying enzymes are available in purified or pure form, and others may become available since they can now be assayed with unmodified tRNAs as substrates. There is hope that within the next few years the function of a number of modified nucleotides will be understood, at least to the first approximation.

In many or all organisms more than one tRNA is found for each of the 20 amino acids. The presence of these so-called isoaccepting tRNAs is the reason for the complexity of the tRNA fractionation pattern that is generally observed. Some isoaccepting tRNAs differ in their nucleotide sequence, others only in the degree of modification. A number of sequence differences can be functionally explained, for instance those related to the degeneracy of the genetic code, to the requirement for a special chain initiator tRNA^{Met}, or to the presence of suppressor tRNAs. Other sequence differences are not understood, such as those outside the anticodon loop, as they were first observed in the two tRNA$^{Ser'}$ species of yeast (9). The simultaneous presence of fully modified and undermodified tRNAs can sometimes be related to fast growth of the organisms or cells. The so-called special-function tRNAs—such as those used in the synthesis of the cell-wall peptidoglycan and, in higher organisms, the special tRNAs of mitochondria and chloroplasts—contribute to the complexity of the isoacceptor patterns.

Studies on the distribution of isoaccepting tRNAs in different organisms and tissues touch on such central questions as regulation, differentiation, growth, and degeneration. Changes in isoacceptor patterns depending on the growth conditions of microorganisms and tissue culture cells have been described. The tRNA population changes when cells are infected with certain bacteriophages or viruses. There are alterations during embryogenesis and differentiation as well as adaptations of the tRNA population to the biosynthesis of specific proteins, with and without hormonal regulation. Particular attention has been paid to the altered tRNAs in malignant cells (55).

Unfortunately the literature on isoaccepting tRNAs is particularly hard to penetrate. When high-resolution column chromatography of tRNA became a standard technique, an avalanche of papers appeared that consisted mainly of elution profiles and lacked control experiments on the degree of aminoacylation, tRNA aggregation, or partial degradation. There are also many papers that include such controls but are purely descriptive; that is, changes of tRNA patterns are reported without any clue as to the involvement of tRNA modification or sequence differences, of organelle tRNA, or of special-function tRNA. It is hoped that some of the studies that deal with biologically or medically important systems will be carried to the next step of sophistication. There are also, of course, many papers that probe deeply into the subject matter. I would like to mention at least one example in which isoaccepting tRNAs and modified nucleotides could be linked to a defined regulatory process: the presence of the sequence U-U instead of Ψ-Ψ in the anticodon loop of a tRNA^{His} is correlated with the derepression of the histidine operon in *Salmonella typhimurium* (56); but again the mechanism of this derepression is not really understood yet. It has been proposed that proteins capable of recognizing partial or complete tRNA structures are involved not only in various steps of the translation process, but also in the transcription (and replication) of nucleic acids; this is likely in the case of the histidine operon and in a few other systems and may be a fairly general phenomenon (57). At present, however, most questions concerning the role of tRNA in the regulation of transcription and translation are wide open, particularly in higher organisms. Answers to these questions are linked to the general progress in the elucidation of regulatory processes and may open up a completely new field in tRNA research. Much has been learned since the last comprehensive review of this complicated field (58), and a new assessment of the situation is timely.

Transfer RNA coming of age—Does this mean tRNA is becoming less interesting? The emblem of the tRNA symposium in Riga 1970 depicts tRNA as an attractive young lady (reprinted in ref. 59). The teenager has matured since and, obviously, has remained very attractive. Some scientists who are interested in solving basic biological problems only to the first approximation have left the tRNA field. Solving problems at second or higher approximations requires refinement of experimental design, new techniques, and combination with other fields of science. Many biochemists who stayed in tRNA reseach have adapted themselves to these requirements. In addition physicists and physical chemists as well as geneticists, virologists, and cell biologists have become interested in tRNA. On the whole there has been a healthy turnover among the tRNA researchers. The total number of people who work on some aspect of tRNA has increased steadily, as is witnessed by the squeeze for space and time at the tRNA symposia and workshops that have become a regular feature in the second decade of tRNA research.

On both the physical and the biological sides of the tRNA field, important problems remain to be solved. The tRNA field has seen, until recently, surprising findings and great achievements. Examples of the first point are the tRNA-like structures at the 3'-end of plant virus RNAs (60) and at least one animal virus RNA (61) and tRNA as primer of reverse transcriptase (62). In the second category the three-dimensional structure of tRNA[Phe] (20, 21) and the synthesis of a tRNA gene (47) must be mentioned. It is a safe prediction that the third decade of tRNA research will see not only the filling in of details, but also some surprises and achievements. One hopes for answers to some of the basic questions concerning the physical and mechanistic problems as well as the biological problems.

From its early stages tRNA research has been an international enterprise, being conducted in East and West, in the Near East as well as in the Far East, in Europe, and in the United States, where several of the key discoveries have been made. There has been a multitude of contacts between researchers, most of them very friendly, some of course involving hard competition. But collaboration has prevailed, not only in the exchange of information and research samples but also in the carrying out of many joint experiments; the tRNA literature has many examples of joint publications from laboratories in two or three countries. The spirit of the first two decades should certainly be carried over into the third.

From the first years of tRNA research, there has been a continuous flow of review articles, most of them in the *Annual Review of Biochemistry* and in *Progress in Nucleic Acid Research and Molecular Biology,* but also some in journals and in books. The state of knowledge at the end of the first decade of tRNA research is mirrored in the volume of the 1966 Cold Spring Harbor Symposium (11). May the present book, compiled at the end of the second decade, serve a similar function.

References

1. Hoagland, M. B., Zamecnik, P. C., and Stephenson, M. L. (1957). Biochim. Biophys. Acta *24*, 215–216.

2. Ogata, K., and Nohara, H. (1957). Biochim. Biophys. Acta *25*, 659–660.

3. Hoagland, M. B. (1960). In Nucleic Acids, vol. 3, E. Chargaff and J. N. Davidson, eds. (New York: Academic Press), pp. 349–408.

4. Zachau, H. G., Acs, G., and Lipmann, F. (1958). Proc. Nat. Acad. Sci. USA *44*, 885–889.

5. Holley, R. W., and Merrill, S. H. (1959). J. Amer. Chem. Soc. *81*, 753.

6. Zachau, H. G., and Feldmann, H. (1965). Prog. Nucl. Acid Res. Mol. Biol. *4*, 217–230.

7. Zachau, H. G., Tada, M., Lawson, W. B., and Schweiger, M. (1961). Biochim. Biophys. Acta *53*, 221–223.

8. Holley, R. W., Apgar, J., Everett, G. A., Madison, J. T., Marquisee, M., Merrill, S. H., Penswick, J. R., and Zamir, A. (1965). Science *147*, 1462–1465.

9. Zachau, H. G., Dütting, D., and Feldmann, H. (1966). Hoppe–Seyler's Z. Physiol. Chem. *347*, 212–235.

10. Madison, J. T., Everett, G. A., and Kung, H. (1966). Science *153*, 531–534.

11. The Genetic Code (1966). Cold Spring Harbor Symp. Quant. Biol. *31*, 409–656.

12. Crick, F. H. C. (1966). J. Mol. Biol. *19*, 548–555.

13. Harriman, P. D., and Zachau, H. G. (1966). J. Mol. Biol. *16*, 387–403.

14. Zachau, H. G. (1969) Angew. Chem. *81*, 645–662; int. ed. *8*, 711–727; partially updated in The Mechanism of Protein Synthesis and its Regulation, L. Bosch, ed. (Amsterdam: North-Holland, 1972), pp. 173–217.

15. Sanger, F., Brownlee, G. G., and Barrell, B. G. (1965). J. Mol. Biol. *13*, 373–398.

16. Simsek, M., Ziegenmeyer, J., Heckman, and RajBhandary, U. L. (1973). Proc. Nat. Acad. Sci. USA *70*, 1041–1045.

17. Randerath, K., Randerath, E., Chia, L. S. Y., Gupta, R. C., and Sivarajan, M. (1974). Nucl. Acids Res. *1*, 1121–1141.

18. Dirheimer, G., Keith, G., Martin, R., and Weissenbach, J. (1976). Proceedings of the International Conference on Synthesis, Structure, and Chemistry of Transfer Ribonucleic Acids and Their Components, M. Wiewiorowski, ed. (Poznan, Poland), pp. 273–290.

19. Nature (1969) *221*, 13.

20. Quigley, G. J., Wang, A. H. J., Seeman, N. C., Suddath, F. L., Rich, A., Sussmann, J. L., and Kim, S. H. (1975). Proc. Nat. Acad. Sci. USA *72*, 4866–4870.

21. Ladner, J. E., Jack, A., Robertus, J. D., Brown, R. S., Rhodes, D., Clark, B. F. C., and Klug, A. (1975). Proc. Nat. Acad. Sci. USA *72*, 4414–4418.

22. Rich, A., and RajBhandary, U. L. (1976). Ann. Rev. Biochem. *45*, 805–860.

23. Lindahl, T., Adams, A., and Fresco, J. R. (1966). Proc. Nat. Acad. Sci. USA *55*, 941–948.

24. Gartland, W. J., and Sueoka, N. (1966). Proc. Nat. Acad. Sci. USA *55*, 948–956.

25. Streeck, R. E., and Zachau, H. G. (1972). Eur. J. Biochem. *30*, 382–391.

26. von der Haar, F., Schlimme, E., and Gauss, D. H. (1971). In Procedures in Nucleic Acid Research, vol. 2, G. L. Cantoni and D. R. Davies, eds. (New York: Harper and Row), pp. 643–664.

27. Cramer, F. (1971). Prog. Nucl. Acid Res. Mol. Biol. *11*, 391–421.

28. Mirzabekov, A. D., Lastity, D., Levina, E. S., and Bayev, A. A. (1971). Nature New Biol. *229*, 21–22.

29. Thiebe, R., Harbers, K., and Zachau, H. G. (1972). Eur. J. Biochem. *26*, 144–152.

30. Hörz, W., Meyer, D., and Zachau, H. G. (1975). Eur. J. Biochem. *53*, 533–539.

31. Ebel, J. P., Giege, R., Bonnet, J., Kern, D., Befort, N., Bollack, C., Fasiolo, F., Gangloff, J., and Dirheimer, G. (1973). Biochimie *55*, 547–557.

32. Smith, J. D. (1972). Ann. Rev. Genet. *6*, 235–256.

33. Roe, B., Sirover, M., and Dudock, B. (1973). Biochemistry *12*, 4146–4154; see also earlier publications of the authors.

34. Yarus, M. (1972). Proc. Nat. Acad. Sci. USA *69*, 1915–1919.

35. Feldmann, H., and Zachau, H. G. (1977). Hoppe-Seyler's Z. Physiol. Chem. *358*, 891–896.

36. Wübbeler, W., Lossow, C., Fittler, F., and Zachau, H. G. (1975). Eur. J. Biochem. *59*, 405–413.

37. Irwin, M. J., Nyborg, J., Reid, B. R., and Blow, D. M. (1976). J. Mol. Biol. *105*, 577–586.

38. Rigler, R., Pachmann, U., Hirsch, R., and Zachau, H. G. (1976). Eur. J. Biochem. *65*, 307–315.

39. Blanquet, S., Dessen, P., and Iwatsubo, M. (1976). J. Mol. Biol. *103*, 765–784.

40. Riesner, D., Pingoud, A., Boehme, D., Peters, F., and Maass, G. (1976). Eur. J. Biochem. *68*, 71–80.

41. Loftfield, R. B. (1972). Prog. Nucl. Acid Res. Mol. Biol. *12*, 87–128.

42. Söll, D., and Schimmel, P. R. (1974). In Enzymes, vol. 10, 3rd ed., L. Boyer, ed. (New York: Academic Press), pp. 489–538.

43. Kisselev, L. L., and Favorova, O. O. (1974). In Advances in Enzymology, vol. 40, A. Meister, ed. (New York: John Wiley and Sons), pp. 141–238.

44. Robertson, J. M., Kahan, M., Wintermeyer, W., and Zachau, H. G. (1977). Eur. J. Biochem. *72*, 117–125.

45. Daniel, V., Beckmann, J. S., Sarid, S., Grimberg, J. I., Herzberg, M., and Littauer, U. Z. (1971). Proc. Nat. Acad. Sci. USA *68*, 2268–2272.

46. Clarkson, S. G., and Kurer, V. (1976). Cell *8*, 183–195.

47. Khorana, H. G., Agarwal, K. L., Besmer, P., Büchi, H., Caruthers, M. H., Cashion, P. J., Fridkin, M., Jay, E., Kleppe, K., Kleppe, R., Kumar, A., Loewen, P. C., Miller, R. C., Minamoto, K., Panet, A., RajBhandary, U. L., Ramamoorthy, B., Sekiya, T., Takeya, T., and van de Sande, J. H. (1976). J. Biol. Chem. *251*, 565–570.

48. Altman, S. (1975). Cell *4*, 21–29.

49. Smith, J. D. (1976). Prog. Nucl. Acid Res. Mol. Biol. *16*, 25–73.

50. Dunn, D. B. (1959). Biochim. Biophys. Acta *34*, 286–288.

51. Feldmann, H., Dütting, D., and Zachau, H. G. (1966). Hoppe-Seyler's Z. Physiol. Chem. *347*, 236–248.

52. Biemann, K., Tsunakawa, S., Sonnenbichler, J., Feldmann, H., Dütting, D., and Zachau, H. G. (1966). Angew. Chem. *78*, 600–601; int. ed. *5*, 590–591.

53. Hall, R. H., Robins, M. J., Stasiuk, L., and Thedford, R. (1966). J. Amer. Chem. Soc. *88*, 2614–2615.

54. Nishimura, S. (1972). Prog. Nucl. Acid Res. Mol. Biol. *12*, 49–85.

55. Symposium on Transfer RNA and Transfer RNA Modification in Differentiation and Neoplasia (1971). E. Borek, ed. Cancer Res. *31*, 591–721.

56. Singer, C. E., Smith, G. R., Cortese, R., and Ames, B. N. (1972). Nature New Biol. *238*, 72–74.

57. Allende, J. E. (1975). Paabs Revista *4*, 343–352.

58. Littauer, U. Z., and Inouye, H. (1973). Ann. Rev. Biochem. *42*, 439–470.

59. Zachau, H. G. (1972). In Functional Units in Protein Biosynthesis, R. A. Cox and A. A. Hadjiolov, eds. (New York: Academic Press), pp. 93–101.

60. Yot, P., Pinck, M., Haenni, A.-L., Duranton, H. M., and Chapeville, F. (1970). Proc. Nat. Acad. Sci. USA *67*, 1345–1352.

61. Salomon, R., and Littauer, U. Z. (1974). Nature *249*, 32–34.

62. Dahlberg, J. E., Sawyer, R. C., Taylor, J. M., Faras, A. J., Levinson, W. E., Goodman, H. M., and Bishop, J. M. (1974). J. Virol. *13*, 1126–1133.

2 GENERAL FEATURES AND IMPLICATIONS OF PRIMARY, SECONDARY, AND TERTIARY STRUCTURE

Brian F. C. Clark

2.1 Biological Activities of tRNA

Now that a three-dimensional structure for one tRNA species and more than 87 primary structures with predictable secondary structures are known, there is interest in considering the functions of tRNA in terms of structure. A summary of the current state of identification of functions or, more precisely, biological activities of tRNAs in procaryotes and eucaryotes is given in table 2.1. Transfer RNA plays a central role in protein biosynthesis (activities 1–9 in table 2.1; see refs. 1, 2), and much more is known about the biochemistry of the processes involved than in any other function. It is therefore more feasible to relate structure and function in this field. Indeed in modern terms tRNA can be considered a molecular interface at which protein and nucleic acid languages impinge on each other. The existence of this molecular type was predicted by Crick (3), who realized that such an adaptor molecule was required during the molecular transmission of genetic information. The additional activities or functions are in general less well defined, so they are listed in table 2.1 for future interest. This chapter cites a limited number of original works; a much more extensive list of references is contained in a recent literature survey by Rich and RajBhandary (4).

In normal protein biosynthesis each tRNA species is charged with an amino acid (activity 1) by an aminoacyl-tRNA synthetase ("activating enzyme"), and then the charged species is carried to the ribosome in the form of a ternary complex made with the elongation factor Tu (EF-Tu) and

GTP (activity 2). In similar fashion the unique tRNA species, the initiator tRNA, a special class of methionine tRNA, is thought to be carried to the ribosome by an initiation factor and GTP (activity 6). The aminoacyl-tRNA is located by an uncharacterized mechanism in the A-site of the ribosome (activity 3) where it decodes mRNA via its anticodon triplet (activity 4). In contrast the initiator tRNA, formylmethionyl-tRNA$_f^{Met}$ in procaryotes and methionyl-tRNA$_f^{Met}$ (sometimes called Met-tRNA$_i$) in eucaryotes, is located in the initiation (I)-site on the small ribosomal subunit (activity 7) for decoding the initiator triplet codon. This site becomes part of the ribosomal P-site. Since the procaryotic Met-tRNA$_f$ has to be formylated, it is also recognized by a special enzyme for this, the transformylase (activity 8).

Table 2.1 Biological activities of tRNA

Protein Biosynthesis

1. Activation of amino acids	Initiation
2. Recognition by EF-Tu	6. Recognition of initiator tRNA by IF
3. Location in A-site	7. Location in I-site (part of P-site)
4. Decoding mRNA	8. Recognition by transformylase
5. Signal for magic spot	
9. Regulation	
a. Repressor	
b. Feedback inhibitor	
c. Suppression	

RNA Metabolism
10. As precursor recognized by cleavase and maturation enzymes
11. Enzymes for making modified bases
12. C-C-A repair enzyme
13. Pep-tRNA degraded by hydrolase
14. Nuclease degradation
15. Reverse transcriptase primer
16. Selection during viral encapsulation
17. Correlation with 3'-end of viral RNA
18. Alteration of *E. coli* endo I specificity

Aminoacyl-tRNA Transferases
19. Cell-wall biosynthesis
20. Membrane components
21. Protein modification

When procaryotic cells are starved for amino acids, an unusual role has been detected for uncharged tRNA (5). The uncharged tRNA is bound to the ribosomal A-site as in mRNA decoding, but it sets off a signal for the formation by the so-called stringent factor of unusual guanosine nucleotide derivatives, ppGpp and pppGpp, called magic spots (activity 5).

A group of assorted roles for tRNA in the regulation of protein biosynthesis has been collected as activity 9. These include bacterial roles as a repressor of the histidine operon (6), a less-defined regulator of amino acid biosynthesis (7), and the well-characterized suppressor of nonsense mutations (8) and a possible role as a feedback inhibitor in yeast (9). In eucaryotes there are also nondefinitive roles relating to evidence on the binding to tryptophan pyrrolase (in *Drosophila;* see ref. 4) and the inhibition of protein synthesis in virally infected animal cells by the degradation of one or more essential tRNA species, a process that seems to accompany interferon production (M. Revel, personal communication).

Another set of phenomena concerning amounts of isoaccepting species existing in different types of cells at various stages in growth or transformation can be classified under a general regulatory role. Obviously restricting the amount of one specific decoding aminoacyl-tRNA will control protein biosynthesis at this point in translation (10, 11). However, despite the large literature on the subject, especially in connection with control of protein synthesis in carcinogenic states, the regulatory role of tRNA has not been clearly characterized (4).

In addition to their role in protein synthesis, tRNAs have activities concerned with a number of reactions conveniently classified as being concerned with RNA metabolism (activities 10–18; refs. 2, 4). The tRNA is trimmed to size and matured from the precursor molecule by a series of still poorly characterized enzymes that are presumably linked to other metabolic roles (activities 10–12). The C-C-A repair enzyme (activity 12), also called tRNA nucleotidyl transferase, certainly repairs tRNAs with incomplete 3'-ends and is probably concerned with maturation as well. If peptidyl-tRNA should fall off the ribosome, there is a peptidyl-tRNA hydrolase (activity 13) that can hydrolyze off the peptide, thus permitting the tRNA to be recycled through protein biosynthesis. Little is known about tRNA turnover, but specific nucleases (activity 14) must be involved.

Recently some interesting properties of eucaryotic tRNAs with regard to virus metabolism have been identified (activities 15–17). Reverse transcriptases from RNA tumor viruses use a specific $tRNA^{Trp}$ or $tRNA^{Pro}$ as a primer during synthesis of virally coded DNA (4). Further, a certain number of selected tRNA species (perhaps 10–15) are incorporated non-covalently into RNA tumor virus particles during encapsulation from the cell membrane (A. E. Smith, personal communication). It is also well established that many viruses, especially plant viruses, have elements of tRNA structure so that their 3'-ends can be charged specifically with an amino acid (activity 17), for example, turnip yellow mosaic viral RNA with valine. Activity 18 is not easily interpreted, but the specificity of bacterial endo-nuclease I for double-stranded cuts in DNA is altered to a nicking property when it binds tRNA.

Finally, a special class of enzymes called aminoacyl-tRNA transferases (4) transfer amino acids from charged tRNAs to a variety of acceptor molecules without the involvement of a decoding mechanism and ribosomes. At least in the case of activity 19, where the amino acids are incorporated into cell walls, a special type of tRNA best characterized in a series of staphylococcal glycine tRNA species is involved (12). These special cell-wall tRNAs do not contain all the constant features of the general cloverleaf structure, presumably excluding those for ribosome binding. However, their primary structures can be arranged in normal cloverleaf structures (13). The acceptor molecule can also be a phospholipid (14) so that the transferred amino acid becomes incorporated into a cell membrane referred to in activity 20. Activity 21 refers to finding that the presently last known acceptor molecule is a finished protein. Thus for unexplained reasons these enzymes transfer amino acids to the N-terminal ends of certain proteins (15) including membrane proteins (16) from specific aminoacyl-tRNAs.

2.2 Subclassification and Generalized Primary Structure of tRNA

The information from 87 different tRNA sequences (primary structures) known in December 1976 and listed in table 2.2 (see refs. 4, 17, 18, 19; Clark (20) plus 10 new structures) has been conveniently incorporated into standard cloverleaf forms as shown in figure 2.1. This remarkable feature of all the primary structures was first proposed by Holley and his colleagues (13) and is based on Watson-Crick base pairing. The simple classification shown in figure 2.1 is based on size (see table 2.2 for species).

Thus we have small and large tRNAs dependent on the size of the extra arm. The 10 new structures are those for Sw Ala$_1$ and Sw Ala$_2$ (K. V. Sprague, O. Hagenbüchle, and M. C. Zuniga, personal communication); Ec Asn (21), Tt Met$_f$ (22), Bl + Rbl Trp$_2$ (M. Fournier, J. Labouesse, G. Dirheimer, and G. Keith, personal communication); Mul Pro (F. Harada,

Table 2.2 Classes of tRNA according to arm sizes and structure correlations

Class 1 (72)

A (40)	4 base pairs in b-stem (D-stem) 5 bases in extra loop III and containing m^7G
(and with A9)	$Ec\ Ala_1$, $Sw\ Ala_1$, $Sw\ Ala_2$, $Ec\ Arg_1$, $Ec\ Asn$, $Ec\ Asp_1$, $Ec\ Gly_3$, $Ec + Sal\ His_1$, $Ec\ Ile_1$, $Ec\ Lys$, $y\ Lys$, $Rbl\ Lys_{2A(2B)}$, $Rbl\ Lys_3$, $Svt\ Lys_4$, $An\ Met_f$, $Ec\ Met_m$, $y\ Met_3$, $Mye + Rbl\ Met_4$, $Ec\ Phe$, $Mp\ Phe$, $Bs\ Phe$, $y\ Phe$, $Wg + Ps\ Phe$, $Rbl + Hvp\ Phe$, $T4\ Pro$, $Ec + su^+\ Trp$, $Ec\ Val_1$, $Ec\ Val_{2A}$, $Ec\ Val_{2B}$, $Bs\ Val_{2A}$, $Mye\ Val_1$
(and with m^1G9 or G9)	$y\ Cys$, $Ec\ Met_{f1}$, $Ec\ Met_{f2}$, $Bsu\ Met_f$, $Tt\ Met_f$, $y\ Met_f$, $Mye + Rbl + St + Xl + Smg + Hup\ Met_f$, $y\ Trp$, $Chi + Bl + Rbl\ Trp_1$
B (7)	4 base pairs in b-stem (D-stem) 5 bases in extra loop III without m^7G
	$y\ Ala_1$, $Tu\ Ala_1$, $y\ Arg_3$, $Hay\ Lys$, $Sf\ Met_f$, $Ec\ Thr$, $Bl + Rbl\ Trp_2$
C (7)	4 base pairs in b-stem (D-stem) 4 bases in extra loop
	$y\ Asp_1$, $Ec\ Glu_1$, $Ec\ Glu_2$, $Ec + Sal\ Gly_1$, $Sta\ Gly$, $Wg\ Gly$, $y\ Gly$
D (18)	3 base pairs in b-stem (D-stem) small extra arm with number of bases (3–5)
	$y\ Arg_2$ (5), $Ec\ Cys$ (4), $Ec\ Gln_1$ (5), $Ec\ Gln_2$ (5), $T4\ Gln$ (5), $y\ Glu_3$ (4), $Ec\ Gly_2$ (4), $T4\ Gly$ (4), $Tu\ Ile$ (5), $Mul\ Pro$, $y\ Thr_{1A}$, $y\ Thr_{1B}$, $y + su^+\ Tyr$ (5), $Tu\ Tyr$ (5), $y\ Val_1$ (5), $y\ Val_{2A}$ (5), $y\ Val_{2B}$ (5), $Tu\ Val$ (3)
Class 2 (15)	3 base pairs in b-stem (D-stem) large extra arm with number of bases (13–21)
	$Ec + Sal\ Leu_1$ (15), $Ec\ Leu_2$ (15), $T4\ Leu$ (14), $y\ Leu_3$ (13), $y\ Leu_4$ (13), $Ec\ Ser_1$ (16), $Ec\ Ser_3$ (21), $T4\ Ser$ (18), $y\ Ser_1$ (14), $y\ Ser_2$ (14), $Rl\ Ser_1$ (14), $Rl\ Ser_3$ (14), $Ec + su_3^+\ Tyr_1$ (13), $Ec\ Tyr_2$ (13), $Bs\ Tyr$ (13)

Abbreviations: su = Suppressor. An = *Anacystis nidulans*. Bl = Beef liver. Bs = *Bacillus stearothermophilus*. Bsu = *Bacillus subtilis*. Chi = Chicken. Ec = *Escherichia coli*. Egc = *Euglena gracilis* chloroplasts. Hay = Haploid yeast. Hup = Human placenta. Mp = Mycoplasma. Mul = Murine leukemia virus. Mye = Myeloma. Nc = *Neurospora crassa*. Ps = *Pisum sativum*. Rbl = Rabbit liver. Rl = Rat liver. Sal = *Salmonella typhimurium*. Sf = *Streptococcus faecalis*. Smg = Sheep mammary gland. St = Salmon testis. Sta = *Staphylococcus*. Stf = Starfish. Svt = Svt 2 cells. Sw = Silkworm. Tt = *Thermus thermophilus*. Tu = *Torulopsis utilis*. Wg = Wheat germ. Xl = *Xenopus laevis*. y = Yeast.

Classes of tRNA

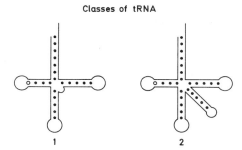

Figure 2.1 Simplified classes of tRNA. The open circle indicates that a Watson-Crick base pair is not always found in this position.

G. Peters, and J. E. Dahlberg, personal communication); y Thr_{1A} and y Thr_{1B} (23), y Val_{2B} (V. D. Axelrod, personal communication); and Bs Tyr (R. S. Brown, J. R. Rubin, H. Guillet, and D. Rhodes, personal communication).

The information in figure 2.2 from the small class 1 sequences has been incorporated into a standard generalized cloverleaf. To obtain this information, 63 of the 72 class 1 species indicated in table 2.2 have been used; clear exceptions based on functions such as Met_f species and Sta Gly to the generalized form were omitted.

The Watson-Crick base pairs shown in figure 2.2 give rise to four double helical stem regions a, b, c, and e; three are closed by nonbase-paired loop regions I, II, and IV. The double helical regions constitute the secondary structure of tRNA, which defines the cloverleaf. However, the functional role of secondary structure is still unclear. Another point of nomenclature illustrated in figure 2.2 is that a stem plus a loop is called an arm. Most of the tRNA cloverleaf forms have remarkably constant regions. There is a phosphate at the 5'-end; but at the 3'-end where the amino acid is attached, there is a common sequence C-C-A. In addition stems a, c, and e contain 7, 5, and 5 base pairs, respectively, and loops II and IV each contain 7 non-base-paired nucleotides.

Stem b contains three or four Watson-Crick base pairs (figure 2.2, open circles in stem b; see also table 2.2), but our knowledge of the three-dimensional structure of tRNA (24, 25) permits us to propose that the non Watson-Crick base pairs occurring in the fourth position of stem b can be accommodated without distortion of the tertiary structure (18). There is then no point in differentiating between tRNAs on the basis of whether this position is a Watson-Crick base pair or not (18). The variable regions are con-

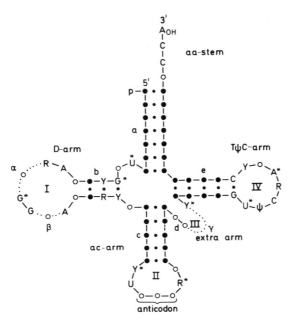

Figure 2.2 Class 1 generalized cloverleaf. Solid circles indicate bases involved in helical stems containing Watson-Crick base pairs. Open circles signify non Watson-Crick base-paired bases in cloverleaf arrangement. Starred nucleosides are positions where modifications of that nucleoside can possibly occur. This nomenclature is based on an original proposal by B. F. C. Clark, T. H. Jukes, and W. E. Cohn.

fined to loops I and III and stem d. For convenience the arms can also be referred to by trivial historical names as shown in figure 2.2; for example, loop I + stem b can be called the D-arm since it usually contains some D-bases; loop II + stem c can be called the anticodon (ac) arm since the loop contains the anticodon; loop III + stem d, the variable finger or extra arm; and loop IV + stem e the TΨC-arm. Stem a is also called the amino acid (aa) stem since this is where the amino acid is attached.

There are also many invariant and semi-invariant nucleotide positions in the generalized structure. The invariant positions are shown by nucleoside letters, whereas the semi-invariant positions are shown by R (signifying a purine nucleoside A or G) or Y (signifying a pyrimidine C or U). The dotted stretches in the diagram are regions of variable length. For example, and including class 2 tRNAs, the extra arm can vary from 3 to 21 nucleotides in length. In contrast the D-loop I is much less variable—from 7 to 10 nucleotides long when the base pair b4, even though not of the Watson-Crick type, is considered part of the D-stem. For convenience the variable dotted regions can be named a, β, and γ.

Until recently an interesting but open question was, What do the invariant and semi-invariant nucleosides represent? It was not possible to decide whether these positions in the primary structure were conserved for metabolic or for structural reasons. Now that a three-dimensional structure has been determined (24, 25), the primary and secondary structural information can be viewed in a new light and we can see a structural role for most of these special nucleosides. A semi-invariant base pair in the D-stem at position b2 (see figure 2.2) was noted only after additional H-bonding interactions between bases and the backbone in the tertiary structure became known (24, 26).

As shown in table 2.2, the class of small tRNAs (class 1) can be conveniently subdivided according to structural characteristics. Class 1 (72 structures) contains four base pairs in the D-stem (stem b) and a small extra loop (three to five nucleotides). Class 2 (15 structures) has only three standard base pairs in the D-stem and a large extra loop of 13 to 21 nucleotides. On the basis of the sequences in the extra loop and standard base pairs in the D-stem, class 1 can be conveniently divided into subclasses A, B, C, and D. Subclass 1A contains five nucleotides, one of which is m^7G, in the extra loop (III), 1B also contains five nucleotides in the extra loop, but now there is no m^7G. Subclass 1C is irregular; its extra loop contains only four nucleotides. And Subclass 1D contains only three Watson-Crick base pairs in the D-stem. The tertiary structure has been determined for yeast tRNA[Phe], which belongs to subclass 1A, but the structure appears relevant to subclass 1B and will probably also accommodate class 1C and class 1D structures (24, 25). For example, a short extra arm, only four nucleotides in length but making all the tertiary interactions, can be built with the excision of the residue U47, which is not involved in the tertiary bonding found for yeast tRNA[Phe] (24, 25).

The list of different structures shown in table 2.2 is somewhat arbitrary since some similar structures such as the minor glutamic tRNA$_1$ (Ec Glu$_1$) and major glutamic tRNA$_2$ (Ec Glu$_2$) are both listed while uncertain sequences and spontaneous mutants showing evidence of gene duplication are not. Ec Arg$_2$, which was included in previous lists (18, 20), has been omitted since it is now agreed that the sequences determined for Ec Arg$_1$ and Ec Arg$_2$ are the same, with Ec Arg$_2$ being correct but containing the modification reported for Ec Arg$_1$ (K. Chakraburtty and S. Nishimura, personal communication).

2.3 Primary Structures According to Biological Source

A simple way to review our state of knowledge regarding tRNA primary structures is given by table 2.3 (see also ref. 19). The table lists known sequences chargeable with each of the 20 essential amino acids for the major sources so far studied. The table enumerates the number of isoaccepting species including different strains and different cell types. Those sequences not yet published are also indicated.

Currently there are known tRNA sequences corresponding to all 20 amino acids, but this is not the case for one biological source. However, in the case of bacteria as source material, tRNA sequences for all amino acids except for Pro are known. Nevertheless, as the table shows, the sequence of bacteriophage tRNA^{Pro} is known. Yeast has been the next most popular source material (27 sequences) after bacteria (37 sequences); mammalian cells are still much less used (12 sequences) largely for reasons of availability and technology. It is generally accepted that there are of the order of 55 different tRNA gene types in a particular cell. So even for the most popular bacterial sources, *Escherichia coli* (30 sequences), there is a long way to go before the complete set of primary structures will be known for one source material. Although there are considerable differences in sequences for different isoaccepting species, evidence so far for several different amino acids suggests that tRNA sequences are largely conserved in different mammalian cell types (see table 2.2 and ref. 19).

2.4 Exceptions to the Generalized Structure

Several exceptions to the generalized tRNA structure in figure 2.2 are shown with their locations in the generalized cloverleaf in figure 2.3. Apart from these, non Watson-Crick base pairs sometimes occur in helical stems. The most frequent non Watson-Crick base pairs occurring in stems are G·U, G·Ψ, and A·Ψ.

In double helical stems G·Ψ is considered equivalent to G·U, and A·Ψ to A·U. When the simplified classification of tRNA structures is used (figure 2.1), then the other non Watson-Crick base pairs that occur in stems are U·U, C·C, A·A, G·A, C·A, Ψ·U, and Ψ·Ψ. No more than one of these unusual base pairs is found in a stem, and there are only a few instances altogether. Two interesting examples of such exceptions are found for position c5 in the ac-stem; here the tRNAs for both y Met$_3$ and Mye + Rbl Met$_4$ contain Ψ·Ψ.

Table 2.3 List of tRNAs of known primary structure from different sources

Charging Amino Acids	Bacteriophage	Bacteria	Yeast	Mammals
Ala		+	+ +	
Arg		+	+ +	
Asn		+		
Asp		+	+	
Cys		×	+	
Gln	+	+ +		
Glu		+ +	+	
Gly	+	+ + + +	+	
His		+		
Ile		+	+	
Leu	+	+ +	+ +	
Lys		+	+ +	× × ×
Met $_f$		+ + + + ×	+	+
Met $_m$		+	+	+
Phe		+ +	+	+
Pro	+			×
Ser	+	+ +	+ +	+ +
Thr		+	+ +	
Trp		+	+	+ ×
Tyr		+ + ×	+ +	
Val		+ + + +	+ + + ×	+

Note: + represents a published isoaccepting species from same and different strains and different cell types. × represents nonpublished sequences.

The exceptions (most are cited in refs. 4 and 19) to the standardized cloverleaf are as follows:

1. Ec tRNA$_1^{His}$ contains eight base pairs in the aa-stem. It is thus longer by one nucleotide at the 5′-end, and the single-stranded region at the 3′-end of the aa-stem contains only C-C-A.

2. Procaryotic initiator tRNAs, tRNA$_f^{Met}$, do not contain a Watson-Crick base pair at position a1 of the aa-stem, for example, C·A in Ec tRNA$_f^{Met}$ (4).

3. The modified U is s^{4}U in most *E. coli* tRNAs but an unmodified U in yeast and mammalian tRNAs.

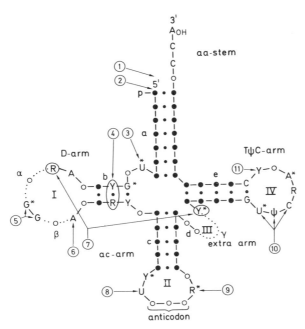

Figure 2.3 Exceptions to the generalized cloverleaf referred to in the text. The locations of the exceptions are given by the arrowheads. The positions with circled numbers are described accordingly in the text.

4. Procaryotic initiator tRNAs contain an A·U instead of Y·R in the standard cloverleaf at position b2.

5. Transfer RNAs that participate in protein biosynthesis have G-G or G*-G at position 5, but for cell-wall biosynthesis, Sta tRNA^{Gly} species have this constant doublet replaced by U-U.

6. The second constant A in the D-loop I is replaced by G in some class 2 tRNAs, for example, Ec tRNA$_1^{Leu}$ and Ec tRNA$_2^{Leu}$, and by D in Ec tRNA^{Cys}.

7. Levitt (27) noticed a common base change in tRNAs with different purines at position 15 (location 7) coordinated with position 48 (numbering as in yeast tRNA^{Phe}). Usually there is a G15·C48 or A15·U48. Exceptions are A15·C48 in Ec tRNA$_3^{Gly}$ and y tRNA$_3^{Glu}$, and G15·G48 in Ec tRNA^{Cys}.

8. The constant U in the anticodon loop is replaced by C in mammalian tRNA$_f^{Met}$ (19).

9. The modified R in the anticodon loop is replaced by a pyrimidine Y (usually C) in cell-wall tRNA^{Gly}, and there is a possible sixth Watson-Crick base pair in the ac-stem.

10. The common sequence G-U*-Ψ-C where U* is T, U, Ψ, or s^{2}T (19) is changed to G-A-U-C in eucaryotic initiator tRNAs, to G-A-Ψ-C in recently determined silkworm alanine tRNAs (Sw Ala$_1$ and Sw Ala$_2$; K. U. Sprague et al., personal communication), and to G-U-G-C in cell-wall tRNA^{Gly} but also in part of wg tRNA^{Gly}.

11. The semi-invariant Y in the TΨC-loop is replaced by A in eucaryotic initiator tRNAs but also in Mye tRNA$_1^{Val}$, Sw tRNA$_1^{Ala}$, and Sw tRNA$_2^{Ala}$.

The most important exceptions to the standard cloverleaf occur in tRNAs that have special functional roles, such as initiator tRNAs and glycine tRNAs involved in cell-wall metabolism. These exceptions are thought to point out positions in the cloverleaf that could be concerned with these special functions. This kind of interpretation may be too simplistic. Nevertheless the special role of procaryotic initiator tRNAs may be related to their exceptional structures at positions 2 and 4 in figure 2.3. Previously eucaryotic initiator tRNAs appeared to be characterized by the special sequence A-U instead of U*-Ψ in TΨC-loop IV (location 10 in figure 2.3). Now that Sw tRNA$_{1+2}^{Ala}$ contain A-Ψ at this position, the special initiating role can be related only to the replacement of U by the constant Ψ. In the case of mammalian initiator tRNAs, C before the anticodon (location 8 in figure 2.3) may play a special role.

Since the Sta tRNA^{Gly} species do not play a role in protein biosynthesis, their lack of constant features should point out parts of the generalized tRNA involved in protein biosynthesis. It is likely that the common sequence of the TΨC-loop missing from Sta tRNA^{Gly} is involved in ribosomal site binding of the tRNA, and there is experimental evidence supporting this proposition (28). The lack of the constant G*-G of the D-loop in the cell-wall tRNAs also implicates this constant feature in protein biosynthesis. However, rather than being a recognition point for some enzyme, it may have a structural role because we know that it is concerned with holding the tertiary structure of yeast tRNA^{Phe} together.

From the lack of a purine after the anticodon in the cell-wall tRNAs we may infer that the purine is a structural requirement concerning proper fitting during the anticodon-codon decoding interaction. However, this feature may also be influenced by the possible sixth Watson-Crick base pair in the ac-stem of cell-wall tRNAs.

In addition to these exceptions to the generalized cloverleaf, there may be one exception to what we have considered possible sizes in the extra arm, loop III. Our knowledge of the tertiary structure permits the ready accommodation of a loop III containing as few as four bases. However, a shorter length such as three reported for Tu tRNA^{Val} (see ref. 17) does not appear possible based on the tertiary structural interactions found in y tRNA^{Phe}. Thus the sequence of Tu tRNA^{Val} can be considered a structural exception. The short (three nucleotides long) extra loop III region previously reported for y tRNA^{Gly} (see ref. 17) has now been shown to contain four nucleotides (M. Yoshida, personal communication).

2.5 Significance of G·U Base Pairs

The wealth of tRNA primary structural information now available permits several interesting analyses of the secondary structure. In particular we have examined the occurrence of non Watson-Crick base pairs such as G·U, U·U, C·A, or G·A in stem regions. Usually only one of these occurs in a given molecule, and where there are more than one, no two are adjacent. It is therefore likely that the RNA double helix can accommodate a single non Watson-Crick base pair without a serious distortion.

The most common non Watson-Crick base pair occurring in tRNA is G·U. Figure 2.4 shows the positions and frequency of occurrence of this base pair in the known class 1 primary structures. The frequency of occurrence of the probably equivalent G·Ψ pair is shown in brackets. So far G·U base pairs have not been found in stem positions a1, b3, c2, c3, and c5 or in e4 and e5. Although G·U base pairs occur frequently and, according to our knowledge of the tertiary structure 4, 5, without gross helix distortion, there appears to be a particularly strong conservation of pure Watson-Crick base pairs in the anticodon stem (c) and the loop end of the TΨC-stem (e). A structural reason for this conservation is probable.

One may ask why G·U base pairs and other non Watson-Crick base pairs occur at all in the stem regions. Clearly a G·U base pair in the middle of a piece of helix requires a certain accommodation of the helix backbone if two H-bonds are to be made. This is therefore a potential point of weakness, or rather irregularity, in such a stem or possibly a special point for enzyme recognition. In particular this could apply to the G·U base pairs in the aa-stem (a).

G·U base pairs at the end of a helix may play a different role. Thus in the 0.25 nm model of yeast tRNA^{Phe} the phosphate of nucleotide 49 in yeast tRNA^{Phe} (at the break in the long double helix formed by the aa-stem stacked on the TΨC-stem; see figure 2.10) is moved from its regular double helical position to allow the adjacent nucleotide in the extra loop to make a

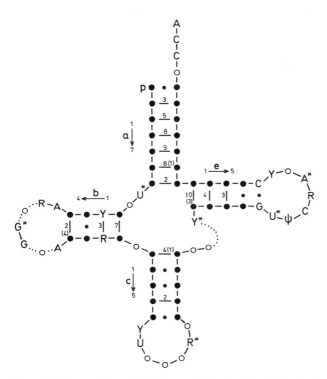

Figure 2.4 The positions of G·U base pairs in the class 1 cloverleaf are indicated by •—• The number of occurrences at each position is given by the numbers with the number of G·Ψ occurrences in parentheses. The arrows give the direction of the standard numbering of stem base pair positions.

nearly right-angle bend with it. A G·U base pair in this position could be very helpful in relieving this tight corner. Perhaps it is significant that the highest frequency of G·U pairs is observed in the base pair position e1; moreover in 12 of 13 cases the G is always on one strand and the U(Ψ) on the other. Similar remarks apply to G·U pairs in position b1 which is close to another sharp bend in the backbone (figure 2.10).

2.6 Position of Ψ

The positions of Ψ occurring in class 1 tRNA structures are shown in figure 2.5 on a standard generalized cloverleaf. Except for the possible structural involvement of the Ψ contained in the common loop IV sequence G-U*-Ψ-C, it seems unlikely that these Ψ positions have significant struc-

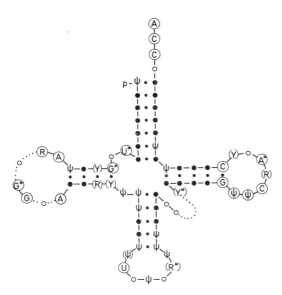

Figure 2.5 Locations for Ψ in the class 1 generalized structure are indicated by Ψ.

tural roles. More likely they are concerned with metabolic roles still largely unidentified, except that the Ψ in the position after R in the anticodon loop II appears to confer repressor activity on bacterial $tRNA_1^{His}$ in connection with regulation of the histidine biosynthetic pathway (6).

2.7 Analysis of Primary Structure and Significance for Function

The extensive data on tRNA nucleotide sequences naturally raises the question of whether a comparison of sequences showing common sequences (sequence homologies) can be interpreted in terms of tRNA functional sites. Different tRNAs recognized by the same enzyme have therefore been compared to determine the common parts of their primary structures. Of course the location of these common parts in the three-dimensional structure is now known, so we can evaluate the results in a more sensible way than before the tertiary structure was known. The sequence homologies can be discussed in terms of several of the activities listed in table 2.1: initiator tRNAs binding to ribosomal sites, common charging of myeloma and *E. coli* methionine tRNA by the appropriate synthetases, the transformylase recognition of initiator tRNAs, and common charging of different phenylalanine tRNAs. Two suitable examples for considering this problem are phenylalanine and methionine tRNA, for which several primary structures from different sources are now known.

Although our early studies comparing the sequences of bacterial methionine tRNAs (tRNAs chargeable with the same activating enzyme) did not permit the assignment of common parts of the cloverleaf structures. detailed studies (29) of the different tRNAs that can be mischarged by yeast phenylalanine tRNA synthetase suggested obligatory common sequences for enzyme recognition. In addition to invariant and semi-invariant regions and the fourth nucleotide from the 3′-end, the short extra loop III and a region of the D-stem were implicated in the recognition studies.

It is difficult to judge whether this line of reasoning based on sequence homology by the criterion of mischarging is valid. For example, we have data on two instances of normal charging by methionyl-tRNA synthetases for which the sequence homology information leads to different conclusions. When the methionine tRNA sequences (Mye Met$_f$ and Mye Met$_4$) recognized by myeloma cell methionyl-tRNA synthetase are compared, regions similar to those previously noted (29) can be suggested for recognition by the mammalian aminoacyl-tRNA synthetase (18). On the other hand, when the same analysis is applied to tRNA sequences recognized by the bacterial methionyl-tRNA synthetase, no such common set of recognition points emerges. This enzyme charges bacterial, yeast, and myeloma tRNA$_f^{Met}$ sequences and bacterial tRNA$_m^{Met}$. When these sequences are compared and the invariants and semi-invariants (circle and brackets) are neglected, then the only common bases that remain are those shown in figure 2.6. This figure shows that the D-stem sequence is unlikely to be recognized by this enzyme, and only the base pair b4 in the D-stem is left as the possibility for recognition. However, yeast tRNA$_3^{Met}$ and Mye tRNA$_4^{Met}$ are not charged by this enzyme, yet they both contain C·G at b4. Further, the common base pairs at a2, a3, a7, and e4 can be rejected since these unchargeable tRNAs, Mye tRNA$_4^{Met}$ and yeast tRNA$_3^{Met}$, contain respectively the pairs a3, e4 and a2, a7 of the common sequence. Thus no common base pairs remain. Finally there is evidence (30) that the common anticodon sequence is not a recognition site. Molecular details of the recognition remain undetermined.

Clearly there are also difficulties in just examining the sequences of initiator tRNAs to explain their interactions with special proteins in initiation. So far no more convincing information relating structures with function than that described for the exceptions has emerged. In addition the yeast, mam-

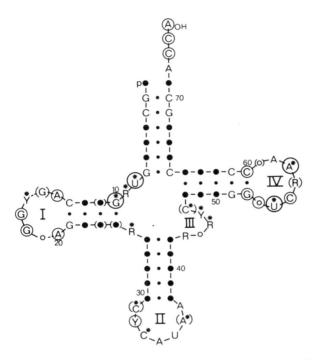

Figure 2.6 Cloverleaf nucleotides common to the sequences of Ec Met$_f$, y Met$_f$, Mye Met$_f$, and Ec Met$_m$, all chargeable by bacterial methionyl-tRNA synthetase. Invariants for the generalized structure of figure 2.2 are shown circled and semi-invariants are in parentheses.

malian, and bacterial tRNA$_f^{Met}$ structures can all be charged by the bacterial aminoacyl-tRNA synthetase, and when charged they can then all be formylated by the bacterial transformylase. When these sequences are compared and when invariances, semi-invariances, and sequences common to activating enzyme recognition are subtracted, the common nucleotides shown in figure 2.7 remain. However, we can subtract base pairs at b3 and c4 because these are present in Mye tRNA$_4^{Met}$ and Mye tRNA$_f^{Met}$, both of which are charged by the myeloma enzyme, so that b3 and c4 could possibly specify synthetase recognition. Of course the transformylase and synthetase could have some recognition points in common. This type of analysis then suggests that the anticodon stem as well as the 3'-end of the tRNA is concerned. Our knowledge of these positions in the known tertiary structure makes it questionable whether the transformylase is large enough for this, so with our limited knowledge of protein-nucleic acid recognition such an analysis does not appear to give a reasonable answer to the problem. I have used the

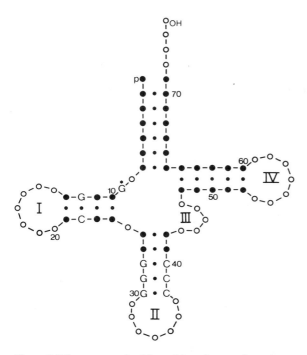

Figure 2.7 Common nucleotide positions for transformylase recognition. Nucleotides common to sequences of Ec Met$_f$, y Met$_f$, and Mye Met$_f$, all formylatable in the charged form by bacterial transformylase. For convenience the invariant and semi-invariant positions shown in figure 2.6 are omitted.

foregoing examples to illustrate that with our present incomplete knowledge of structure and similar experience in comparing protein sequences, we now realize that analysis of recognition in terms of simple sequence analogy is naive and unlikely to be fruitful and that tertiary structural knowledge is essential to discuss function.

2.8 Functional Significance of the Tertiary Structure of Yeast tRNA[Phe]

After tRNA was first crystallized (31), several laboratories sought a species that would give crystals suitable for an X-ray crystallographic analysis to high resolution (4, 32, 33). A systematic study of many species from *E. coli* and yeast yielded suitable crystals of an orthorhombic form and a monoclinic form of yeast tRNA[Phe]. The monoclinic form had a smaller unit cell (34, 35) than the related orthorhombic form (36) and was better ordered

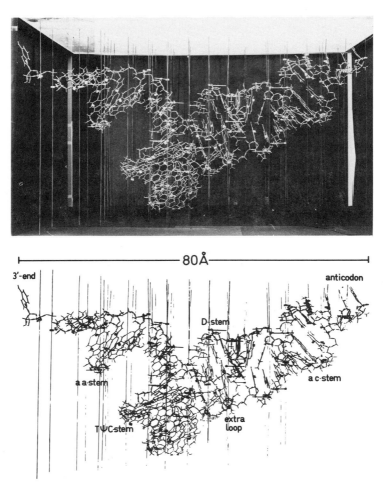

Figure 2.8 *Top*: Photograph of a 0.25 nm model of yeast tRNA[Phe] (26) built at Aarhus University under the guidance of Dr. J. E. Ladner. *Bottom*: Silhouette of the model with labeling of parts referred to in the text. The distance shown of 80 Å is a projection. Actually on the model the distance from the anticodon to the 3′-terminal A is about 85 Å.

than a different orthorhombic form obtained earlier by Cramer and his colleagues (37). The three-dimensional structure of the yeast tRNA[Phe] in the monoclinic form has been solved to 0.25 nm resolution in Cambridge by the method of isomorphous replacement (26), using five heavy atom derivatives. Independent structural determinations using the orthorhombic form (38) and the monoclinic form (39) have yielded the same structure in broad outline. A photograph of the Kendrew skeletal model built to fit the electron density and a silhouette key to structural parts are shown in figure 2.8. The second stage of the X-ray crystallographic analysis to 0.25 nm resolution (26) resolved some ambiguities in the structure. The chain tracing of the ribose-phosphate backbone was completed with certainty, and many more features of the molecular stereochemistry have been revealed. In addition to the base pairs of the cloverleaf arrangement of yeast tRNA[Phe] (figure 2.9), many other interactions have been deduced (see figure 2.10) that fix the tertiary structure of yeast tRNA[Phe] and show how the cloverleaf of figure 2.9 is folded.

From the X-ray studies we now have a picture of the ordered complexity of a folded RNA molecule, a complexity as great as that of a protein. An interesting by-product is a detailed picture of the stereochemistry of a G·U base pair in a double helical stem, where the pairing is that predicted by the wobble hypothesis (40).

Accounts of the 0.3 nm tertiary structure focused on the interactions between bases. Now an extensive network of H-bonds, including ribose and phosphate groups as well as bases, has been traced; almost half the bonds involve 2'-OH groups of riboses as acceptors, donors, or both (26, 38). A prominent set of H-bonds stabilizes the T-join made in the structure by the two long double helices that meet approximately at right angles. Another set reinforces the links between bases at the bends of the D-loop and TΨC-loop.

In the 0.3 nm model all sugar conformations were assumed to be C3'-endo, as found in RNA double helices. However, in the more detailed model building occasioned by fitting the 0.25 nm map, the pucker of ten of the sugars was changed to the C2'-endo conformation (26). Although the resolution of the electron density map is still not sufficient to show the shape of a sugar unequivocally, the conformation can be deduced from the restrictions imposed on it by the relative dispositions of the two phosphate groups on either side and of the base that emanates from it. A good example is that of nucleotide m[1]A58 whose sugar is C2'-endo in the 0.25 nm model. In the 0.3 nm model the phosphate and base densities were well fitted, but

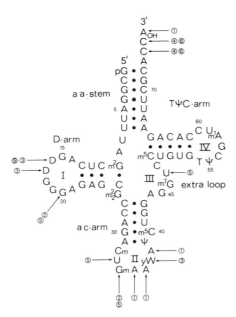

Figure 2.9 Primary structure of y tRNA[Phe] arranged in the cloverleaf form. Also shown is the chemical reactivity of y tRNA[Phe]. The arrows indicate points of chemical reaction. Reagents used were (1) perphthalic acid, specific for A-residues, (2) kethoxal for exposed G-residues, (3) Na BH reduction, (4) methoxyamine (44), (5) carbodiimide (44), and (6) I_2/TlCl$_3$ (45; see also refs. 18, 40, 46).

the C3′-endo ribose was not. Fitting the ribose density moved the base of m[1]A58 too close to its base-paired partner T54, which was well fixed. After the change the C2′-endo sugar resided in its proper density, and the base pair was comfortably made.

The topography of the model deduced from the crystallographic study agrees with a companion study of the chemical reactivity of yeast tRNA[Phe] in solution (41). There is thus no reason to presume that the structure in the crystal departs significantly from that in solution.

I shall summarize some of the structural features of the 0.3 nm and 0.25 nm tRNA models that appear to have functional significance.

The molecule depicted in the model (figure 2.8 and schematically in figure 2.10) can be described in terms of three major skeletal substructures: (1) the long double helix formed by the stacking of the aa-stem and TΨC-stem on top of each other; (2) the augmented D-helix forming the central part or "thorax" of the molecule, consisting of the D-stem, augmented laterally by interactions with the short "stretcher" region 8–9, with the extra loop III

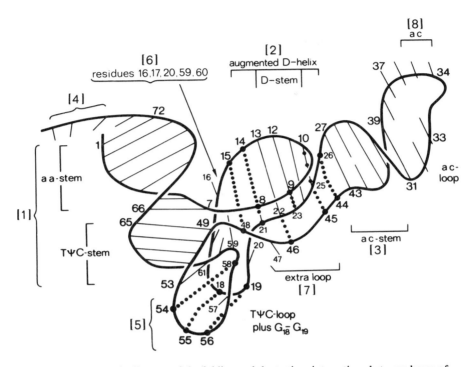

Figure 2.10 A schematic diagram of the folding and the tertiary interactions between bases of y tRNA[Phe] (adapted from Ladner et al., ref. 26). The ribose-phosphate backbone is represented by a continuous line. Base pairs in the double helical stems are represented by long lightly weighted lines, and nonpaired bases by shorter lines. Base interactions additional to those in the cloverleaf formula are indicated by dotted lines. Substructures referred to in the text are given by numbers in brackets.

and with a part of the D-loop; (3) the anticodon stem tilted by about 20° to the D-stem and apparently hinged to it by hydrogen bonds between nucleotides A44 and m$_2^2$G26 and between G45 and m^{2}G10. The molecule has been described as L-shaped (25), but the two major long double helical stretches, that is, substructure 1 and substructure 2 + 3, are arranged as the letter T (24). Thus the T-form of tRNA is a tertiary structural arrangement of the secondary structure represented by the cloverleaf form.

To these skeletal substructures are joined five regions with established, or potential, functional roles. (4) At one end of the long double helix is the 3'-end C-C-A to which the amino acid is attached. (5) At the other end the invariant TΨC-loop is tightly folded and interacts with a part (constant G*-G) of the D-loop in a complex set of interactions, such as those that have been interpreted in detail in the electron density map at 0.25 nm resolution (26). Briefly, U59 stacks on the pair G15·C48; T54 is base-paired with m¹A58 in a reverse Hoogsten type; G57 intercalates between G18 and G19, giving rise to a stack of four purines 58, 18, 57, and 19, interacts with the backbone of the D-loop C56, and forms a Watson-Crick base pair with G19; and G18 appears to interact with Ψ55 and with the T Ψ C-backbone on either side of it (26). The interaction of the T Ψ C-loop with the invariant G18 and G19 probably masks the potential of the T-Ψ-C-G sequence for its proposed role of binding to the ribosomal A-site during peptide bond formation (28). (6) The variable regions in the D-loop may form a discrimination site (or part of one) for enzymes to distinguish between tRNAs (not necessarily the aminoacyl-tRNA synthetase recognition). The two sections a and β of D (see figure 2.2) are near each other and form a surface patch on the molecule. They each contain a variable number (one to three) of nucleotides and, either on their own or together with the nearby nonbase-paired residues 59 and 60, could well provide a general enzyme discrimination site. The variable extra loop (or at least a part of it, such as U47 in tRNA[Phe]), which is region 7, may also play such a role in some cases. Finally the decoding function of tRNA is provided for by region 8, the anticodon loop, containing the anticodon. This loop contains a stack of five bases on the 3′-side; the other two bases C32 and U33 are also stacked separately on the end of the anticodon stem (24) but with a sharp bend between U33 and Gm 34 (26). It is possible that the loop conformation changes during the translocation of the tRNA.

An interesting observation on the modified purine nucleoside found following the anticodon still does not have a clear-cut functional or structural interpretation. Nishimura (42) pointed out an apparent connection between this particular modified base (location 9 in figure 2.3) and the type of first base in the codon that the particular tRNA decodes. For example, tRNAs containing ms²i⁶A decode codons with U as the first base, and tRNAs containing t⁶A decode codons with A as the first base. In other words this implies a connection between the modified base and the third position of the anticodon. Whether the connection is due to a structural specification or to a metabolic pathway linkage is not known.

2.9 Generality of Yeast tRNA[Phe] Tertiary Structure

The structural integrity of the molecule as a whole depends on the central part at which elements of four chains come close together and a number of base triples or pairs are found. These additional base pairs or triples are stacked or intercalated so as to make their H-bonds inaccessible to water molecules. The tertiary structure is further stabilized by the stacking of non-paired bases on each other or on base pairs as, for example, in the anticodon loop. Indeed about 90 percent of the bases can be said to be internal as hydrophobic groups are internal in proteins.

The elucidation of the tertiary structure of yeast tRNA^{Phe} naturally leads one to ask whether other tRNA primary structures can be folded in the same way. In particular can other bases be substituted into the base pair or base triple positions of yeast tRNA^{Phe} so as to give equivalent interactions? Most changes of sequence in class 1 can indeed be compensated by concomitant changes elsewhere, so that local geometry in the three-dimensional structure remains essentially unchanged (24, 43). These changes are listed in table 2.4.

Levitt (27) first noticed a coordinated base change at positions 15 and 48 (tRNA^{Phe} numbering, see figure 2.9) by examining tRNA primary structures. However, the pair is not of the Watson-Crick type (24); moreover it is asymmetric since only a purine R is found in position 15. Structural considerations (43) require that such an asymmetry form a reverse Watson-Crick base pair. G·C can be replaced by A·U to give the same disposition of

Table 2.4 Correlated base changes in class 1 tRNAs

	Direction of Chains
Base pair at positions 15–48	
G15:C48	parallel
A15:U48	parallel
Base triple at positions 13–22–46	
C13:G22:m⁷G46	antiparallel/antiparallel
U13:A22:A46	antiparallel/antiparallel
U13:G22:G46	antiparallel/antiparallel
Base triple at positions 12–23–9	
U12:A23·A9	antiparallel/parallel
G12:C23·G9	antiparallel/parallel
U12:A23·m¹G9	antiparallel/parallel
Base pair at 26–44	
A26·G44	antiparallel
G*26:A44	antiparallel

Note: Dots represent H-bonds.

H-bonds in this nonstandard base pair. Both y $tRNA_3^{Glu}$ and Ec $tRNA_1^{Gly}$ provide interesting exceptions to this structural correlation; they contain A15·C48 instead of the usual semi-invariant A15·U45 or G15·C48. Further, the recently determined Ec $tRNA^{Cys}$ contains G15·G48.

Class 1 is subdivided into 1A and 1B in table 2.2 based on the two nucleoside positions 46 and 9 (in the numbering for $tRNA^{Phe}$). Position 46 has either m^7G or not (G or A), and this is the basis for the subdivision into the two groups 1A and 1B. However, the base triple 46·22·13 can be made equivalent in these subclasses, with nearly the same disposition of glycosyl bonds (24, 43). Nucleoside 9 is G, m^1G, or A; here again a base triple 9·23·12 can be made for all members of subclasses 1A and 1B, though the number of hydrogen bonds is not the same in all cases (24, 43). The coordinated base changes are summarized in table 2.3.

The model is perhaps also directly relevant to sequences of class 1C. In the model it is possible to remove the exposed nucleotide U47 and still bridge the gap between 46 and 48 by the phosphate-sugar linkage, maintaining all the tertiary interactions. Class 1C could have this type of structure; any base triples equivalent to 13·22·46 would now have to be made in a somewhat different way, but they would still have to preserve the relative disposition of the pieces of backbone being bridged by the base·base interactions. By an extension of the argument the model is also relevant for class 1D. If a base pair, albeit nonstandard, can be fitted into position b4, an equivalent augmented D-helix might be made without disrupting the pattern of the four chains that come together to make it. Hence it is supposed that the yeast $tRNA^{Phe}$ structure is common to all class 1 structures. We have no firm tertiary structural information about class 2 structures, although there is recent speculation (44) about such structures.

2.10 Solution Structure of tRNA

The bases of yeast $tRNA^{Phe}$ that react with a carboiimide reagent and with methoxyamine in solution have been identified (45). This information has been combined with other results from the literature to produce the composite picture (18, 41) of base accessibility in yeast $tRNA^{Phe}$ shown in figure 2.9. For the chemistry of the reactions indicated in figure 2.9 and for additional reactions of other tRNA species, see a recent review by Brown (47). The bases that react chemically can be correlated with exposed positions in the three-dimensional structure of tRNA (compare figures 2.9 and 2.10). Those that do not react are either in the double helical regions or are involved in maintaining the tertiary structure. These results confirm the usual structural assumptions made about reactivity in chemical studies and hence support the extension of such work to other tRNA structures. Further, since the chemical studies are carried out on tRNAs in solution rather than in the crystalline state, they also support the contention that there is no significant change in the conformation of yeast tRNA^{Phe} on crystallization.

This picture for chemical reactivity is similar to that obtained earlier for the bacterial initiator tRNA, tRNA$_f^{Met}$ (48), which also belongs to class 1. It is therefore reasonable to propose that the overall tertiary structure of bacterial initiator tRNA is similar to that described for yeast tRNA^{Phe}. The special function of *E. coli* tRNA$_f^{Met}$ is probably explained by a subtle recognition of an initiation factor for locating it in the correct ribosomal site rather than in a different folding of the tRNA. Recent physical chemical studies, especially those using the NMR technique, have identified the H-bonds involved in the tertiary interactions between bases (49). These studies also imply that the same structure exists in solution as in the crystal form.

2.11 Structural Changes during Function

Although we now have strong evidence that the tRNA conformation is the same in solution as in the crystal form, it is unclear whether the structure changes its conformation for its role in protein biosynthesis. Indeed it is likely that the conformation does change and is effected by the many different interactions that the tRNA undergoes with other protein and nucleic acid components. There is evidence that the binding of the aminoacyl-tRNA to the ribosome requires the exposure of sequence T-Ψ-C-G for complementary binding to the ribosomal 5S RNA (28). Should this be really the case, a change in conformation of the tRNA is necessary from the structure represented in figure 2.10 where the common sequence is unavailable for further base pairing because of its involvement in the tRNA tertiary structural design. Presumably the change in conformation would be affected by the elongation factor EF-Tu or ribosomal proteins. Conclusive information about these possibilities is not yet available.

The anticodon loop is another place where conformational changes may occur during protein synthesis. The hingelike region of the structure where the anticodon arm is attached to the rest of the structure (see figure 2.10 around residues 24 and 45) could provide a point for directing conformational change in the anticodon loop or for conformational change concomitant with movement of the tRNA from the A-site to the P-site (2) during protein synthesis. As can be seen in figure 2.10 the two bases following the anticodon to the 3'-side are stacked on the anticodon. There is a break in the regular stacking to the 5'-side of the anticodon, but the two preceding pyrimidines are stacked on each other and on the anticodon stem. This structure is similar to that originally proposed by Fuller and Hodgson (50) from model-building studies. However, the stacking of the anticodon, with somewhat more overlapping of bases than in a regular RNA double helix, appropriately exposes the anticodon bases for interaction with a mRNA codon. We still need much information about the way that the ribosomal environment ensures this specificity and stability.

Results from experiments testing the strength of binding of specific complementary oligonucleotides to regions of the anticodon loop indicate that the anticodon can be stacked on the 5'-side of the anticodon at least on the next pyrimidine (51). This result was also obtained for tRNA$_f^{Met}$(52), which does not contain a modified base after the anticodon. The structural role of rare base W in y tRNA^{Phe} is still unclear. Thus in solution, but in the presence of a complementary oligonucleotide, the conformation of the crystal structure could be changed. Clearly the possibility also holds for the natural interaction with mRNA.

More surprising is the lack of conclusive evidence concerning the possible difference in conformations between uncharged and charged tRNA. The literature contains evidence both in favor of and against a change in conformation. The evidence in favor suggests that the methods producing negative results are not capable of measuring the subtle changes that may occur. Large structural rearrangements that would be readily detectable do not seem to occur. To gain more insight into these putative conformational changes we need information on the different tRNA-protein complexes formed during the steps of protein biosynthesis from chemical, X-ray diffraction, and other physical chemical studies.

2.12 tRNA-like Structures Chargeable with Specific Amino Acids

As stated earlier, tRNAs with primary structures containing obvious exceptions to the generalized cloverleaf pattern usually have special functional roles such as involvement in polypeptide chain initiation, cell-wall metabolism, and possibly RNA virus replication. It should be emphasized that it may be dangerously misleading to consider exceptions to the standard primary and secondary structural pattern to be directly related to these func-

tional roles. Clearly knowledge of the spatial structures of these functionally different tRNAs is necessary before the exceptions to common sequence
positions can be adequately interpreted. So far this knowledge is not forth-
coming, so that this discussion of sequence is, of necessity, not definitive.
The common feature of the tRNA-like structures to be discussed in this section is that they are all chargeable with a specific amino acid by the appropriate aminoacyl-tRNA synthetase. What defines the chargeability of a
tRNA or tRNA-like structure remains unknown. We can also surmise that
there are other tRNA-like structures, defined by other biological properties
of tRNA, that are thus far outside our range of knowledge.

The best-known tRNA function other than its role in polypeptide chain
elongation is in polypeptide chain initiation. In both eucaryotic and procaryotic cells a special methionine tRNA acts as a chain initiator. In procaryotic cells the special Met-tRNA $_f$ acting as a chain initiator is formylated, whereas in eucaryotic cells it is not. The difference apparently is due
to the lack of the formylating enzyme transformylase in the eucaryotic cytoplasm, since the eucaryotic Met-tRNA $_f$ can be formylated in vitro with the
procaryotic transformylase. Primary and secondary structures of procaryotic initiator tRNAs appear to be characterized by the lack of a Watson-
Crick base pair at the first position (a1) of the amino acid stem (see also
under exceptions 2 and 4 earlier) and by an A·U instead of Y·R at the second Watson-Crick base pair (b2) in the D-stem. Until very recently eucaryotic initiator tRNAs were thought to be specially characterized by the
sequence G-A-U-C instead of the more usual G-T-Ψ-C in the TΨC-loop
(exception 10 earlier). However, a major alanine tRNA from silkworm (53)
contains the sequence G-A-Ψ-C in this region. Unless this tRNA is an initiator, which seems unlikely, the only special characteristic of a general eucaryotic initiator may be U instead of the constant Ψ in the TΨC-loop. In
addition, in the special case of mammalian initiator tRNA the constant U
just before the anticodon is replaced by C.

As mentioned earlier, the well-established example of biological activity
listed under aminoacyl-tRNA transferases in table 2.1 (activity 19) involves
a set of tRNA-like structures characterized by chargeability with glycine.
These structures from staphylococcal sources have had their sequences determined (12, 54) and can be arranged in the cloverleaf form but contain
several exceptions to the generalized cloverleaf shown in figure 2.2. As far
as is known, these tRNA-like structures are used in the cell to transfer glycine to acceptors in cell-wall components and are not used to decode mRNA
on the ribosome. Thus the features defining these roles in protein biosyn-

thesis may be obviously lacking in the cell-wall tRNA^{Gly} sequences. Again, without a knowledge of the tertiary structure an analysis of the primary structural differences may not be meaningful. However, some clear differences in the cell-wall tRNA-like cloverleaf forms from the generalized cloverleaf form stand out (see also under exceptions 5 and 9). The type of tertiary interactions between the D-loop and the TΨC-loop described for the yeast tRNA^{Phe} tertiary structure cannot be made for the cell-wall tRNA^{Gly}. Instead of the standard G*-G in the D-loop and G-U*-Ψ-C in the TΨC-loop the cell-wall tRNA^{Gly}s contain U-U and G-U-G-Y. Perhaps once again the replacement of the standard Ψ is meaningful. In addition the structure of the anticodon loop of tRNA^{Gly} must be abnormal to prevent its interfering with decoding by normal tRNA^{Gly}; a sixth Watson-Crick base pair is possible at the bottom of the anticodon stem, and there is also a pyrimidine Y and not a modified purine R* next to the 5'-end of the anticodon.

In recent years apparently normal eucaryotic tRNA structures have been shown to have new, not completely defined properties (activities 15 and 16 in table 2.1). These tRNAs appear to play an important role in virus metabolism (55, 56). Discovery of tRNAs of host origin within various oncogenic RNA viruses led to study of these tRNAs to determine whether they play a role in virus biogenesis. Not all tRNA species are contained in the virion, and apparently their relative concentrations in the virion differ from those in the host cell. This implies that during the formation of virion the tRNA species are selected in a specific, nonrandom manner. Thus besides the usual 70S RNA genome, purified virions contain a mixture of 4S RNAs containing tRNAs, 5S RNA, and 7S RNA. The isolated 70S RNA still contains a mixture of tRNAs which by sequential heating can be liberated from the complex. There is some indication that tRNAs may be involved in determining the structure of the genome. But the most tightly bound of these turns out to be a pure species of tRNA which serves as a primer for the enzyme reverse transcriptase to make a DNA copy of the viral RNA genome.

So far there are two well-documented cases of this finding in which the specific tRNA primer has been sequenced, namely tRNA^{Trp} in avian sarcoma virus (57, 58) and tRNA^{Pro} in murine leukemia virus (J. E. Dahlberg, personal communication; 56). The most obvious peculiarity in the cloverleaf forms of these two tRNAs is the sequence G-Ψ-Ψ-C instead of G-U*-Ψ-C in the TΨC-loop region. Whether this feature has significance for recognition by the reverse transcriptase enzyme is not yet known. It is of interest that the tRNA^{Trp} species characterized by the primary structure determined for the avian sarcoma virus is the only tryptophan-accepting species detected in chicken cells. So far the tRNA primer has been shown to interact at its 3'-end with the template 35S component of the RNA genome. This interaction probably involves as little as one-fourth of the length of the tRNA molecules. In spite of much recent progress the presence of most of the tRNAs inside a tumor virus remains an open question.

The last tRNA-like structures or elements of tRNA structure to be discussed here are related to biological activity 17 of table 2.1. A number of RNA viruses—mostly plant viruses—have been shown to be chargeable at their 3'-ends with specific amino acids by aminoacyl-tRNA synthetases. The plant viruses turnip yellow mosaic virus (59) and eggplant mosaic virus (60) are chargeable with valine, tobacco mosaic virus with histidine, and each of the four brome mosaic virus RNAs with tyrosine (62). The picornaviral RNAs from Mengovirus and encephalomyocarditis virus can also be charged with histidine (63) and serine (64), respectively, but less efficiently. There is also evidence that RNA bacteriophages $Q\beta$, MS2, and R17, and RNAs coded by T-even DNA bacteriophages, contain nonchargeable tRNA elements of structure within their genomes (55). Although the biological significance of the chargeability of these viral RNAs still eludes us, there are some experimental results hinting at possible involvement in protein biosynthesis and viral replication. For example, charged turnip yellow mosaic virus RNA or tobacco mosaic virus RNA interacts in vitro with plant elongation factor. Moreover, since bacterial elongation factor Tu has been implicated as a component of $Q\beta$ replicase (65), these structures may be involved in viral RNA replication.

More information on these tRNA-like structures within viral genomes has come recently from sequence analyses of the viral RNAs. Secondary structural arrangements into a cloverleaf has been attempted for the known sequences with varying degrees of success, none completely satisfying. The secondary structure of the 3'-end regions of turnip yellow mosaic virus RNA and possibly eggplant mosaic virus RNA (66, 67) resemble tRNA fairly well, but the secondary structure of the 3'-end of tobacco mosaic virus is less easily folded to resemble a tRNA cloverleaf (68). Finally, reasonable base pairing schemes with the 3'-end, 161 bases long, identical in each of the four brome mosaic virus RNAs, yield at best a very distorted cloverleaf form (69). In the last case it is known that these charged brome mosaic virus RNAs will not serve as donors of amino acids during polypeptide chain elongation in vitro (69). It has also been established that the charged valyl-turnip yellow mosaic virus RNA contains a structure different from host-cell valine tRNAs (55), in contrast to the situation described earlier for reverse transcriptase primer tRNA. Another important feature of the elements of tRNA structure so far sequenced is the lack of minor (modified) bases. Evidently minor bases are not involved in specifying enzyme recognition of these tRNA-like elements. With respect to this topic it should be noted that the sequenced 3'-end fragment of turnip yellow mosaic virus RNA was obtained by RNase P (a tRNA processing enzyme) cleavage of the virus RNA (66).

In summary all that can be said is that some elements of tRNA secondary structure, such as identifiable arms and loops, can be observed in the known plant viral sequences, but they are not joined in a common, clearly defined way. The proper interpretation of these teasingly interesting facts must await further knowledge of the tertiary structural arrangement of these elements of tRNA structure.

Acknowledgment

I am grateful for the continuing cooperation of and collaboration with Dr. A. Klug's group and for support by the University of Aarhus, Danish Natural Science Research Council grant No. 511-3820, and NATO research grant no. 893.

References

1. Lipmann, F. (1969) Science *164*, 1024–1031.

2. Clark, B. F. C. (1974). In Companion to Biochemistry, A. T. Bull, J. R. Lagnado, J. O. Thomas, and K. F. Tipton, eds. (London: Longman Group), pp. 1–86.

3. Crick, F. H. C. (1957). Biochem. Soc. Symp. *14*, 25.

4. Rich, A., and RajBhandary, U. L. (1976). Ann. Rev. Biochem. *45*, 805–860.

5. Pedersen, F. S., Lund, E., and Kjeldgaard, N. O. (1973). Nature New Biol. *243*, 13–15.

6. Lewis, J. A., and Ames, B. (1972). J. Mol. Biol. *66*, 131–142.

7. Allende, J. E. (1975). Paabs Revista *4*, 343–361.

8. Smith, J. D. (1972). Ann. Rev. Genet. *6*, 235–256.

9. Bell, J. B., Gelugne, J. P., and Jacobson, K. B. (1976). Biochim. Biophys. Acta *435*, 21–29.

10. Smith, D. W. (1975). Science *190*, 529–535.

11. Sharma, O. K., Beezley, D. W., and Roberts, W. K. (1976). Biochemistry *15*, 4313–4318.

12. Roberts, R. J. (1972). Nature New Biol. *237*, 44–46.

13. Holley, R. W., Apgar, J., Everett, G. A., Madison, J. T., Marquisee, M., Merrill, S. H., Penswick, J. R., and Zamir, A. (1965). Science *147*, 1462–1465.

14. Gould, R. M., and Lennarz, W. J. (1970). J. Bacteriol. *104*, 1135–1144.

15. Scarpulla, R. C., Deutch, C. E., and Soffer, R. L. (1976). Biochem. Biophys. Res. Commun. *71*, 584–589.

16. Kaji, H., and Rao, P. (1976). FEBS Letters *66*, 194–197.

17. Barrell, B. G., and Clark, B. F. C. (1974). Handbook of Nucleic Acid Sequences (Oxford: Joynson Bruvvers).

18. Clark, B. F. C., and Klug, A. (1975). Proceedings of the Tenth FEBS Meeting, vol. 39, F. Chapeville, and M. Grunberg-Manago, eds. (Amsterdam: North Holland/American Elsevier), pp. 183–206.

19. Dirheimer, G., Keith, G., Martin, R., and Weissenbach, J. (1976). Synthesis, Structure and Chemistry of Transfer Ribonucleic Acids and Their Components, Proceedings of the International Conference, Dymaczewo, Sept. 1976. Polish Academy of Sciences, Poznan, pp. 273–290.

20. Clark, B. F. C. (1977). Prog. Nucl. Acid Res. Mol. Biol. *20*, 1–19.

21. Ohashi, K., Harada, F., Ohashi, Z., Nishimura, S., Stewart, T. S., Vögeli, G., McCutchan, T., and Söll, D. (1976). Nucl. Acids Res. *3*, 3369–3376.

22. Watanabe, K., and Nishimura, S. (1975). Protein Nucleic Acids and Enzymes (Japan) *20*, 181.

23. Weissenbach, J., Kiraly, I., and Dirheimer, G. (1976). FEBS Letters *71*, 6–8.

24. Robertus, J. D., Ladner, J. E., Finch, J. T., Rhodes, D., Brown, R. S., Clark, B. F. C., and Klug, A. (1974). Nature *250*, 546–551.

25. Kim, S. H., Suddath, F. L., Quigley, G. J., McPherson, A., Sussman, J. L., Wang, A. H. J., Seeman, N. C., and Rich, A. (1974). Science *185*, 435–440.

26. Ladner, J. E., Jack, A., Robertus, J. D., Brown, R., Rhodes, D., Clark, B. F. C., and Klug, A. (1975). Proc. Nat. Acad. Sci. USA *72*, 4414–4418.

27. Levitt, M. (1969). Nature *224*, 759–763.

28. Erdmann, V. A., Sprinzl, M., and Pongs, O. (1973). Biochem. Biophys. Res. Commun. *54*, 942–948.

29. Williams, R. J., Nagel, W., Roe, B., and Dubock, B. (1974). Biochem. Biophys. Res. Commun. *60*, 1215–1221.

30. Bruton, C. J., and Clark, B. F. C. (1974). Nucl. Acids Res. *1*, 217–221.

31. Clark, B. F. C., Doctor, B. P., Holmes, K. C., Klug, A., Marcker, K. A., Morris, S. J., and Paradies, H. H. (1968). Nature *219*, 1222-1224.

32. Sigler, P. B. (1975). Ann. Rev. Biophys. Bioeng. *4*, 477–527.

33. Brown, R. S., Clark, B. F. C., Coulson, R. R., Finch, J. T., Klug, A., and Rhodes, D. (1972). Eur. J. Biochem. *32*, 130–134.

34. Ladner, J. E., Finch, J. T., Klug, A., and Clark, B. F. C. (1972). J. Mol. Biol. *72*, 99–101.

35. Ichikawa, T., and Sundaralingam, M. (1972). Nature New Biol. *236*, 174–175.

36. Kim, S. H., Quigley, G., Suddath, F. L., McPherson, A., Sneden, D., Kim, J. J., Weinzierl, J., Blattmann, P., and Rich, A. (1972). Proc. Nat. Acad. Sci. USA *69*, 3746–3750.

37. Cramer, F., von der Haar, F., Holmes, K. C., Saenger, W., Schlimme, E., and Schulz, G. E. (1970). J. Mol. Biol. *51*, 523–530.

38. Quigley, G. J., Wang, A., Seeman, N. C., Suddath, F. L., Rich, A., Sussman, J. L., and Kim, S. H. (1975). Proc. Nat. Acad. Sci. USA *71*, 3711–3715.

39. Stout, C. D. Mizuno, H., Rubin, J., Brenner, T., Rao, S. T., and Sundaralingam, S. (1976). Nucl. Acids Res. *3*, 1111–1123.

40. Crick, F. H. C. (1966). J. Mol. Biol. *19*, 548–555.

41. Robertus, J. D., Ladner, J. E., Finch, J. T., Rhodes, D., Brown, R. S., Clark, B. F. C., and Klug, A. (1974). Nucl. Acids Res. *1*, 927–932.

42. Nishimura, S. (1972). Prog. Nucl. Acids Res. Mol. Biol. *12*, 49–85.

43. Klug, A., Ladner, J. E., and Robertus, J. D. (1974). J. Mol. Biol. *89*, 511–516.

44. Brennan, T., and Sundaralingam, M. (1976). Nucl. Acids Res. *3*, 3235–3251.

45. Rhodes, D. (1975). J. Mol. Biol. *94*, 449–460.

46. Batey, I., and Brown, D. M. (1975). Mol. Biol. Rep. *2*, 65–72.

47. Brown, D. M. (1974). Basic Principles in Nucleic Acid Chemistry, P. O. P. T'so, ed., vol. 2 (London: Academic Press), ch. 1.

48. Chang, S. E. (1973). J. Mol. Biol. *75*, 533–547.

49. Robillard, G. T., Tarr, C. E., Vosman, F., and Berendsen, H. J. C. (1976). Nature *262*, 363–369.

50. Fuller, W., and Hodgson, A. (1967). Nature *215*, 817–821.

51. Pongs, O., and Reinwald, W. (1973). Biochem. Biophys. Res. Commun. *50*, 357–363.

52. Uhlenbeck, O. C. (1972). J. Mol. Biol. *65*, 25–41.

53. Sprague, K. U., Hagenbüchle, O., and Zuniga, M. C. (1977). Cell *11*, 561–570.

54. Roberts, R. J. (1974). J. Biol. Chem. *249*, 4787–4796.

55. Haenni, A. L., Bénicourt, C., Teixeira, S., Prochiantz, A., and Chapeville, F. (1975). Proceedings of the Tenth FEBS Meeting, vol. 39, F. Chapeville and M. Grunberg-Manago, eds. (Amsterdam: North Holland/American Elsevier), pp. 121–131.

56. Waters, L. C., and Mullin, B. C. (1977). Prog. Nucl. Acid Res. Mol. Biol. *20*, 131–160.

57. Harada, F., Sawyer, R. C., and Dahlberg, J. E. (1975). J. Biol. Chem. *250*, 3487–3497.

58. Waters, L. C., Yang, W.-K., Mullin, B. C., and Nichols, J. L. (1975). J. Biol. Chem. *250*, 6627–6629.

59. Pinck, M., Yot, P., Chapeville, F., and Duranton, H. M. (1970). Nature *226*, 954–956.

60. Pinck, M., Generaux, M., and Duranton, H. M. (1974) Biochimie *56*, 423–428.

61. Oberg, B., and Philipson, L. (1972). Biochem. Biophys. Res. Commun. *48*, 927–932.

62. Hall, T. C., Shih, D. S., and Kaesberg, P. (1972). Biochem. J. *129*, 969–976.

63. Salomon, R., and Littauer, U. Z. (1974). Nature *249*, 32-34.

64. Lindley, I. J. D., and Stebbing, N. (1977). J. Gen. Virol. *34*, 177–182.

65. Blumenthal, T., Landers, T. A., and Weber, K. (1972). Proc. Nat. Acad. Sci. USA *69*, 1313–1317.

66. Silberklang, M., Prochiantz, A., Haenni, A. L., and RajBhandary, U. L. (1977). Eur. J. Biochem. *72*, 465–478.

67. Briand, J. P., Richards, K. E., Bouley, J. P., Witz, J., and Hirth, L. (1976). Proc. Nat. Acad. Sci. USA *73*, 737–741.

68. Guilley, J., Jonard, G., and Hirth, L. (1975). Proc. Nat. Acad. Sci. USA *72*, 864–868.

69. Dasgupta, R., and Kaesberg, P. (1977). Proc. Nat. Acad. Sci. USA *74*, 4900–4904.

3   BIOSYNTHESIS
OF tRNA

Sidney Altman

3.1 Organization of Transfer RNA Genes

3.1.1 Escherichia coli

In *E. coli* the intracellular concentrations of some isoaccepting species of tRNA are much higher than others. One may ask whether the genes for the major species of tRNA are clustered together and whether their transcription is controlled by one single promoter. These questions illustrate some issues to be resolved in a consideration of tRNA biosynthesis and the requirement for information concerning the organization of tRNA genes in the *E. coli* chromosome. Such information can yield clues to the expected products of transcription of tRNA genes and the regulation of the appearance of the various species of tRNA.

Both genetic and biochemical approaches can be applied to the study of the organization of tRNA genes. The first method depends on the isolation of mutants in tRNA genes that confer certain recognizable phenotypes on the cells carrying the mutations. For example, it is possible to isolate mutants containing single nucleotide changes in the anticodons of some tRNAs that enable these tRNAs to interact with nonsense codons and prevent premature termination of mutant mRNAs. Clones derived from such mutant cells are recognizable as nonsense suppressor strains. Missense suppressors can be similarly isolated (1, 2, 3). Once several such mutant strains are available, it is possible to map the individual mutations to see, for example, whether mutant tRNA genes lie together or are scattered around the *E. coli* chromosome. Since the tRNA genes that have been mapped do not all lie together (see figure 5.1; ref. 4) there is no reason to anticipate that the products of tRNA gene transcription should be made in equal quantities under the same transcriptional controls or that there could be very long transcripts containing many tRNA sequences.

Biochemical studies of tRNA genes involve the hybridization of mature tRNA species to isolated fragments of DNA from transducing phages carrying suppressor genes or to restriction endonuclease fragments of DNA from *E. coli*. The DNA fragments can be of various sizes, and the kinetics of hybridization of the tRNA to the fragments allow calculation of the number of copies of tRNA genes present in any preparation of fragments of a given size. These biochemical studies have shown that some tRNA genes are clustered in groups at various places on the genetic map (5). These clustered sequences could be transcribed as polycistronic RNA molecules, and this possibility has been confirmed by direct analysis of gene transcripts (6–10). Hybridization of bulk tRNA to DNA fragments can also give the average number of tRNA genes per organism, which is about 60 in *E. coli*.

Perhaps one of the most interesting discoveries about tRNA gene organization is that certain tRNA gene sequences are located within some or all of the 6–10 ribosomal RNA cistrons in *E. coli* (11, 12). Transfer RNA sequences have been mapped by hybridization to DNA restriction fragments and lie between the 16S and 23S sequences (figure 3.1). To date $tRNA_{1b}^{Ala}$, $tRNA_2^{Glu}$, and $tRNA_1^{Ileu}$ have been identified in rRNA cistrons. A transcript is made of the whole cistron, which is then cleaved; this posttranscriptional processing probably starts before transcription is completed. Since there are 6–10 such ribosomal RNA cistrons per *E. coli* genome, it is clear that the synthesis of an equal (or smaller) number of tRNA species must be coordinately controlled with the synthesis of rRNA. Why this should be the case is not known, but apparently it is important that these species are produced in the same abundance as the rRNA species in *E. coli*.

The genetic and biochemical methods described have also been used for the identification and mapping of the tRNA genes coded by bacteriophages T4 and T5 (13–20). In the T4 phage all eight tRNA genes are clustered together on the phage chromosome (15–17, 20) (figure 3.1). These may be transcribed as one long transcript (21), but such a complete transcript has not yet been demonstrated. Smaller transcription fragments have been isolated; so if there is a single long gene transcript, processing must begin before transcription is completed.

Figure 3.1 Organization of rRNA transcription unit of *E. coli* and tRNA genes in bacteriophage T4. *Top*: The organization of an *E. coli* rRNA transcription unit. Transfer RNA$_2^{Glu}$ hybridizes to a DNA restriction fragment in the spacer region between the 16S and 23S sequences. The rRNA transcription unit analyzed had been transposed to a transducing phage. The total length of the unit shown is about 5000 nucleotide pairs (Lund et al., ref. 11). *Bottom*: The organization of T4 tRNA genes as determined by hybridization mapping of the various RNA species to DNA restriction fragments. The abbreviations refer to the tRNA species found in this cluster, and the relative size of the region can be judged from the size of the tRNA genes. In addition, there are RNA species of unknown functions (bands 1 and 2) encoded in this region (Velten et al., ref. 21).

3.1.2 Eucaryotes

Since it is difficult to obtain suppressor strains and transducing phages corresponding to those used in the analysis of the *E. coli* genome, the study of the organization of eucaryotic tRNA genes has proceeded primarily along biochemical lines. However, several suppressor genes have been identified and mapped in yeast, and at least one has been shown to be a structural gene for a tRNA (23). Hybridization kinetics of DNA from several organisms, but most notably yeast, *Drosophila melanogaster* and *Xenopus laevis,* indicate that there are about 350, 750, and 8000 copies of tRNA genes respectively in these organisms and that the gene organization is clearly different from that of procaryotes (24–26). In eucaryotes the tRNA gene sequences appear to have adjacent nontranscribed spacer regions up to ten times the size of the tRNA sequence itself. In situ hybridization studies in *D. melanogaster* have shown that tRNA genes are not localized on one chromosome or in one area of several chromosomes (25). The wide distribution of tRNA genes and the possible function of the spacer regions between them, so different from procaryotic organization, pose questions to be answered. Presumably the gene structure is closely related to the transcriptional control mechanisms operating in higher organisms.

3.1.3 Size of tRNA Gene Transcripts

Promoters and terminators   The steps and mechanisms in the control of transcription of DNA and the processing of the RNA transcripts into functional RNA molecules are currently under extensive investigation (27, 28). A comparatively well-understood system at present is the biosynthesis of tRNA, encompassing the transcription of tRNA genes and the regulation and processing of primary transcripts.

Although we know that in some cases tRNA genes are clustered together, in both procaryotes and eucaryotes, we cannot predict whether these tRNAs are transcribed singly or consecutively as a single transcript or monocistronic RNAs since the size of the transcription product is governed by the location of promoter sites in the DNA (sites to which RNA polymerase can bind strongly and initiate transcription) and transcription termination sites. Hybridization studies with mature tRNAs may reveal that a eucaryotic tRNA gene consists of a "structural" region and a large spacer region, but no prediction can be made about the size of the original gene transcript. Nevertheless the characteristics of the gene transcripts (precursor molecules) identified so far do seem compatible with a picture of an economical biological system. Clusters of procaryotic tRNA genes appear to be transcribed as polycistronic RNAs and the transcripts of both procaryotic and eucaryotic genes not having other contiguous (within 100 nucleotides or so) tRNA sequences are usually not much longer than the individual, mature tRNA sequence (29–31). Since the data are obtained from transcripts as they are extracted from cells, these conclusions must be viewed with reservation especially in relation to the sites of transcription termination in vivo.

Pulse-labeling of tRNA precursors   Transfer RNA precursor molecules were originally identified in eucaryotic cells (31). It was shown that pulse-labeled species of RNA, with sizes estimated by mobility in polyacrylamide gels as about 4.5S (90–115 nucleotides), could be converted in vivo into 4S molecules, and on conversion these RNA molecules acquired more of the nucleotide modifications characteristic of mature tRNA. Further, the 4.5S species could be converted to 4S molecules and simultaneously partially modified by crude cell extracts in vitro. It has also been shown in various organisms that these 4.5S molecules can hybridize to DNA enriched for tRNA genes. The 5'-termini of these 4.5S RNAs, when made in vitro in isolated nuclei, contain pppPu, the nucleotide tetraphosphate found at the site of initiation of transcription (32–34). Thus excellent circumstantial evidence exists showing that these molecules are tRNA precursor molecules. Although the presence at the 5′-termini of pppPu indicates a real initiation site for transcription, no information is available concerning the natural sites of termination of transcription. The isolated precursor molecules are heterogeneous in size with respect to the numbers of extra nucleotides on both sides of the mature tRNA sequence (31, 35). Although some processing has probably occurred in vivo before extraction of these primary transcripts, it is very likely that the natural promoter and terminator sites are not far from the ends of the tRNA sequences. This picture is consistent with the known organization of eucaryotic tRNA genes and indicates that the spacer sequences are indeed spacers and are not transcribed.

The use of mutants that retard the posttranscriptional processing of the gene transcripts has allowed the positive identification of tRNA precursors in procaryotes (29, 30). Mutations, in either the processing enzymes (36, 37) or the substrates themselves (38–43), are indispensable because $E.$ $coli$ precursor species are extremely labile. Precursors of many sizes have been isolated; some have pppPu at their 5′-termini. The average length of an $E.$ $coli$ tRNA cistron, that is, a single mature tRNA plus extra transcribed sequences, appears to be about 100–125 nucleotides. Thus a polymeric precursor containing five tRNA sequences may be 600 or more nucleotides long.

Results from transcription in vitro of a Φ80 transducing phage DNA that carries two tRNA^{Tyr} genes (Φ80su$_3^+$su$^-$) indicate that posttranscriptional processing occurs extremely quickly at the 3′-ends of tRNA sequences and in long intercistronic regions. The gene transcript obtained from this phage DNA in vitro is about 600 nucleotides long (44–46). The intercistronic region is about 120 nucleotides long. Attempts to isolate this long transcript in vivo have failed in spite of genetic and physiological manipulations (47). Only one monocistronic precursor can be isolated; it is about 130 nucleotides long with very few extra nucleotides at the 3′-terminus.

Chemical and enzymatic studies of the tRNA^{Tyr} have revealed that the promoter regions of the genes for tRNA$_1^{Tyr}$ and tRNA$_2^{Tyr}$ are identical (45, 48, 49). This has been achieved in experiments involving hybridization of a chemically synthesized DNA fragment of the tRNA^{Tyr} gene to the complementary DNA strand of a transducing phage carrying either to the tRNA$_1^{Tyr}$ or the tRNA$_2^{Tyr}$ gene. DNA polymerase is then used to elongate the fragment. The nucleotides adjacent to the hybrid double-stranded fragment can be identified by suitable labeling regimes during the polymerization reaction. The nucleotide sequence adjacent to the 5′-termini of both genes shows a twofold rotational symmetry which may be characteristic of strong RNA polymerase binding sites. No other tRNA gene promoter region has been characterized in such detail, but it will be interesting to compare promoter sequences for the genes for other species of tRNA when these become available. The nature of the extra nucleotide sequences found in tRNA precursor molecules and the ways these relate to the regulation of synthesis of these molecules are the subject of continued investigation.

Precursor sequences   Although tRNA precursor molecules from eucaryotic cells have half-lives of about one hour and can be isolated readily, the nucleotide sequence of an individual species has not yet been determined. The discussion of nucleotide sequences is thus restricted to those of precursors isolated from *E. coli*. These precursors (aside from those coded for by T4) have been isolated from strains defective either in having enzymes that cleave the transcripts to make molecules of mature tRNA size or in having mutations in the substrates themselves that affect the affinity of the processing enzymes for these molecules. If there is a mutation in one of the enzymes required for processing all tRNA precursors, an accumulation of precursors to all cell tRNAs is observed. An illustration of the use of such mutants in the isolation of the $tRNA_1^{Tyr}$ precursor in particular is shown in figure 3.2. It is also possible to isolate individual precursor species by affinity chromatography (refs. 50, 51; see legend to figure 3.3). In mutants in which the ability of enzymes to interact with individual substrates is affected, the accumulation of only those individual species is observed. Some T4 precursors have a half-life of several minutes and can be isolated without special ancillary techniques (52–55). As soon as radiochemically pure species of tRNA precursors were obtained, nucleotide sequence analysis became feasible.

The nucleotide sequences of several precursor species are shown in figures 3.3 and 3.4. In addition to these complete sequences, partial information is available concerning several other species (58, 59). Aside from the $tRNA_1^{Tyr}$ and $tRNA_2^{Tyr}$ precursors none of the "extra" sequences are identical. Since there do not appear to be different processing enzymes for different precursors, these enzymes must recognize structural features shared by all these RNA molecules. Such common features could clearly be those possessed by the tRNA moieties. If the extra sequences do play an important but so far unrecognized role in transcription regulation, we might expect them to be conserved throughout evolution. It is therefore important to characterize at

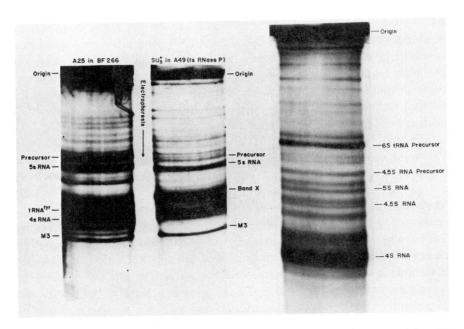

Figure 3.2 Autoradiographs of acrylamide gel separations of RNA species extracted from *E. coli* under different conditions for tRNA precursor separation. *Left panel*: RNA extracted from a host normal in RNase P function infected with Φ80 su$_3$-A25. The A25 mutation slows the action of RNase P on the tRNA precursor. *Central panel*: RNA extracted from host cell temperature-sensitive for RNase P function, infected at the restrictive temperature with Φ80 su$_3$. Note the underproduction of mature tRNA and the many bands in the upper portion of the gel. These bands represent multimeric tRNA precursors. *Right panel*: RNA extracted from cells as in the center panel except that the cells are uninfected. Thus there is no enrichment for any one particular tRNA species. Note again the many unlabeled bands throughout the gel that are mostly tRNA precursor species (Altman et al., ref. 40; and Bothwell et al., ref. 57).

least one eucaryotic precursor tRNA to a corresponding known tRNA species from *E. coli* to see whether the extra sequences are the same.

3.2 Transition to Final Product

3.2.1 Endoribonucleases

The most striking difference between precursor tRNAs and their mature tRNA products is their relative sizes. Nucleotides must be removed from both ends of most precursor molecules to yield the mature species. Removal of the 5′-proximal nucleotides to generate the usual 5′-phosphate adjacent to the beginning of the mature tRNA sequence is accomplished by the endoribonuclease RNase P (36, 37, 39, 56). This enzyme, which cleaves phosphodiester bonds to yield 3'-OH and 5'-phosphate groups, always cleaves any particular precursor at exactly the right place to generate the correct 5'-terminus of the mature tRNA. Since the primary sequences adjacent to the cleavage site are different in all precursor species, the enzyme must recognize some aspect of precursor tertiary structure (57).

The endoribonuclease RNase P was first discovered through the observation that incubation of the precursor to tRNA$_1^{Tyr}$ with a crude extract of E. coli resulted in the removal of the 41 extra nucleotides from the 5'-end of the mature tRNA (39). This cleavage reaction was subsequently used as the assay for the purification of RNase P. It appears that RNase P is the enzyme responsible for generating the correct 5'-terminus of every tRNA whose precursor, monocistronic or polycistronic, has been investigated. The extra sequences at the 3'-end of the mature tRNA are removed by 3'- to 5'-exonuclease activity.

The processing of polycistronic precursors may require a second endoribonuclease that can cleave the precursor in intercistronic regions (58, 59). The identification of such an enzymatic activity on polycistronic precursors has taken advantage of the existence of temperature-sensitive mutants of E. coli which appear to be defective in RNase P function at high temperatures (36, 37). These mutants were isolated by the use of suppressor phenotypes that could not be expressed at high temperatures in the mutant cells. The characterization of the mutant cells has shown accumulation of at least 35 precursor tRNA species at high temperatures, indicating that only one enzyme, RNase P, is needed for the processing of the 5'-termini of most (and probably all) tRNAs in E. coli. At high temperatures polycistronic precursors accumulate in these mutant cells. When these large precursors are then exposed to extracts of the mutant cells, previously incubated at high temperature to destroy the temperature-sensitive RNase P function, many of the polycistronic precursors are cleaved into smaller pieces containing one, two, or more (but always fewer than in the original substrate) tRNA sequences (58, 59). However, none of the products of this reaction have the 5'-terminus of a mature tRNA. Thus in extracts in E. coli in which RNase P has been inactivated, another endoribonuclease is present that can cleave polycistronic precursors in the region between tRNA sequences. This enzymatic activity, identified independently by Sakano and Shimura (58) and by Schedl et al. (59), has been named RNase O or P$_2$. Whether these two differently named activities are indeed the same enzyme

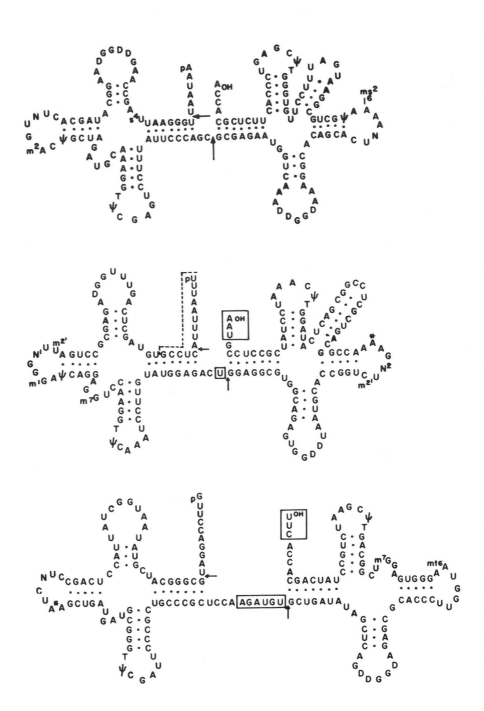

or whether they differ from another *E. coli* endoribonuclease, RNase III (60, 61), has not been shown. RNase III is known to be involved in the processing of rRNA cistrons, presumably through the attack of double-stranded RNA structures. However, under certain ionic conditions this enzyme can cleave single-stranded regions in RNA (61); and since it shares some properties with those attributed to the putative RNases P_2 and O, they may all be the same activity, but this remains to be proved. To rigorously specify whether intercistronic cleavage of polycistronic tRNA precursors is required for their biosynthesis, it will be necessary to examine mutants defective in the presumed RNase P_2 (O, III) function. In any case it is instructive to examine the nature of the cleavage of large precursors, by this activity and to consider how it might play a role in tRNA biosynthesis.

We may first ask why large precursors persist at high temperature in cells with a thermosensitive RNase P function if the second endoribonuclease is present and functioning. The answer appears to lie in the conformational specificity of RNase P_2 (O, III).

It has been shown that the large precursor tRNAs are not uniformly susceptible to RNase P or P_2 action when incubated in vitro with partially purified samples of these enzymes. To be processed completely some polycistronic precursors may require cleavage by the different RNases in a defined order, for instance first by RNase P_2 into mono- and dicistronic precursors (58, 59). A variety of precursors are isolated from the thermosensitive RNase P strains, and some of the precursors do not contain pppPu at the 5'-terminus but may be, for example, monocistronic; thus it is assumed that these have already been cleaved by RNase P_2 and accumulate only because RNase P action is blocked. Ordered cleavage events in vitro have been demonstrated under conditions that do not necessarily duplicate those found in vivo, and appropriate control experiments regarding the suscepti-

Figure 3.3 Nucleotide sequences of dimeric tRNA precursor molecules. The upper sequence is that of the Gln-Leu precursor encoded by T4. The arrows pointing toward the sequence represent RNase P cleavage sites. The middle sequence is that of the T4 Pro-Ser precursor and the lower one that of the Gly-Thr precursor of *E. coli*. Nucleotides at the 3'-termini not found in the mature tRNAs are boxed. The arrows indicate the RNase P cleavage sites. In the T4 Pro-Ser precursor, the RNase T_1-generated fragment that is resistant to RNase P cleavage is outlined by the dotted line (Guthrie, ref. 55; Seidman et al., ref. 53; and Chang and Carbon, ref. 9). The Gly-Thr precursor was isolated with aid of an *E. coli* mutant thermosensitive for RNase P function and the transposition of the tRNA genes to a transducing phage. The T4 precursors can be isolated without any special genetic manipulation, but growth of the bacteriophage in an *E. coli* host thermosensitive for RNase P function does enhance the yield of the precursor.

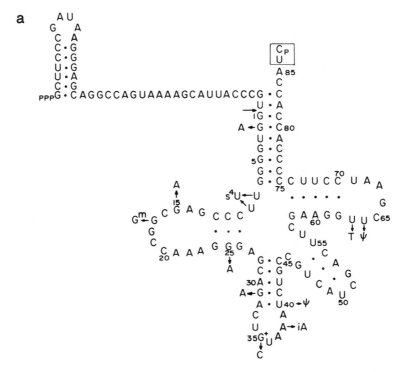

Figure 3.4 Nucleotide sequences of monomeric tRNA precursor molecules of *E. coli*. (a) Nucleotide sequence of the tRNA$_1^{Tyr}$ precursor molecule showing modifications found in the mature molecule, the position and nature of various mutations decreasing tRNA yield (see table 3.1) and the anticodon su$_3^+$ mutation (C35) that does not alter tRNA yield. High yield of precursor to tRNA^{Tyr} su$_3^+$ is possible only if an *E. coli* host thermosensitive for RNase P function is used as a host for Φ80 su$_3^+$. The arrow pointing in is at the site of RNase P cleavage and extra 3'-terminal nucleotides are boxed (Altman and Smith, ref. 39). The 5' extra nucleotides are identical in the precursor to tRNA$_2^{Tyr}$. (b) Nucleotide sequences of three monomeric tRNA precursors isolated by affinity chromatographic methods from an RNase P thermosensitive host. These methods involve chromatography of the RNA-containing cell extract either on a tRNA-cellulose column, in which the immobilized tRNAs have an anticodon complementary to that of tRNA^{Phe} or tRNA$_2^{Glu}$, or on a borate-substituted DEAE cellulose column to which tRNAs containing the modified base Q absorb selectively. The arrows point to the sites of RNase P cleavage. *Left*: Precursor to tRNA^{Phe} (50). *Right*: Precursor to tRNA$_2^{Glu}$ (50). *Below*: Precursors to tRNA^{Asp}; that is, four species were isolated (51). No extra nucleotides were found at the 3'-termini of these molecules. The mature tRNA sequences are drawn with modifications intact.

b

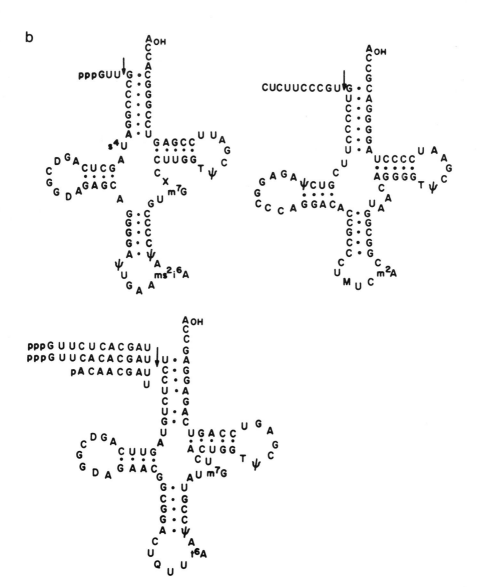

bility of substrates have not always been done. An illustration of the kinds of product generated by either RNase P or RNase P_2 (O, III) from a polycistronic precursor is shown in figure 3.5. A combination of action of the enzymes, in different order for different precursors, may be needed to generate the monocistronic products with correct 5'-termini. Presumably exonuclease action at the 3'-termini of these products would complete the trimming. Such a scheme seems plausible if RNase P_2 (O, III) is required in the biosynthesis of precursors. It is equally likely that RNase P and 3'- to 5'-exonuclease action are sufficient to complete tailoring events.

RNAase P Monomers

RNAase P_2 Monomers

Figure 3.5 Schematic plan for steps in the processing of large *E. coli* tRNA precursors. The top part of the diagram shows the hypothetical action of RNase P on a pentameric precursor molecule. In this case it is suggested that the first cleavage splits the large molecule into two parts, and subsequent cleavages generate all the correct 5'-termini. The resulting monomeric precursors must still have their 3'-termini trimmed. The lower part of the diagram shows the hypothetical action of RNase P_2 on the large precursor. This latter enzymatic activity cleaves in intercistronic regions generating fragments that must be further acted on by RNase P and a 3'-terminus-trimming enzyme. If RNase P_2 does indeed play a role in normal tRNA precursor maturation, one can envision processing of large precursor molecules occurring through a combination of the steps shown in the upper and lower parts of the diagram. That is, cleavages by RNase P might be preceded or followed by cleavages by RNase P_2, each cleavage facilitating the next in the sequence. Different large precursor molecules might be processed by a different combination and ordered sequences of RNase P and P_2 action (Schedl et al., ref. 59).

The specificity of RNase P is partly understood, and this enzyme is not guided by primary sequence information alone. All tRNA precursors have in common the conformation of their tRNA moieties. It has been shown by direct chemical experimentation in at least one case and by inference from the results of mutant studies in other cases that the conformation of the tRNA sequences in the precursor molecules is very similar to that of the mature tRNAs. Thus RNase P can be expected to recognize some aspect of tRNA conformation when it binds tightly to its precursor substrates. Mutants that have lost suppressor activity have been isolated for various suppressing tRNAs. In these mutants the anticodons are unchanged, but a new mutation has been identified elsewhere in the molecule that reduces the rate at which the precursor is processed by RNase P (table 3.1). (Some of these mutations alter 3'-exonuclease processing which may be a prerequisite for RNase P cleavage.) These mutations occur in virtually every part of the tRNA sequence except the aminoacyl stem (where additional complications may arise), and presumably they affect the three-dimensional conformation of the molecule. In many cases additional mutations can revert the phenotype of the mutation to allow improved interaction with RNase P (table 3.2). Not all these latter revertants can be easily understood in terms of restoring a "normal" tRNA conformation because the three-dimensional conformation of tRNA is complex and it is not easy to predict how the structure changes in three dimensions.

With certain mutant forms of tRNA$_1^{Tyr}$ the rate of cleavage of the precursor by RNase P is greatly reduced in vitro although it is not yet known whether the binding or catalytic step is affected. In no case has the site of cleavage been changed although the rate of cleavage may be much below normal. One can understand the unaltered specificity of cleavage by assuming that the enzyme always binds to a unique three-dimensional configuration provided by the tRNA moiety of the molecule. The site where the enzyme binds tightly to the RNA is probably physically distant from the catalytic site of the protein. The active site could be separated from a possible tight binding site, the junction of the two helical segments of the tRNA moiety, by a distance corresponding to the seven base pairs of the aminoacyl stem. Since we expect this distance to be invariant in all tRNAs, the enzyme because of its own architecture always makes the cleavage at the right place in all its substrates.

Recently it has been shown that RNase P can cleave other RNA molecules besides tRNA precursors. These molecules, the precursor to a stable 4.5S RNA species in *E. coli* (of unknown function) and an RNA coded for by

Table 3.1 Location in the tRNA moieties of precursor molecules of mutations that affect processing events

tRNA Precursor	Mutation	Function Affected
E. coli Tyr $_1$	G15 →A15	RNase P cleavage
	G25 →A25	RNase P cleavage
	G31 →A31	RNase P cleavage
	G31 →U31	Probably RNase P cleavage
	G46 →A46	Probably RNase P cleavage
	G1 →A1	Not known: no precursor identified
	G2 →A2	Not known: no precursor identified
	C(−4)→U(−4)[a]	Not known: enhanced tRNA production
T4 *Gln*[b]-Leu	C11→U11	Nucleotide modification; partial loss of RNase P and C-C-A repair at Gln moiety
	G40 →A40	Nucleotide modification; partial loss of RNase P and C-C-A repair at Gln moiety
	C62 →U62	Nucleotide modification; partial loss of RNase P and C-C-A repair at Gln moiety
T4 Gln-*Leu*[b]	C72 →U72	Nucleotide modification; RNase P cleavage of both moieties
T4 Pro-*Ser*[b]	G67 →A67	Nucleotide modification; 3′ → 5′ exonuclease at Ser terminus
	C70 →U70	Nucleotide modification; 3′ →5′ exonuclease at Ser terminus
	C75 →U75	Nucleotide modification; 3′ →5′ exonuclease at Ser terminus

Source: Data for this table taken from references 39–41, 53, and 55.
[a] This mutation is four nucleotides from the 5′-terminus of the mature tRNA sequence and is in the 5′ "extra" sequence.
[b] Italicized moiety contains the mutation noted.

bacteriophage Φ80 (62, 63), do not contain tRNA-like structures so the basis of enzyme-substrate recognition in these cases must be different in some respects from that for the tRNA precursors. At least one other *E. coli* ribonuclease, RNase III, has also been shown to use various kinds of nucleotide sequence and conformational information in the selection of cleavage sites in RNA.

Table 3.2 Second-site mutations of tRNA$_1^{Tyr}$ that restore normal biosynthesis

Original Mutation	Second-Site Mutation
G31 → U31	C16 → U16
	C45 → U45
	C41 → A41
G31 → A31	C16 → U16
	C45 → U45
	C41 → U41
G25 → A25	C11 → U11
G46 → A46	C54 → U54
G1 → A1	C81 → U81
G2 → A2	C80 → U80

Source: Data for this table taken from references 39, 40, and 83.
Note: In addition to the mutations shown, the anticodon of tRNA^{Tyr} molecules has the C35 mutation that makes it an amber suppressor. This mutation does not affect tRNA biosynthesis.

3.2.2 Exoribonucleases

The role of exoribonucleases has been clearly demonstrated in the trimming of extra nucleotides from the 3′-end of tRNA precursors. Such trimming was first shown using the precursor to tRNA$_1^{Tyr}$: at least four of the 3′-terminal nucleotides can be removed by an activity present in crude extracts of *E. coli* (39). Subsequently more carefully controlled experiments have demonstrated that this enzymatic activity is exonucleolytic in nature (44). Several exoribonucleases are found in *E. coli* including RNase II, a 3′- to 5′-exonuclease that might be responsible for this action (64). At least one other activity, identified by Bikoff and Gefter (65, 66) and demonstrated to be distinct from RNase II, performs the same exonucleolytic function as RNase II on precursors. There may exist in *E. coli* several exonucleases that can perform the precursor trimming in vitro, while in vivo only one of these may perform this particular function under normal conditions. Another good candidate is the so-called BN exonuclease which is absent from an *E. coli* strain called BN. In this *E. coli* strain T4 tRNA precursors cannot be processed to mature tRNAs because extra nucleotides at the 3′-termini are not removed, the normal terminating nucleotides C-C-A are not added, and the RNase P function is severely impeded (67, 68). These data have suggested a sequence of five distinct steps in the maturation of some T4 tRNAs (figure 3.6). The BN mutation, though not perhaps in the structural

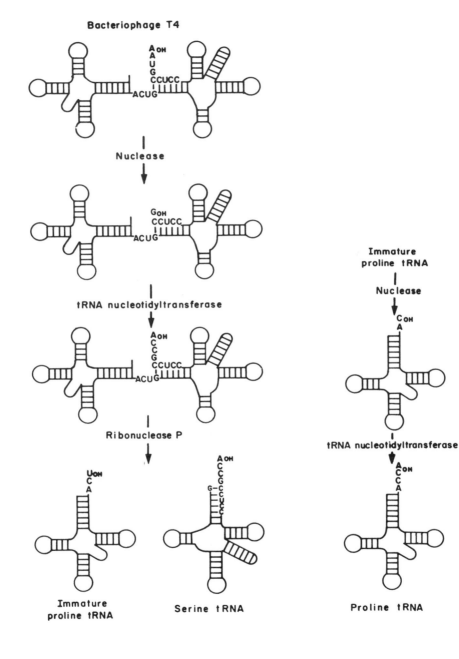

Bacteriophage T4

Nuclease

tRNA nucleotidyltransferase

Ribonuclease P

Immature
proline tRNA

Serine tRNA

Immature
proline tRNA

Nuclease

tRNA nucleotidyltransferase

Proline tRNA

gene for an exonuclease, clearly prevents the action of an exonuclease responsible for the trimming of at least some T4 precursors. It is necessary to determine whether this exonuclease is identical to those labeled as RNase II, Q, or P_3.

Consistent with the notion that precursor structure is the determining factor specifying the action of processing enzymes, several mutations located in the TΨCG-loop of T4 tRNA^{Ser} that have been isolated block processing by the 3'-exonuclease (table 3.1). There is another, less interesting role for 3'-exonucleases in tRNA biosynthesis, namely a degradative role. The 5'-proximal fragments released by RNase P cleavage of precursors must be degraded to mononucleotides. This function can be performed by 3'- to 5'-exonucleases.

3.2.3 Transfer RNA Nucleotidyltransferase (Action at the -C-A-A$_{OH}$ Terminus)

All mature tRNAs contain the 3'-terminal sequence -C-A-A$_{OH}$, yet some of the T4 tRNA gene transcripts investigated do not have this sequence intact (53, 55). An enzyme that can add the terminal Cs and A to transcripts lacking these nucleotides or to mature tRNAs from which C-C-A$_{OH}$ has been enzymatically removed in vitro has been purified from both *E. coli* and rat liver. The enzyme is designated the cca enzyme (69). Before these nucleotides are added to tRNA gene transcripts, nucleotides found in the transcript beyond the 3'-terminus of the base-paired acceptor stem must be removed by exonuclease action. In addition to the obvious need for the cca enzyme to synthesize the necessary sequence on pretrimmed tRNA gene transcripts, it has been postulated that this enzyme plays a role in the repair of mature tRNAs. Using suitably labeled nucleotides and/or tRNAs, one can demonstrate the constant turnover in vivo of the 3'-terminal adenosine of mature tRNAs. So in a sense the cca enzyme is a repair enzyme (70).

Figure 3.6 Five steps in the processing of T4-encoded Pro-Ser tRNA precursor. The first step is trimming of the 3'-terminus to remove nucleotides not found in the mature tRNA^{Ser} sequence. This terminus is then repaired by the cca enzyme (nucleotidyl transferase), and subsequently RNase P cleaves the precursor at the sites indicated. Trimming and repair of the tRNA^{Pro} 3'-terminus then occurs (Seidman et al., ref. 53). For other tRNA precursor molecules already having the sequence -C-C-A encoded in the tRNA genes, the requirement for 3'-terminus processing prior to RNase P cleavage does not seem to be absolute (7, 40, 53). Nucleotide modification is occurring at all times after gene transcription. For at least one *E. coli* large precursor (58) and certain mutant precursors (77), some modifications appear in stoichiometric amounts only after RNase P cleavage has occurred.

E. coli cells lacking or having reduced levels of the cca enzyme will not support the normal production of mature T4 tRNAs since the C-C-A$_{OH}$ sequence is added very slowly if at all to these molecules. Some T4 tRNAs accumulate, but they lack the C-C-A terminus in such a strain. Precursor molecules to which C-C-A has not been added also accumulate (71). This is evidence that cleavage of certain tRNA precursors by RNase P might require the presence of the C-C-A sequence for efficient action by the endoribonuclease. This expectation has been confirmed in vitro by Schmitt et al. (73) who compared the rate of RNase P cleavage of precursor molecules to T4 tRNA^{Ser} and T6 tRNA^{Ser}, which are identical in sequence except at their 3'-termini. In vitro the discrimination between the two species is not as rigorous as the in vivo results suggest. Experiments in vitro are usually done with the enzyme present in great excess and with concentrations of carrier tRNA that are probably lower than in vivo. These lower concentrations are important because the ability of RNase P to discriminate between precursors having or lacking the C-C-A terminus depends on the concentration of mature tRNA, which is an inhibitor of RNase P in the reaction mixture (72, 73). Nevertheless one can conclude that the presence of the C-C-A sequence in a tRNA precursor does affect the cleavage rate of RNase P, though not its specificity.

The cca enzyme may also play a role in T4 phage maturation (74). Phage grown in the *E. coli* cca mutant produce only 10 percent of the normal phage progeny when grown subsequently on normal hosts. It appears that DNA replication and/or packaging is impeded in the second-cycle infection. This result suggests that perhaps tRNA is packaged in phage heads and plays some role in directing normal phage DNA replication. In phage produced in cells with reduced levels of cca enzyme, the packaged tRNA is defective and thus supports aberrant DNA replication in the next round of replication. Of course the cca enzyme may play a more direct role in normal T4 DNA replication.

3.2.4 Nucleotide Modification

Besides the enzymatic events that determine the size of tRNA gene products and the addition of -C-C-A$_{OH}$to the 3'-terminus of gene transcripts without this sequence encoded in the DNA, maturation of tRNA requires the enzymatic modification of certain nucleotides to produce the so-called rare or minor nucleotides illustrated in figure 7.1 (75, 76).

Transfer RNA gene transcripts are made by RNA polymerase using the four ribonucleotides A, C, U, and G. The nucleotides are then modified with an intact polymer as substrate. Analysis of precursor molecules along the pathway to maturation shows that they become more highly modified as they approach the size of mature tRNA species. The nucleotide modification reactions can be performed in vitro using as substrates either chemically or naturally undermodified tRNA or partially processed precursor molecules. Enzymes that have been partially purified can carry out many of the known nucleotide modifications in vitro. The nature of the enzymatic mechanisms catalyzing many of these events is not fully known.

For our purposes it is important to consider the role nucleotide modifications might perform in governing the maturation of the tRNA gene product. The presence or absence of nucleotide modifications in tRNA precursor does not appear to affect the size maturation of the tRNA gene transcripts (77). Several modification-deficient mutants of *E. coli* have been characterized, but from none is any particular tRNA gene product missing (though the function of certain mature tRNAs may be affected). Since the full range of modification-deficient mutants (all possible generally occurring modifications) is not available, we must rely on experiments in vitro to indicate whether modification is necessary for processing. Indeed tRNA synthesized enzymatically in vitro can be processed to tRNA-sized gene products in the absence of any modifying enzymes (44). Various enzymatic modifications of specific nucleotides in tRNA can also be carried out using cell extracts and precursor made in vitro (78).

With precursor tRNAs made in vivo it has been shown that RNase P cleavage can occur normally without the full complement of T, Gm, i^6A, and s^4U in the substrate (77). Further, increasing the level of some of these modifications enzymatically did not alter RNase P cleavage rates in subsequent tests. An analysis of several different precursors has led, however, to the notion that certain modifications are carried out at a faster rate than others on precursor tRNAs (58, 77). It also appears that some modifications such as Gm appear only on the mature tRNA of certain species in vivo. That T4 precursors are usually isolated fully modified may be explained by their very long lifetime in vivo (which may be due to slow cleavage by endonucleases before the action of 3'-trimming activities). Thus even modification reactions that proceed at a relatively slow rate on precursors have time to go to completion.

3.2.5 Transfer RNA Gene Transcript Processing in Eucaryotes
Eucaryotic tRNA precursors are generally smaller than their procaryotic analogs, and no more than one tRNA sequence has been found per precur-

sor (31). The enzymology of processing appears similar to that in *E. coli* except for the function of RNase P$_2$ (35). It has long been established that cytoplasmic extracts of eucaryotic cells, specifically mammalian cells, yeast, and silkworms, contain an activity capable of reducing the molecular weight of precursor species to that of tRNA size (31). The cleavage sites have not been firmly identified, nor has the accuracy of cleavage of these enzymes on eucaryotic substrates been determined.

Enzyme-substrate specificity of tRNA processing enzymes seems to lie in structural conformation. Thus if we assume that these structures have been preserved throughout evolution, a procaryotic tRNA precursor exposed in vitro to a eucayotic cell extract should be accurately and efficiently cleaved. This approach, initially using the tRNA$_1^{Tyr}$ precursor from *E. coli*, has demonstrated the pressure of RNase P-like activities in mammalian cells (35), chick embryo thigh muscle, and silkworms (E. J. Bowman, R. Garber, and S. Altman, unpublished experiments). In every case an endoribonuclease activity capable of generating the correct 5'-terminus of the mature tRNA$_1^{Tyr}$ sequence has been identified; and in each case 3'-exonucleolytic activity has also been identified. The general characteristics of the endoribonucleases are similar to those of RNase P from *E. coli*. The experiments with the *E. coli* precursor have been instructive in identifying these activities and in demonstrating the conservation of such enzymatic mechanisms. It has also been shown that these partially purified eucaryotic enzymes, specifically from KB cells and silkworms, can process eucaryotic precursor tRNAs. It has not yet been possible to show that the RNase P-like activity in mammalian cells (KB tissue culture cells) actually created the correct 5'-termini of homologous tRNAs because no individual tRNA precursor species has been isolated in radiochemically pure form from these cells. Unsuccessful attempts to isolate a single tRNA precursor species have utilized two-dimensional gel electrophoresis, affinity chromatography on tRNA-cellulose, and borate-substituted DEAE cellulose columns (R. Koski, R. Heimer, and S. Altman, unpublished experiments). However, similar efforts with silkworm extracts have yielded some success.

When silkworms (or isolated silk glands) are labeled radioactively with a short period label during the fifth instar just before silk fibroin synthesis, precursor tRNAs can be easily isolated by phenol extraction and electrophoresis in acrylamide gels (refs. 34, 79, 80; figure 3.7). Two-dimensional gel electrophoresis yields ^{32}P-labeled precursor species according to the criterion that a high degree of homology exists between their fingerprints and those of certain mature tRNA species (34). These individual precursor species are excellent substrates for a partially purified, RNase P-like activity isolated from silkworm glands using the *E. coli* tRNA$_1^{Tyr}$ precursor as substrate in purification assays. Proof that the cleavage occurs at the 5'-termini of the mature tRNA sequences awaits the total sequence analysis of individual precursor substrates and their cleavage products. As in extracts from KB cells and chick embryo thigh muscle, a 3'-exonucleolytic activity can also be identified in silkworm extracts. The RNase P-like activities identified in vitro from eucaryotic sources have not been proved identical to the enzymes used in vivo in tRNA processing, but it seems probable that this will be the case.

3.3 Control of Gene Expression

3.3.1 Regulating the Amount of Gene Product
The previous sections have focused on the nature of tRNA gene transcripts and the ways in which they are processed by the intracellular machinery after they are made. Now we shall discuss the regulation of the amount of gene product made in the cell. Regulation can occur at two levels: through control of the amount of transcript made and through control of the amount of transcript actually processed into a usable molecule by the cell. The latter control is exemplified in a discussion of posttranscriptional processing enzymes and their substrate specificity. Transfer RNA molecules in which the secondary and/or tertiary structure has been altered by mutation are not produced in normal amounts in growing cells because the processing steps are delayed. This control can be exerted at the level of 3'-exonuclease action, C-C-A addition, or RNase P cleavage. The increase in rates of processing leads not only to the accumulation of precursor, but also to larger amounts of gene transcript being degraded during the waiting time. There appears to be constant competition for precursor molecules between enzymes that recognize these RNAs as unstable molecules prone to complete degradation and the normal processing enzymes. Normally the processing enzymes work so efficiently that the bulk of the transcribed products are made into mature products. However, any mutation that decreases the rate of correct substrate cleavage allows a precursor to have an effectively longer half-life during which it is susceptible to degradation. Such a competitive situation can explain why rates of transcription of ribosomal and tRNA genes are too fast to account for the amount of final product observed in cells (81). Any small adjustment in degradative enzyme function would alter the amount of final product produced.

a

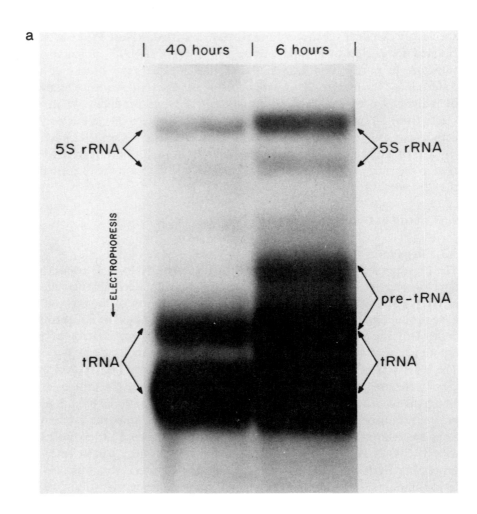

b

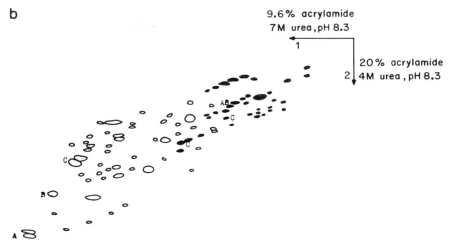

9.6% acrylamide
7M urea, pH 8.3
1

20% acrylamide
2 4M urea, pH 8.3

Figure 3.7 Identification of tRNA precursors from the silkworm *Bombyx mori*. (a) Autoradiogram of a polyacrylamide gel (10 percent acrylamide; pH 8.3) separation of ^{32}P-labeled RNA extracted from *B. mori* silkglands. The label was administered for the times indicated to animals on the third day of the fifth instar. The relative positions of precursor tRNA species, mature tRNA, and 5S RNA are indicated. (b) Composite tracing of autoradiograms of ^{32}P-labeled low molecular weight RNA species separated by two-dimensional gel electrophoresis. Separation in the first dimension was in 9.6 percent acrylamide, 7M urea, pH 8.3 and in the second dimension in 20 percent acrylamide, 4M urea, pH 8.3. The open circles represent mature tRNA species, and the filled ones represent precursor tRNA species (species appearing only in short-labeled preparations). The lettered species have been identified as precursors to certain mature tRNAs (labeled with the same letter) by direct fingerprint and nucleotide sequence analysis. AB: Two spots in autoradiograms each denoting a mixture of precursors to tRNA$_{2a}^{Ala}$ and tRNA$_{2b}^{Ala}$. The corresponding mature species are A: tRNA$_{2a}^{Ala}$ (a doublet in which the species differ only by a nucleotide modification); and B: tRNA$_{2b}^{Ala}$. C: Precursors and mature species of tRNA$_1^{Gly}$. The two precursors represent different stages of processing.

The control of transcription itself of necessity encompasses the mechanism of RNA polymerase and the accessibility of template DNA to that enzyme. In *E. coli* the rate of transcription of tRNA and rRNA genes seems to be tightly coupled to the cell's growth rate, but the coupling mechanism is not understood. The coupling mechanism may be similar to the stringent-relaxed control system, where the transcription of tRNA and rRNA genes is related to the intracellular amino acid concentration and mediated through the action of ppGpp and pppGpp (magic spots I and II). These agents could decrease the tight binding of RNA polymerase to promoter regions, but the way in which tRNA gene transcription is directly regulated is not known (27, 81).

3.3.2 Factors Affecting RNA Polymerase-Promoter Interaction

The affinity of RNA polymerase for different promoter regions may be an intrinsic property of the enzyme itself or may be determined by the state of template DNA at any particular time. For example, all other things equal, the transcription of tRNA genes for two different tRNA species might be governed by promoters of different strength so that under no circumstances could the concentration of these two tRNA species ever be identical in the cell. On the other hand, the accessibility of the promoter sites themselves can vary, depending in eucaryotes for instance on the state of the chromatin and attendant protein.

The binding of RNA polymerase to various promoter regions is also controlled by "sigma" factors, proteins that can associate reversibly with the core RNA polymerase and greatly enhance the binding of the enzyme to promoter sites rather than to random sites in *E. coli* DNA (27, 81). The mechanism of enhanced promoter binding is exploited by nature to control the temporal expression of various genes. For example, after T4 infection of *E. coli,* not only do alterations in RNA polymerase core subunits occur that enhance the affinity of the enzyme for T4 DNA, but various additional regulatory proteins are made, at different times after infection, that control the affinity of the new core enzyme for promoters of genes to be expressed at those specified times. Similar mechanisms may be operative during sporulation of *B. subtilis.*

The DNA template in eucaryotes may not be uniformly accessible to RNA polymerase due to the nonuniform architecture of the chromosome (different chromosomal proteins bound at different places). The unfolding of local chromosomal structure may be affected during various developmental events by hormone action, acetylation of histones, or the phosphorylation of certain proteins. In eucaryotes an additional control over transcription has been introduced through the diversity of the RNA polymerase enzymes themselves. Of the three different RNA polymerases only one (RNA polymerase III) seems to be used for the transcription of tRNA genes (82). The controls exerted over RNA polymerase action in *E. coli* could also in principle apply to the eucaryotic RNA polymerase III. Fine tuning of the amounts of tRNAs needed for specific protein synthesis, however, may be more easily governed by posttranscriptional alterations of specific tRNA through nucleotide modification of specific tRNA and tRNA precursor degradation.

3.4 Conclusion

Examination of tRNA biosynthesis in organisms as diverse as *E. coli* and man reveals a familiar theme. The basic enzymological processes, once solved in the simple organisms, seem to be preserved throughout evolution. By this we mean not only the means of making or breaking phosphodiester bonds or modifying nucleotides, but also the stereospecificity of these interactions in terms of enzyme substrate recognitions. Thus tRNA structures are essentially the same in *E. coli* and in human cells; similarly the specificity of an enzyme such as RNase P remains unchanged in its recognition requirement for a tRNA-like conformation in the two organisms. What has changed or in a sense become more complex during evolution is the nature of control of tRNA gene expression. The accessibility of RNA polymerase to eucayotic tRNA genes, their transcription as monocistronic units, their more complex modifications and infinite variations in isoacceptor and nucleotide modification levels during developmental processes appear to be responses to the need for intricate control mechanisms for modulating the expression of all genes in higher organisms. The conclusion from tRNA biosynthesis studies is similar to that from studies of DNA and protein synthesis: the basic mechanisms in procaryotes and eucaryotes are similar; the controls are more complex.

Acknowledgments

I wish to acknowledge the many contributions to work in my laboratory and reported in this chapter made by A. Bothwell, E. Bowman, R. Garber, R. Koski, and B. Stark. Research in my laboratory was supported by a grant GM-19422 from the United States Public Health Service.

References

1. Smith, J. D. (1972). Ann. Rev. Genet. *6*, 235–256.

2. Hartman, P. E., and Roth, J. R. (1973). Adv. in Genetics, E. W. Caspari, ed. (New York: Academic Press), pp. 1–105.

3. Berg, P. (1973). Harvey Lectures (New York: Academic Press).

4. Bachmann, B. J., Low, K. B., and Taylor, A. L. (1976). Bacteriol. Revs. *40*, 116–167.

5. Brenner, D. J., Fournier, M. Y., and Doctor, B. P. (1970). Nature *227*, 448–451.

6. Schedl, P., Primakoff, P., and Roberts, J. (1974). Brookhaven Symp. Biol. *26*, 53–76.

7. Schedl, P. (1975). Ph.D. dissertation, Stanford University, Stanford, Calif.

8. Ikemura, T., Shimura, Y., Sakano, H., and Ozeki, H. (1975). J. Mol. Biol. *96*, 69–86.

9. Chang, S., and Carbon, J. (1975). J. Biol. Chem. *250*, 5542–5555.

10. Ilgen, C., Kirk, L. L., and Carbon, J. (1976). J. Biol. Chem. *251*, 922–929.

11. Lund, E., Dahlberg, J. E., Lindahl, L., Jaskunas, R., Dennis, P. P., and Nomura, M. (1976). Cell *7*, 165–177.

12. Lund, E., and Dahlberg, J. E. (1977). Cell *11*, 247–262.

13. Daniel, V., Sarid, S., and Littauer, U. Z. (1970). Science *167*, 1682–1688.

14. Scherberg, N. H., and Weiss, S. B. (1970). Proc. Nat. Acad. Sci. USA *67*, 1164–1171.

15. Wilson, J. H., Kim, J. S., and Abelson, J. (1972). J. Mol. Biol. *71*, 547–556.

16. McClain, W. H., Guthrie, C., and Barrell, B. G. (1972). Proc. Nat. Acad. Sci. USA *69*, 3703–3707.

17. Wilson, J. H., and Abelson, J. N. (1972). J. Mol. Biol. *69*, 57–73.

18. Chen, M. J., Locker, J., and Weiss, S. B. (1976). J. Biol. Chem. *251*, 536–547.

19. Hunt, C., Hwang, L. T., and Weiss, S. B. (1976). J. Virol. *20*, 63–69.

20. Velten, J., Fukuda, D., and Abelson, J. (1976). Gene *1*, 93–106.

21. Kaplan, D. A., and Nierlich, D. P. (1975). J. Biol. Chem. *250*, 934–938.

22. Gilmore, R. A., and Mortimer, R. K. (1966). J. Mol. Biol. *20*, 307–311.

23. Piper, P. W., Wasserstein, M., Engback, F., Kaltoft, K., Celis, J. E., Zeuthen, J., Liebman, S., and Sherman, F. (1976). Nature *262*, 757–761.

24. Feldmann, H. (1976). Nucl. Acids Res. *3*, 2379-2386.

25. Grigliatti, T. A., White, B. N., Tener, G. M., Kaufman, T. C., Holden, J. J., and Suzuki, D. T. (1973). Cold Spring Harbor Symp. Quant. Biol. *38*, pp. 461–474.

26. Clarkson, S. G., and Birnstiel, M. L. (1973). Cold Spring Harbor Symp. Quant. Biol. *38*, pp. 451–459.

27. Travers, A. (1976). Nature *263*, 641–646.

28. Chamberlin, M., and Losick, R., eds. (1976). RNA Polymerases (Cold Spring Harbor, N.Y.: Cold Spring Harbor Press).

29. Altman, S. (1975). Cell *4*, 21–29.

30. Schaefer, K., and Söll, D. (1974). Biochemie *56*, 795–804.

31. Burdon, R. H. (1971). Prog. Nucl. Acid Res. Mol. Biol. *11*, 33–79.

32. Marzluff, W. F., Jr., Murphy, E. C., and Huang, R. C. C. (1974). Biochemistry *13*, 3689–3696.

33. Marzluff, W. F., Jr., and Huang, R. C. C. (1975). Proc. Nat. Acad. Sci. USA *72*, 1082–1086.

34. Garber, R. L., Siddiqui, M. A. Q., and Altman, S. (1978). Proc. Nat. Acad. Sci. USA, in press.

35. Koski, R. A., Bothwell, A. L. M., and Altman, S. (1976). Cell *9*, 101–116.

36. Schedl, P., and Primakoff, P. (1973). Proc. Nat. Acad. Sci. USA *70*, 2091–2095.

37. Sakano, H., Yamada, S., Ikemura, T., Shimura, Y., and Ozeki, H. (1974). Nucl. Acid Res. *1*, 355–371.

38. Altman, S. (1971). Nature New Biol. *229*, 19–21.

39. Altman, S., and Smith, J. D. (1971). Nature New Biol. *233*, 35–39.

40. Altman, S., Bothwell, A. L. M., and Stark, B. C. (1974). Brookhaven Symp. Biol. *26*, 12–25.

41. Smith, J. D. (1974). Brookhaven Symp. Biol. *26*, 1–11.

42. McClain, W. H., Barrell, B. G., and Seidman, J. G. (1975). J. Mol. Biol. *99*, 717–732.

43. McClain, W. H., and Seidman, J. G. (1975). Nature *257*, 106–110.

44. Daniel, V., Beckmann, J. S., Grimberg, J. I., and Zeevi, M. (1976). In Control of Ribosome Synthesis, N. Kjeldgaard and O. Maaloe, eds. (Copenhagen: Munksgaard), pp. 268–278.

45. Sekiya, T., Contreras, R., Kupper, H., Landy, A. and Khorana, H. G. (1976). J. Biol. Chem. *251*, 5124–5140.

46. Kupper, H., Contreras, R., Landy, A., and Khorana, H. G. (1975). Proc. Nat. Acad. Sci. USA *72*, 4754–4758.

47. Ghysen, A., and Celis, J. C. (1974). Nature 29, 418–421.

48. Sekiya, T., and Khorana, H. G. (1974). Proc. Nat. Acad. Sci. USA 71, 2978–2982.

49. Loewen, P. C., Sekiya, T., and Khorana, H. G. (1974). J. Biol. Chem. 249, 217–226.

50. Vogeli, G., Grosjean, H., and Söll, D. (1975). Proc. Nat. Acad. Sci. USA 72, 4790–4794.

51. Vogeli, G., Stewart, T. S., McCutchan, T., and Söll, D. (1977). J. Biol. Chem. 253, 2311–2318.

52. Guthrie, C., Seidman, J. G., Altman, S., Barrell, B. G., Smith, J. D., and McClain, W. H. (1973). Nature New Biol. 246, 6–11.

53. Seidman, J. G., Barrell, B. G., and McClain, W. H. (1975). J. Mol. Biol. 99, 733–760.

54. Barrell, B. G., Seidman, J. G., Guthrie, C., and McClain, W. H. (1974). Proc. Nat. Acad. Sci. USA 71, 413–416.

55. Guthrie, C. (1975). J. Mol. Biol. 95, 529–547.

56. Robertson, H. D., Altman, S., and Smith, J. D. (1972). J. Biol. Chem. 247, 5243–5251.

57. Bothwell, A. L. M., Stark, B. C., and Altman, S. (1976). Proc. Nat. Acad. Sci. USA 73, 1912–1916.

58. Sakano, H., and Shimura, Y. (1975). Proc. Nat. Acad. Sci. USA 72, 3369–3373.

59. Schedl, P., Roberts, J., and Primakoff, P. (1976). Cell 8, 581–599.

60. Robertson, H. D., Webster, R. E., and Zinder, N. D. (1968). J. Biol. Chem. 243, 82–91.

61. Dunn, J. J. (1976). J. Biol. Chem. 251, 3807–3814.

62. Bothwell, A. L. M., Garber, R. L., and Altman, S. (1976). J. Biol. Chem. 251, 7709–7716.

63. Bothwell, A. L. M., and Altman, S. (1975). J. Biol. Chem. 250, 1460–1463.

64. Singer, M. F., and Tolbert, G. (1965). Biochemistry 4, 1319–1330.

65. Bikoff, E. G., and Gefter, M. L. (1975). J. Biol. Chem. 250, 6240–6247.

66. Bikoff, E. K., LaRue, B. F., and Gefter, M. L. (1975). J. Biol. Chem. 250, 6248–6255.

67. Seidman, J. G., Schmidt, F. J., Foss, K., and McClain, W. H. (1975). Cell 5, 389–400.

68. Seidman, J. G., and McClain, W. H. (1975). Proc. Nat. Acad. Sci. 72, 1491–1495.

69. Deutscher, M. P. (1973). Prog. Nucl. Acid Res. Mol. Biol. *13*, 51–92.

70. Deutscher, M. P., and Evans, J. A. (1977). J. Mol. Biol. *109*, 593–597.

71. Deutscher, M. P., Foulds, J., and McClain, W. H. (1974). J. Biol. Chem. *249*, 6696–6699.

72. Stark, B. C. (1977). Ph.D. dissertation, Yale University, New Haven, Connecticut.

73. Schmidt, F. J., Seidman, J. G., and Bock, R. M. (1976). J. Biol. Chem. *251*, 2440–2445.

74. Morse, J. W., and Deutscher, M. P. (1976). Biochem. Biophys. Res. Comm. *73*, 953–959.

75. Söll, D. (1971). Science *173*, 293–299.

76. Nishimura, S. (1972). Prog. Nucl. Acid Res. Mol. Biol. *12*, 49–85.

77. Schaefer, K., Altman, S., and Söll, D. (1973). Proc. Nat. Acad. Sci. USA *70*, 3626–3630.

78. Ciampi, M. S., Arena, F., Cortese, R., and Daniel, V. (1977). FEBS Letter *77*, 75–82.

79. Chen, G. S., and Siddiqui, M. A. Q. (1973). Proc. Nat. Acad. Sci. USA *70*, 2610–2613.

80. Tsutsumi, K., Majima, R., and Shimura, K. (1974). Jap. J. Biochem. *76*, 1143–1145.

81. Kjeldgaard, N., and Maaloe, O., eds. (1976). Control of Ribosome Synthesis (Copenhagen: Munksgaard).

82. Roeder, R. G. (1976). In RNA Polymerases, M. Chamberlain and R. Losick, eds. (Cold Spring Harbor, N.Y.: Cold Spring Harbor Press) pp. 285–329.

83. Smith, J. D. (1976). Prog. Nucl. Acid Res. Mol. Biol. *16*, 25–73.

4 TRANSFER RNA FUNCTION IN PROTEIN SYNTHESIS: RIBOSOME (A-SITES AND P-SITES) AND mRNA INTERACTIONS

O. Pongs

4.1 Introduction

The linear nucleotide sequence of mRNA is translated into the amino acid sequence of protein on the ribosome (for review, see refs. 1–3). This process is mediated by aminoacylated tRNA species, which interact with the 50–60 proteins and the three ribonucleic acid species that make up the ribosome (4–6) and are required for the complex process of protein synthesis.

In the course of translation the codons in mRNA are read sequentially by aminoacyl-tRNAs (aa-tRNA) (7, 8). Messenger RNAs are read in the 5'- to 3'-direction, and proteins are synthesized from the amino terminal toward the carboxyterminal amino acid (9). The growing polypeptide chain is also attached to tRNA (peptidyl-tRNA) (10). Each mRNA contains a nucleotide sequence that specifies the initiation of protein synthesis (11), followed by the sequence that is actually translated and then the sequence that specifies the termination of translation. In addition mRNAs may contain so-called noncoding sequences, whose function is still unknown.

Each initiation sequence contains at its 5'-end an initiator codon, in general either A-U-G or G-U-G. This specifies N-formyl methionine as the N-terminal amino acid of the growing polypeptide chain (12). Thus all proteins originally contain the same N-terminal amino acid (13), and there is a specialized initiator tRNA, $tRNA_f^{Met}$, that recognizes the initiation codon (12). Because the genetic code is degenerate, in most cases more than one nucleotide triplet codes for a particular amino acid. One might expect the occurence of possible triplet sequences in a given mRNA sequence to be random. The recent elucidation of several mRNA sequences clearly demonstrates that this is not the case. Of a given group of triplet sequences specifying one amino acid, very often one particular codon is predominantly used. The choice of codons is apparently characteristic for the given mRNA and varies from one mRNA to another (14). This means that the choice of codons is a characteristic of the individual gene and not of the organism as a whole. The degeneracy of the genetic code has been exploited in the sequences of individual mRNAs and is apparently not a useless biological peculiarity.

The multiplicity of codons for a given amino acid is accompanied by a multiplicity of tRNAs; but they do not correlate with each other in a particular fashion (15). Transfer RNAs of different primary structure, which are charged with the same amino acid, are designated isoaccepting tRNAs. However, there is no trivial relationship between the number of isoaccepting tRNAs and the degree of codon degeneracy. In some cases isoaccepting tRNAs read the same codon, and in other cases several codons are read by the same aminoacyl-tRNA.

Each termination sequence of mRNA begins with one of the three possible nonsense or terminator codons, U-G-A, U-A-A, or U-A-G (16). Termination sequences are usually recognized by the ribosome itself and are not read by a tRNA (17). This means that tRNA molecules are involved in the initiation and elongation steps of the translation but do not function in the termination of protein synthesis. Mutational events can generate a termination signal inside the normal mRNA reading frame, thus causing a premature termination of protein synthesis. As a kind of repair mechanism, special tRNA molecules have evolved that can read and thereby suppress premature termination signals (18). These tRNAs are called nonsense suppressor tRNAs.

Thus tRNA has three functions in the translation of mRNA into protein on the ribosome: the initiation of peptide synthesis, the elongation of polypeptide chains, and the attachment of the growing polypeptide chain to the ribosome. All functions are mediated by the ribosome at the sites where tRNA molecules are bound, the fidelity of codon-anticodon, mRNA-tRNA interactions, is controlled, peptide bonds are formed, and the mRNA is advanced sequentially (three nucleotides at a time). The most thoroughly investigated system is that of the *E. coli* ribosome.

The *E. coli* ribosome is designated by its sedimentation coefficient as a 70S particle (MW 2.7×10^6) and consists of two dissociable subunits, denoted 30S (MW 0.9×10^6) and 50S (1.8×10^6) (see review in ref. 2). The 30S subunit consists of 21 proteins, designated S1–S21 (S for small subunit) and 16S RNA (MW 0.55×10^6). The sequence of 16S RNA is essentially known. The 50S subunit consists of 34 proteins, designated L1–L34 (L for large subunit), and two ribonucleic acids, 5S RNA and 23S RNA. The complete sequence of 5S RNA has been determined (19). Protein synthesis is conveniently divided into three primary processes: chain initiation, chain elongation, and chain termination. The overall initiation process is described by

$$30S + 50S + mRNA + fMet\text{-}tRNA_f^{Met} + GTP$$
$$\xrightarrow{IF1, IF2, IF3} 70S\cdot mRNA\cdot fMet\text{-}tRNA_f^{Met} + GDP + Pi. \qquad (4.1)$$

Formation of the initiation complex requires the catalytic participation of the three initiation factors IF1, IF2, and IF3 and the hydrolysis of GTP.

The overall elongation process is described by

$$70S\cdot pep_n\text{-}tRNA\cdot mRNA + aa\cdot tRNA + 2GTP$$
$$\xrightarrow{EF\text{-}Tu, EF\text{-}Ts, EF\text{-}G} 70S\cdot pep_{n+1}\text{-}tRNA\cdot mRNA + tRNA + 2GDP + 2Pi, \qquad (4.2)$$

where pep_n is the growing peptide chain consisting of n amino acid residues. In the first elongation step n would be one and pep_n would be formylmethionine. Elongation comprises three separate and consecutive steps, which are catalyzed by the three elongation factors EF-Tu, EF-Ts, and EF-G with concomitant hydrolysis of two equivalents of GTP. The sequence of codon-directed aa-tRNA binding, peptide bond formation, and translocation is outlined in scheme 4.1.

The function of EF-Tu is to provide additional points of interaction for the codon-directed aa-tRNA binding to the ribosome by forming a complex with GTP, aa-tRNA, and the ribosome (20). GTP is then hydrolyzed and EF-Tu is released from the ribosome in the form of EF-Tu·GDP.

Since EF-Tu binds only to charged aminoacyl-tRNAs, its specificity assures that only charged aminoacyl-tRNAs, as long as they are available, are bound to the ribosome in response to the appropriate codon. Peptide bond formation occurs apparently without any requirement for additional factors or GTP in the so-called peptidyltransferase center of the ribosome (21). After peptide bond formation, newly formed pep_{n+1}-tRNA is "translocated" in order to allow another aa-tRNA to bind in response to the next codon. Translocation involves release of the tRNA that carried the peptidyl moiety before peptide bond formation, movement of the tRNA that carries the pep_{n+1} moiety after peptide bond formation, and movement of the

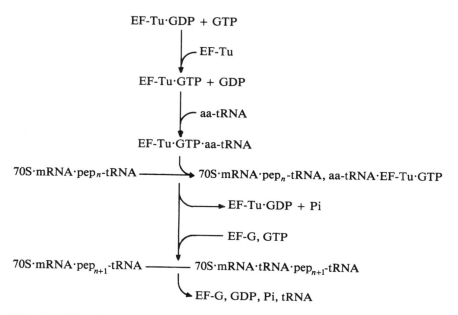

EF-Tu·GDP + GTP

EF-Tu

EF-Tu·GTP + GDP

aa-tRNA

EF-Tu·GTP·aa-tRNA

70S·mRNA·pep$_n$-tRNA ⟶ 70S·mRNA·pep$_n$-tRNA, aa-tRNA·EF-Tu·GTP

EF-Tu·GDP + Pi

EF-G, GTP

70S·mRNA·pep$_{n+1}$-tRNA ⟶ 70S·mRNA·tRNA·pep$_{n+1}$-tRNA

EF-G, GDP, Pi, tRNA

Scheme 4.1 Elongation sequence

mRNA in accordance with the reading frame of three nucleotides per codon. This complicated operation requires the catalytic participation of EF-G and the hydrolysis of two equivalents of GTP.

The termination process is described by

70S·mRNA·protein·tRNA
$$\xrightarrow{\text{RF-1 or RF-2}} \text{protein} + 70S + mRNA + tRNA, \tag{4.3}$$

where the ester linkage between protein and tRNA is cleaved and the remaining 70S·mRNA·tRNA complex dissociated. Termination requires a termination factor, which is either RF-1, in response to the nonsense codons U-A-A or U-A-G, or RF-2, in response to the nonsense codons U-A-A or U-G-A (17).

4.2 Peptide Chain Initiation

Peptide chain initiation involves the assembly of the following components on the ribosome: the initiation signal of mRNA, initiator tRNA, initiation factors, and GTP (scheme 4.1). The sequence of events leading to the formation of a functional initiation complex is known in broad outline.

The initiation signal of mRNA very often consists of a sequence of 3 to 7 nucleotides located about 10 bases to the 5'-side of the initiator codon (11) and, of course, the initiator codon itself. The former sequences apparently base-pair with a complementary sequence near the 3'-terminus of 16S rRNA (see figure 4.1; refs. 11, 22). The variability in the length of these sequences as well as their distance from the initiator-codon is probably important for the formation and the stability of the mRNA-ribosome interaction. This base-pairing mechanism is probably used to achieve fine control over the relative efficiency of initiation at different cistrons of polycistronic mRNA. Base pairing between mRNA and the 3'-end of 16S rRNA might be an important component of initiation site selection and mRNA binding to ribosome (11). However, this rRNA-mRNA complementarity is not a necessary prerequisite for mRNA-ribosome interaction, since mRNA containing no additional sequence on the 5'-side of the initiation codon A-U-G has been isolated (23). Moreover, the A-U-G triplet itself can direct initiation of a polypeptide chain. The interaction of initiator tRNA with the initiation codon is obviously the major event, which leads to the initiation of protein synthesis. It has even been proposed that initiator tRNA binds to ribosomes before mRNA in order to help in the recognition of the initiation signal of mRNA (24, 25).

Initiator tRNAs are highly specialized with respect to their function in peptide chain initiation (26). fMet- or Met-tRNA$_f^{Met}$ recognizes A-U-G in the starting position but in no other position of the mRNA coding sequence. They do not take part in peptide chain elongation. Though there are no striking differences between the primary structures of initiatior tRNAs and those of the others, two important features have been observed. The 5'-terminal nucleotide of procaryotic initiator tRNA is not complementary to its adjacent nucleotide near the 3'-terminus, in contrast to all other tRNAs that have been sequenced (26). This non-base-paired terminus of initiator tRNA is very important for its interaction with initiation factor 2 (IF2). It is apparently also the signal that allows elongation factor Tu to discriminate against initiator tRNA in the elongation process (27).

Eucaryotic initiator tRNA has another feature that makes it slightly different from all other tRNAs (often it is the small difference that counts). It does not contain pseudouridine, which is common to all other tRNAs, in the ribothymidine loop (28). In contrast to eucaryotic initiator tRNA, procaryotic initiator tRNA is not only acylated with methionine, but is also

mRNA (5')   ⌒‾‾‾‾‾A-G-G-A-G-G-U⌒ A-U-G-
                         | | | | | |
16SrRNA (3')   A-U-U-C-C-U-C-C-A-

Figure 4.1 Possible base pairing between mRNA initiation site and ribosomal 16S RNA.

formylated. However, there are exceptions to this rule. Some procaryotes can grow under conditions where their initiator tRNA is not formylated (29). In vitro both fMet-tRNA$_f^{Met}$ and Met-tRNA$_f^{Met}$ act as initiators of polypeptide chains, whereas Met-tRNA$_M^{Met}$ does not (30). These observations suggest that certain characteristics of the tRNA structure rather than formylation, determine its specificity as polypeptide chain initiator. Formylation of initiator tRNA, however, is not absolutely dispensable. Binding experiments with formylated and nonformylated initiator tRNA to 70S ribosomes rather than 30S subunits showed that initiation requires a formylated initiator tRNA (30). This suggests that formylation of initiator tRNA is necessary when initiation of polypeptide synthesis takes place on dissociated 70S ribosomes.

Binding of fMet-tRNA$_f^{Met}$ to ribosomes is promoted by three initiation factors in the presence of mRNA and GTP (see equation 4.1). It is believed that when polypeptide synthesis is terminated, or shortly thereafter, ribosomes dissociate into their subunits. This process is required for the re-formation of the initiation complex. The role of IF3 in this process seems to be to prevent reassociation of dissociated 70S ribosomes rather than to actively dissociate 70S ribosomes into their subunits. IF3 shifts the equilibrium between undissociated and dissociated ribosomes toward the latter by binding to the 3'-terminus of 16S RNA of the 30S subunits, which then cannot reassociate with 50S subunits (31, 32). This mechanism ensures that the concentration of free 30S subunits is increased and that the binding of mRNA and of initiator tRNA is promoted. IF2 is involved in binding initiator tRNA to the small subunit initiation complex. Free IF2 can bind GTP and fMet-tRNA$_f^{Met}$ loosely. It is not certain whether this weak complex is an intermediate in peptide chain initiation. The exact sequence in mRNA/ribosome/initiator tRNA/IF2 complex formation is still disputed. Most likely initiator tRNA and IF2 bind to the small subunit before they interact with mRNA. If initiator tRNA binds to the small subunit before mRNA, then initiator tRNA must be actively involved in the selection of the initiation signal and the "right" A-U-G must be selected from the mRNA sequence. Such a mechanism is most probable for initiation in eucaryotic systems according to current experimental evidence (24). The role of IF1 in these processes is not well understood. Apparently it helps stabilize the small subunit initiation complex, which consists of 30S subunit, mRNA, initiator tRNA, IF2, and IF3 (33).

Several ribosomal proteins of the small subunit are involved in binding initatior tRNA, mRNA, and the initiation factors. The results summarized here are based on several lines of evidence provided by RNA-protein and protein-protein cross-linking studies, by inhibition studies with specific antibody fragments (Fab) against individual ribosomal proteins, and by affinity-labeling techniques (for reviews see refs. 3, 34, 35). Special initiation sequences on the 5'-side of the initiation codon A-U-G of mRNA can potentially base-pair with a complementary sequence of the 3'-terminus of 16S RNA (11, 22). Further, IF3 can be cross-linked to the 3'-terminus of 16S RNA (32). This indicates that the 3'-terminus of 16S RNA is a hot spot in initiation. However, the picture that takes shape from further studies is not as straightforward as one would hope. Ribosomal proteins S1 and S21 as well as IF2 can also be cross-linked to the 3'-terminus of 16S RNA (36). Since the technique of cross-linking was identical in all cases, it is puzzling that four different proteins could be cross-linked to the 3'-terminus of 16S RNA. This finding could mean either that the 3'-terminus of 16S RNA is very flexible and not rigidly packed into the 30S subunit or that the 30S subunit can undergo various conformational changes, which can put any of the four proteins close to the 3'-terminus. The latter possibility is more likely since ribosomes acquire different conformations according to their functional states during protein synthesis (37). The proximity of protein S1 to the initiation factors is also indicated by protein-protein cross-linking studies. Both IF2 and IF3 can be cross-linked to S1 with dimethyl suberimidate (38). IF2 can also be cross-linked with proteins S11 and S13, and IF3 with proteins S11, S13, and S14. This result again indicates the complexity of the initiation site. A further protein-protein cross-linking result is of considerable importance when one considers the topography of the ribosomal initiation site. When tetranitromethane is used as a cross-linking reagent, a cross-linked trimer can be isolated, consisting of proteins S11, S18, and S21 (39). Thus close to the 3'-terminus we find IF2, IF3 and ribosomal proteins S1, S11, S12, S13, S14, S18, and S21.

A few models of the arrangement of 30S ribosomal proteins have been constructed. Two recent models have been based on the three-dimensional interpretation of electron-microscopic pictures of 30S subunit-antibody complexes in two different laboratories (40, 41). Though there are some discrepancies between the two models, they are in agreement with respect to

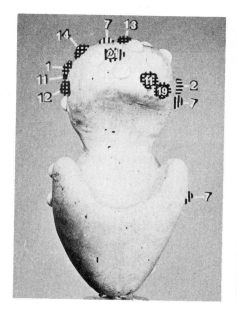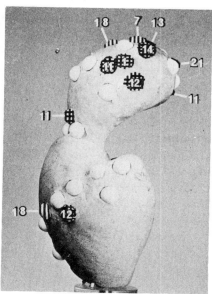

Figure 4.2 Initiation factor binding sites on the 30S subunit. This is a plasticine model of the 30S subunit as derived from electron-microscopic studies. Shading indicates antibody binding sites to proteins, which have been cross-linked to IF2 (horizontal shading), to IF3 (vertical shading), and to both (crosshatched shading).

the shape of the ribosomal subunit and the location of the various ribosomal proteins in the subunit.

Figure 4.2 shows such a topographical model of the location of antibody binding sites, obtained with IgG molecules specific for the proteins, which have been cross-linked to IF2 and IF3. Most of the antibody binding sites of these cross-linked proteins are situated close together on the 30S subunit. These data support preceding conclusions. Proteins S11, S12, and S14 are required for subunit association. The antibody binding sites of these proteins are within areas of the 30S particle, which are in contact with the large subunit. This observation is certainly relevant to an understanding of the function of IF3 as "antiassociation" factor. As long as this site is occupied by IF3, the 30S subunit apparently cannot reassociate with 50S subunits.

Do the ribosomal proteins have any function in initiator tRNA binding? Reconstituted 30S particles deficient in either protein S12, S13, or S21 almost completely lose the capacity to form the IF2-dependent A-U-G/ fMet-tRNA$_f^{Met}$/30S complex (42). However, it is very difficult to assign a direct role to a particular protein by experiments in which a single component is omitted. Specific antibody fragments (Fab) against the ribosomal proteins S3, S10, S14, S19, and S21 strongly inhibit fMet-tRNA$_f^{Met}$-binding. Antibodies against proteins S1, S2, S5, S6, S12, S13, and S20 are less inhibitory, although their effect is distinct. The remaining antibodies (against S4, S7, S8, S9, S11, S15, S16, S18) are weak inhibitors or have no effect (for reviews see refs. 3, 34). It may be surprising that antibodies against so many proteins interfere with the formation of an initiator tRNA complex. However, the outcomes of these inhibition studies can occur for a variety of reasons. Besides a considerable stereochemical effect, which the antibody fragment might impose on the 30S subunit, the formation of an initiation complex requires binding of several macromolecules to the 30S subunit: mRNA, initiator tRNA, and initiation factors. Interference with the binding of any one prevents complex formation. Thus we are left with the vague notion that the ribosomal proteins are involved in initiation complex formation in one way or another. But it is impossible to identify their exact function in this complex formation.

Fortunately another source of information enables us to sort out some of the proteins. Affinity-labeling experiments with initiation complexes, which contain chemically reactive A-U-G derivatives, yield the following result. When an A-U-G derivative carries a chemically reactive group on the 5′-side, cross-linking occurs to proteins IF3, S1, and S18 (43). When the A-U-G derivative carries a chemically reactive group on the 3′-side, cross-linking occurs only to protein S18. These data make it possible to locate these proteins at defined sites in the initiation complex, and they support the protein-protein cross-linking results. The result of another affinity labeling with an fMet-tRNA$_f^{Met}$ derivative that carries a chemically reactive group at the methionine residue is also pertinent here. In this case cross-linking occurs to proteins S2 and S14 (44).

Thus we have the following topographical picture of the involvement and function of initiation factors and ribosomal proteins during the interaction of mRNA and initiator tRNA on the 30S subunit. At the 3′-terminus of 16S RNA, IF3 and S1 are involved in binding mRNA. Both proteins are located near the 5′-side of the initiation codon A-U-G. Since addition of S1 stimulates mRNA binding, it may be involved in the base pairing between the 3′-terminus of 16S RNA and complementary mRNA sequences 5′ to the initiation codon. IF2 and IF3 are close to S1 and IF3 is also close to S14, which seems to bind the acceptor end of initiator tRNA. S18 occupies the site where the initiation codon is bound since it is found on both sides of it.

A peptide analysis of the A-U-G/S18 cross-link has actually shown that it is the N-terminal part (19 amino acids long) of S18 that becomes cross-linked to A-U-G (45). Protein S11 is located in the neighborhood of protein S18 according to cross-linking data (39). Protein S11, in turn, is again close to or at the binding site of IF2 and IF3 (see figure 4.2).

An important ribosomal protein is S12. This protein is a major component in ribosomal control of cistron specificity and codon-anticodon interaction. Proteins S12 and S11 are important for the fidelity of mRNA translation and hence for the fidelity of codon-anticodon interaction (4, 42). Affinity-labeling experiments with A-U-G derivatives have shown that these two proteins are located close to the 5'-side of the initiation codon binding sites. However, A-U-G can only be cross-linked to proteins S12 and S11 in th absence of fMet-tRNA$_f^{Met}$ and of initiation factors (46). Further, for cross-linking to occur ribosomes must be pretreated to force them into a conformation unfavorable for initiation complex formation. This suggests that these two proteins may be indispensable for codon-anticodon interaction and may be located near the site of the interaction.

These results may lead one to ask how all these proteins can be neighbors and , moreover, be involved in the same ribosomal function—the binding of A-U-G and of initiator tRNA. A solution to this apparent paradox is suggested by recent electron-microscopic studies in which the binding of specific antibody fragments to ribosomal proteins of the small (and large) subunit has been visualized (40, 41). Since the 30S subunit is an asymmetrical kidney-shaped particle, electron micrographs of these studies can be used to construct a topological map of the 30S subunit (figures 4.2 and 4.3). Several striking results have been obtained from these studies. First, some ribosomal 30S proteins are not globular, but of extended conformation. Some of them (proteins S4 and S18) run all the way through the 30S subunit. Considering their molecular weight and the dimensions of the subunit, these extended proteins probably have few internal tertiary interactions. Another striking feature can be seen from the data summarized in figure 4.3. The proteins, which have been found to be involved in initiation complex formation, are all located in the head of the 30S subunit close to the bottleneck, where it apparently gets very crowded, suggesting that this site of the 30S subunit represents the mRNA and tRNA binding sites or domains.

When the 30S initiation complex has been formed, it associates with the 50S subunit to form the 70S initiation complex; IF3 is released simultaneously. The formation of the 70S initiation complex is also accompanied

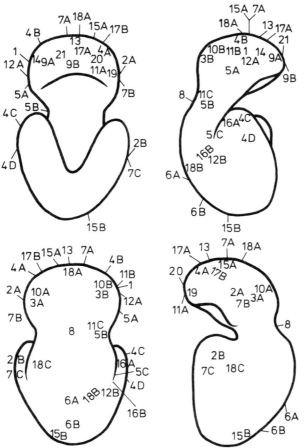

Figure 4.3 Three-dimensional model of the 30S subunit. Locations of antibody binding sites for the 21 ribosomal proteins are indicated by using their respective numbers. A–D designate multiple attachment sites of an antibody specific to one protein. The four forms are related to each other by rotation through an angle of approximately 90° along the vertical axis.

by the hydrolysis of GTP, whereupon IF2 and IF1 are released (37, 47). These processes apparently do not result in a major reorientation of fMet-tRNA$_f^{Met}$. When the inhibition of fMet-tRNA$_f^{Met}$ binding to 70S ribosomes (rather than to 30S subunits) was measured by antibody-binding experiments, the results were similar to those obtained with 30S subunits (48). Any major rearrangement would have been observed in these experiments. Protein S21 probably plays an important role in the formation of the 70S initiation complex by replacing IF3 at the 3′-terminus of 16S RNA. Protein

S21 is not present on native 30S subunits but on monosomes or polysomes in the process of translating mRNA. Addition of protein S21 inhibits initiation complex formation (49). Thus a switch from absence to presence of protein S21 in ribosomes is correlated with the functional status of the ribosome, that is, a switch from initiation to elongation in protein syntheis.

Of all the various ribosomal proteins involved in mRNA-tRNA binding to the 30S subunit, four have been found to be important for subunit assoction. These proteins are S11, S12, S14, and S18. Proteins S11 and S12 even bind to 23S RNA of the 50S sunbunit (50). This affinity to 23S RNA is certainly important for the mechanism of subunit association. IF2 and parts of initiator tRNA, especially that carrying the formylmethionine moiety, interact with the ribosomal proteins of the 50S subunit. These proteins are L7/L12 for IF2 (51) (L7 and L12 differ only in the acylation of the N-terminal serine of L7) and L2/L27 for the 3'-terminus of fMet-tRNA$_f^{Met}$ (52, 53). Affinity-labeling experiments with initiator tRNA derivatives have shown that a part of 23S RNA is near the 3'-terminus of initiator tRNA (54). It is not clear whether there is direct contact between the unpaired 3'-terminal C-C-A- sequence of tRNA and a possible complementary sequence of 23S RNA. After the 70S initiation complex has been formed and IF2 has been released (for which GTP hydrolysis is actually not necessary), fMet-tRNA$_f^{Met}$ becomes reactive toward the antibiotic puromycin (for a review see ref. 55). This reaction leads to the formation of formylmethionyl-puromycin and eventually to the release of discharged initiator tRNA. Since charged tRNA during the ribosome cycle is not always sensitive to the action of puromycin, one differentiates between puromycin-sensitive and puromycin-insensitive tRNA binding sites on the ribosome. The puromycin-sensitive site is usually referred to as the P-site and the puromycin-insensitive as the A-site. The A-site is referred to as the aminoacyl-tRNA binding site. However, the *functional* terms of P-site and A-site should not be confused with tRNA binding sites as physical entities. Any conclusion on physical tRNA binding sites should be checked against this background.

We have seen that the final step in initiation complex formation makes the initiator tRNA reactive toward puromycin. The puromycin reaction actually simulates peptide bond formation, and it is therefore widely used in studies on this part of ribosome function. For the progress of protein synthesis, puromycin-sensitivity of fMet-tRNA$_f^{Met}$ means that the ribosome is now ready to accept the first aminoacyl-tRNA in response to the codon, adjacent to A-U-G on the 3'-side. Thereafter the first peptide bond is formed by the ribosome and the elongation/translation cycle is on its way.

4.3 Aminoacyl-tRNA Binding

Aminoacyl-tRNA is bound to the ribosome as a ternary complex between EF-Tu (MW 47,000), GTP, and the aminoacyl-tRNA itself, of which the anticodon base-pairs with the codon in mRNA in an antiparallel fashion. One aminoacyl-tRNA may donate aminoacyl residues to the growing polypeptide chain in response to several related codons (in agreement with the degeneracy of the genetic code). The relationship between such codons is as follows. There is an apparent requirement for a Watson-Crick complementarity between the central nucleotide of the codon and the central nucleotide of the anticodon, as well as between one of the outside nucleotides (the first or the third) of the codon and the corresponding nucleotide of the anticodon. The relationship between the third nucleotide of the codon and the corresponding nucleotide in the anticodon apparently obeys less strict rules of complementarity, as enuciated by Crick in his wobble hypothesis (15). The specificity of codon recognition by aminoacyl-tRNA is affected by the presence of modified and hypermodified nucleotides which in some tRNAs are part of the anticodon and in others are adjacent to the anticodon. A fairly well-established rule for the occurrence of modified bases 3'-adjacent to the anticodon seems to be the following: Hydrophilic modifications are found adjacent to U-A base pairs between the third nucleotide of the anticodon and the first nucleotide of the codon. Hydrophobic modifications are found adjacent to A-U base pairs between the third nucleotide of the anticodon and the first nucleotide of the codon. The meaning of this apparent regularity in modifications adjacent to the anticodon is still obscure. There are some indications that the modifications may be involved in codon-anticodon recognition. Complementary trinucleotides do not associate detectably in solution. So, we may ask, how does the anticodon on the aminoacyl-tRNA bind to the messenger triplet on the ribosome? Moreover, what makes this interaction so specific that the error frequency in translation is as low as 1 in 10^4 amino acids?

Part of the answer lies in the structure of the tRNA-anticodon and in the structure of the anticodon-codon (tRNA-mRNA) complex. Trinucleotides complementary to the anticodons of tRNA associate with the latter in solution at low temperatures. More dramatically, pairs of tRNA molecules with complementary anticodons bind with an association constant as high as 10^6M (56). This equilibrium constant is 10^6 times greater than expected due to the very slow dissociation reaction of these complexes. The entropy of interaction is not very different from that expected for interactions between trinucleotides, but the enthalpy change (-25 kcal) is about 10 kcal larger than expected (57). This is apparently at the heart of the enhanced anti-codon-anticodon interaction, and very probably stacking interactions are responsible for it. The relevance of these studies for mRNA-tRNA inter-action is obvious. Properly stacked (helical) configurations can account for high binding constants and hence for part of the observed selectivity and fidelity in codon-anticodon recognition.

The Y-base of yeast tRNA^{Phe} can be easily removed (58). It is thus possible to study the influence of a modified base 3'-adjacent to the anti-codon on base pairing. The results of binding studies with complementary triplets show that excision of the Y-base leads to a substantial reduction in the enthalpy of association (59). This finding demonstrates the importance of a modified base 3'-adjacent to the anticodon for a strong anticodon-co-don interaction. On the other hand many data suggest that ribosomes are actively involved in an mRNA-tRNA recognition and play an active role in the fidelity of translation (60, 61, 62).

An unusual mutant form of tRNA^{Trp} can translate the nonsense codon U-G-A and the cysteine codon U-G-U as well as its normal codon U-G-G. This suppressor tRNA unexpectedly contained a normal anticodon loop as well as a single base substitution in the D-stem. The properties of this mutant tRNA suggested that tRNA selection on the ribosome involves both the anticodon and other domains in the tRNA molecule (61). A coupling be-tween codon-anticodon interaction and a conformational change of tRNA was proposed to account for the effects of D-stem alterations on the codon specificity of the mutant tRNA^{Trp}.

Studies with ribosomal mutants have shown that ribosomes significantly influence codon-anticodon interactions. Ribosomal strA mutants restrict mistranslation (63); ram mutations reverse the restrictions imposed by the strA mutations and allow increased levels of ambiguous translation (64). Added to ribosomes in vitro, the drug streptomycin induces a high degree of misreading (65). This means that proper mRNA-tRNA interactions on the ribosome are disturbed by streptomycin such that the ribosome often selects a "wrong" tRNA. The target of streptomycin action is the ribosome and not tRNA or any other component of the cell-free extract necessary for polypeptide synthesis (62). This means that streptomycin stimulates the ribosome to bind a tRNA that is normally not accepted in response to a given codon. The existence of this ribosomal screening for the appropriate tRNA (60) is probably linked to the presence of the modified nucleotide adjacent to the anticodon. Transfer RNA_f^{Met} from which the Y-base has been excised is poorly bound by ribosomes. In the presence of streptomycin, however, this $tRNA^{Phe}$ is recognized by ribosomes almost as well as normal $tRNA^{Phe}$. This indicates that streptomycin affects a ribosomal mechanism that can discriminate between tRNA molecules for the presence or absence of the modified Y-base. The degree of modification of the base 3'-adjacent to the anticodon strongly influences the ability of the tRNA to transfer amino acids into polypeptide (66). Mutants of $tRNA^{Tyr}$ in which the 3'-adjacent adenosine was not or was only partially modified exhibit very poor or, respectively, significantly less affinity to ribosomes than normal $tRNA^{Tyr}$.

Inhibition experiments with specific antibodies show that antibodies against proteins S8, S9, S11, and S18 interfere with EF-Tu–dependent, poly(U)-directed Phe-tRNA binding (48). These antibodies do not interfere with initiator tRNA binding. This indicates that on the 30S subunit the binding sites for $fMet-tRNA_f^{Met}$ and peptidyl-tRNA are not identical. This is a strong indication that ribosomes do have two tRNA binding sites, and this in turn is supported by affinity-labeling experiments with codon analogs. The reaction pattern obtained with A-U-G derivatives in the presence of $Met-tRNA_f^{Met}$ is different from that obtained in the presence of $fMet-tRNA_f^{Met}$ (43).

An investigation of the topographical model depicted in figure 4.3 shows that 30S proteins involved in aminoacyl-tRNA binding are located on the head and on the dorsal side of the subunit around the neck region. This is distinct from the domain on the head, where we find the proteins involved in $fMet-tRNA_f^{Met}$ binding. However, none of these data support a direct involvement of the cited proteins in aminoacyl-tRNA binding.

The situation is clearer in case of the 50S subunit, where the 3'-terminus of aminoacyl-tRNA is bound. This binding site is the heart of the ribosome. At this site the peptide bond is formed by a transfer of the peptide chain to the a-amino group of the amino acid of aminoacyl-tRNA. This site is accordingly called the peptidyltransferase center of the ribosome. Chloramphenicol inhibits the binding of 3'-terminal fragments of aminoacyl-tRNA to ribosomes (67). Affinity-labeling studies with chloramphenicol derivatives, N-substituted Phe-tRNA^{Phe}s, and reconstitution experiments all locate protein L16 at the site of the peptidyltransferase center, which binds the 3'-terminal C-C-A of aminoacyl-tRNA and the aminoacyl-tRNA (for a review see ref. 68). Thus protein L16 is intimately involved in peptide bond formation on the ribosome. Partial reconstitution experiments suggest that the nearest neighbors to protein L16 are proteins

L6 and L11; protein L11 is important for an active peptidyltransferase center in the ribosome (69) and has been identified as the thiostrepton binding protein (70). Thiostrepton is a drug that inhibits EF-G–dependent GTP hydrolysis, which is necessary for translocation (71).

Several lines of evidence suggest that an interaction between the 5S RNA of the 50S subunit and the ribothymidine loop of tRNA is important for aminoacyl-tRNA binding (for a review see ref. 72). The sequence T-Ψ-C-Pu is common to all tRNAs except eucaryotic initiator tRNAs and the tRNAs needed in cell-wall synthesis. Inspection of the three-dimensional structure of yeast tRNA^{Phe} shows that this sequence is at the corner of the molecule, about 70 Å from the 3′-terminal adenosine, which carries the aminoacyl moiety. Though the T-Ψ-C-Pu sequence is on the surface at this corner of the tRNA molecule, it is not readily available for base pairing with complementary bases (for a review see ref. 26). This is apparent from an inspection of the crystal structure as well as from complementary oligonucleotide binding studies, which have shown that complementary oligonucleotides do not bind to this sequence of tRNA. An important stabilizing factor for the interaction of the ribothymidine loop with the dihydrouridine loop at this corner is the presence of a magnesium ion between these two loops. Recent studies, however, have shown that this particular magnesium ion readily exchanges with a manganese ion. This result suggests a flexibility in this structure that could be exploited by the ribosome.

As the T-Ψ-C-Pu sequence is common to tRNAs, so the complementary sequence G-A-A-C is common to 5S RNAs (72). Though complementary oligonucleotides do not bind to the T-Ψ-C-Pu sequence of tRNA, the reciprocal experiments with 5S RNA give positive results (73). Oligonucleotide binding experiments show that the tRNA fragment T-Ψ-C-G binds specifically to 5S RNA. Binding of the fragment T-Ψ-C-G to ribosomes inhibits binding of aminoacyl-tRNA as well as formation of magic spots. Moreover T-Ψ-C-G can mimic tRNA (74). Under appropriate conditions it is possible to induce ribosomes to synthesize magic spots in the presence of mRNA and T-Ψ-C-G, a reaction for which tRNA is usually necessary. These data suggest that the binding to the ribosome frees the T-Ψ-C corner of the aminoacyl-tRNA molecule for an interaction with 5S RNA via complementary base pairing.

A complex of proteins L5, L18, and L25 with 5S RNA has been isolated, which exhibits GTPase activity in vitro (72). This enzymatic activity is inhibited by fusidic acid and thiostrepton (the latter bound exclusively by protein L11). This observation brings us back to the peptidyl transferase center.

Thiostrepton has only one binding site on the ribosome. This means that the 5S RNA/protein complex, which is responsible for GTPase activity, should be close to protein L11. Usually GTP hydrolysis during protein synthesis is coupled to factor-dependent processes on the ribosome like EF-Tu-dependent aminoacyl-tRNA binding, IF2-dependent initiator tRNA binding, and EF-G-dependent tRNA translocation. The three factors bind to the same site on the ribosome. Proteins L7/L12 have been identified to constitute this site. One would expect from the mechanism of protein synthesis that a topographical examination of the 50S subunit should reveal that proteins L7/L12 are close to protein L16 as well as to the 5S RNA/protein complex.

Figure 4.4 summarizes immunoelectron microscopy data on the topographical location of these proteins in the 50S subunit. It can be seen that proteins L7/L12, located at the central protuberance of the 50S subunit at multiple sites, are indeed close to proteins L11 and L16, whereas the 5S RNA–binding proteins L5, L18, and L25 have been found farther from the middle of the 50S subunit, at the edge formed by the seat and the lower anterior surface. The proximity of protein L2 to protein L16 in the peptidyltransferase center is particularly interesting since it influences binding of 5S RNA/protein complexes to 23S RNA. Protein L2 is also part of the peptidyltransferase center and is close to the 3'-terminal adenosine of bound peptidyl-tRNA.

Protein L16 binds toward the 5'-end of 23S RNA and protein L2 toward the 3'-end of 23S RNA (75). This indicates that both parts of 23S RNA contribute to the structure of the peptidyltransferase center and possibly to its function as well. Recall that two-thirds of the mass of the ribosome consists of RNA and only one-third of protein. For some time it was believed that ribosomal RNA merely served as a matrix for the assembly of ribosomal proteins. Lately it has become increasingly clear that this view is distorted. Ribosomal RNA plays an active role in ribosomal functions. The 3'-terminus of 16S RNA is involved in initiation and possibly in termination of protein synthesis, and 5S RNA interacts with the T-Ψ-C corner of the aminoacyl-tRNA molecule; in addition, 16S RNA and 23S RNA are constituents of the ribosomal tRNA binding sites.

Irradiation of tRNA[Val] leads to a cross link with 16S RNA (76), whereas derivatization of aminoacyl-tRNA at the a-amino group of the amino acid leads to cross links between the N-substituted tRNAs and 23S RNA (68). As the roles of the 3'-terminus of 16S RNA in initiation and that of 5S RNA in aminoacyl-tRNA binding are fairly well understood, it is not clear whether

Figure 4.4 Functional sites on the 50S subunit. Frontal view of a 50S subunit model, indicating antibody binding sites relevant to ribosomal functions of the 50S subunit.

the cross-linking data point toward comparably important interactions between 16S and/or 23S RNA and aminoacyl-tRNA. The base-pairing mechanisms in these processes rely on common sequences that occur in either mRNA or tRNA. One is left to speculate whether similar mechanisms might operate between 16S and/or 23S RNA and aminoacyl-tRNA. One possibility is that the 3′-terminal C-C-A sequence common to all tRNAs is recognized on the ribosome via a base-pairing mechanism similar to mRNA recognition or to interaction between T-Ψ-C-Pu and 5S RNA.

The peptidyltransferase center of the ribosome consists of proteins L16, L11, and L2. Close by are proteins L7/L12, the binding site of EF-Tu,

EF-G and IF2. The 5S RNA–binding proteins are farther, as indicated in Figure 4.4. Protein L16 is near the 3′-terminal adenosine of aminoacyl-tRNA. Thus it should be at the site where peptidyl transfer takes place. Once the growing peptide chain has been transferred to the inbound amino-acyl-tRNA, it becomes the new peptidyl-tRNA. This peptidyl-tRNA is then recognized by EF-G, which translocates the peptidyl-tRNA with the consumption of energy provided by hydrolysis of GTP. After translocation the aminoacyl-tRNA binding site is again available so that the next codon of the mRNA can be read.

4.4 Peptidyl-tRNA Binding to Ribosomes

After translocation, peptidyl-tRNA is located in the P-site of the ribosome, which by definition is puromycin-sensitive. Since fMet-tRNA$_f^{Met}$ is also bound in a puromycin-reactive site, it seems that both initiator tRNA and peptidyl-tRNA bind in the same domain of the ribosome, although the actual sites may not be completely identical. Not much is known about the peptidyl-tRNA binding site on the 30S subunit. However, it certainly over-laps with the binding site of fMet-tRNA$_f^{Met}$. N-substituted phenylalanyl-tRNA^{Phe} and methionyl-tRNA$_f^{Met}$ analogs have been constructed as affinity labels of the 50S P-site in the peptidyltransferase center (for a review see ref. 68). These labels react mainly with proteins L2 and/or L27. Thus these two proteins are thought to be located close to the binding site of the aminoacyl moiety in the P-site of the peptidyltransferase center of the ribosome. The affinity-labeling studies with N-substituted peptidyl-tRNA analogs have been extended through preparation of a set of bromoacetyl-peptidyl-rRNAs of increasing chain length. These experiments yield a defined reaction pattern that allows construction of a kind of chemical map for proteins L2, L27, L33a, and L24 (77). The quantitative reaction pattern can be traced as a function of distance. Thus the four proteins in order of increasing distance from the 3′-end of the peptidyl-tRNA are L2, L27, L33, and L24 (see also figure 4.4). These data might seem to imply that the growing poly-peptide chain is not freely movable but to a certain extent protected by ribo-somal proteins. This would ensure that a growing protein is not released into solution before a certain length of the polypeptide has been synthe-sized, which allows for a stabilization of the structure via secondary and tertiary interactions.

How different are the peptidyl-tRNA and aminoacyl-tRNA binding sites? Can two tRNAs bind to the ribosome at the same time? The data discussed in the preceding sections show that there are indeed many differences between the binding sites for peptidyl- and aminoacyl-tRNA. First, different proteins are involved in each site. Second, aminoacyl-tRNA interacts with 5S RNA, whereas peptidyl-tRNA does not. Nevertheless, there appears to be some overlap between the two binding sites. Antibodies against proteins S1, S3, S10, S14, S19, S20, and S21 interfere with fMet-tRNA$_f^{Met}$ binding as well as with aminoacyl-tRNA binding (48). Proteins S2, S3, and S14 stimulate both types of tRNA binding (49, 78). An inspection of the topographical map of the ribosomal 30S proteins (shown in figures 4.2, 4.3) reveals that proteins associated with P-site functions are located on the head of the subunit and that proteins with A-site functions occur both on the head and on the dorsal side of the subunit around the neck region. The proximity of proteins in the head of the subunit is certainly responsible for the observation that, the tRNA binding sites in some experiments exhibit characteristics of a hybrid rather than a twin site.

Binding of streptomycin to the ribosome can induce misreading of codons in the A-site as well as the release of fMet-tRNA$_f^{Met}$ from the P-site (47, 55). If only one streptomycin binding site is assumed, these data suggest that there is only one, hybrid, codon-anticodon binding site, which according to demand functions as either the A-site or the P-site. Recent affinity-labeling studies with streptomycin analogs, however, show that the assumption of one streptomycin binding site is incorrect (79). There are, in fact, two binding sites. This now easily explains the effect of streptomycin on aminoacyl-tRNA and initiator tRNA binding. This leads us to the important question of the mechanism of protein synthesis. Is the codon-anticodon interaction between peptidyl-tRNA and mRNA in the A-site (before translocation) disruped by translocation? Unfortunately there is no solid evidence that peptidyl-tRNA still interacts with mRNA after translocation.

When a chemically reactive A-U-G analog is covalently bonded to protein S12, such ribosomes can bind fMet-tRNA$_f^{Met}$ to a puromycin-sensitive as well as to a GTP-sensitive site (46). This ability suggests that codon-anticodon interaction indeed takes place in the peptidyl-tRNA binding site of the ribosome, in turn implying that streptomycin displaces fMet-tRNA$_f^{Met}$ in the P-site by distorting or displacing the codon A-U-G. When an initiation complex has been formed in the P-site, in which A-U-G is covalently bonded to the ribosome, it is no longer sensitive to the action of streptomycin (43). Since A-U-G was irreversibly bound in the labeling experiment, streptomycin can no longer act on it. Thus these data support the idea that peptidyl-tRNA is still bound to mRNA in the P-site of the ribosome. This point is important for understanding how tRNA actually moves within the ribosome.

4.5 Movement of tRNA within the Ribosome

Since the structural information about the ribosome is not as detailed as that for tRNA, any model of tRNA movement in the ribosome must be based on several assumptions. The first is that the tRNA molecules do not undergo gross conformational changes. This assumption does not rule out small changes during binding and translocation such as an altered folding of the ribothymidine loop, which interacts with 5S RNA in the A-site but not in the P-site. Further, we assume that all tRNA molecules interact with ribosomes in the same manner, determined by common features of their nucleotide sequences. The structure of tRNA^{Phe} as seen in crystals is thought to remain unchanged in solution, when tRNA is biologically active. Further assumption are as follows: Anticodons of the two tRNA molecules in the A-site and the P-site both interact with adjacent codons on the mRNA at the same time. The codon-anticodon base pairs remain stacked and have the geometry known from RNA double helices. The polynucleotide backbone is not stretched out. The final assumption is that the peptidyl chain is transferred from the tRNA in the P-site directly to the aminoacyl-tRNA in the A-site and that a ribosomal protein does not function as an intermediate transfer site.

According to these assumptions two adjacent tRNAs should form complementary hydrogen bonds with adjacent codon triplets without steric interference with each other. This bonding, however, cannot be accomplished easily, since the 3'-ends of aminoacyl-tRNA and of peptidyl-tRNA have to be so close together that they allow direct peptidyl transfer. If the tRNA molecules were extended nucleotide chains, simultaneous binding to mRNA would separate both 3'-termini by an enormous distance. Since tRNAs are bent molecules and can be accommodated on mRNA in such a way that the T ψ C corners are far apart, they can indeed interact with adjacent codons and at the same time have their 3'-termini together. Such an arrangement, however, implies that mRNA turns a corner or is kinked between the two codons, as illustrated in figure 4.5. Such an arrangement would resolve another problem. If two tRNAs are simultaneously interacting with adjacent nucleotide triplets, the tRNA anticodons must be no

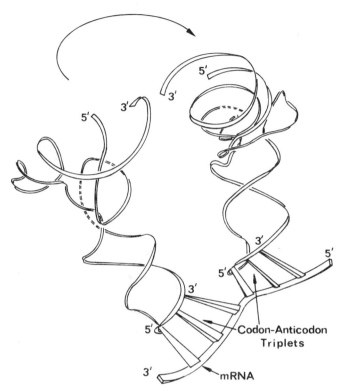

Figure 4.5 Base pairing between mRNA and tRNA on the ribosome. Two tRNAs are arranged by a rotatory movement (indicated by the arrow) such that their 3′-termini are close together and their anticodon base pairing with mRNA is not disrupted. This arrangement requires a bending of mRNA as indicated.

farther apart than 10 Å. Since diameter of the tRNA molecule is approximately 20 Å near the anticodon, a kink or corner between the two codons would mean that the two tRNA molecules bind from different directions to the adjacent codons. In this manner they would not interfere with each other. The three nucleotide pairs in the codon-anticodon interactions would remain stacked, and no base stacking would occur between the two triplet pairs.

According to this model translocation is largely a rotatory operation on tRNA, during which the ribosome turns tRNA around the corner between adjacent codons in the mRNA.

Each translocation step moves a tRNA from the A-site to the P-site, and at the same time the mRNA strand advances one triplet. This simultaneous

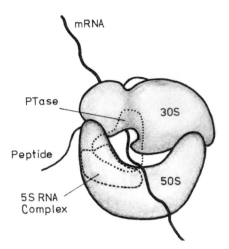

Figure 4.6 General model of the 70S ribosome. This model tentatively shows how mRNA runs through a cavity and where the tRNA molecules should enter and leave the ribosome. Dotted areas indicate the domains of ribosomal functional sites.

mRNA advancement could be promoted by the ribosomal machinery. Another possibility is that the codon-anticodon interaction is not disrupted during the tRNA translocation. It follows from this possibility that the mRNA advancement in the ribosome is a sole consequence of tRNA translocation and that it is not directly carried out as a separate operation by the ribosome.

Recent nucleotide sequence analysis of a frameshift suppressor tRNA supports this latter possibility. This particular suppressor tRNA has an extra base in its anticodon loop (80). The action of the suppressor tRNA strongly suggests the movement of four bases in the mRNA strand instead of the usual three. This could not occur if ribosomal machinery were responsible for moving mRNA three nucleotides at a time. It is therefore more likely that the mRNA movement is regulated by the tRNA movement, which is carried out by the ribosome during translocation. Topographical data, obtained by electron-microscopic studies, give some idea of where mRNA and tRNA are located in the ribosome (40, 3). According to these data mRNA proceeds through a channel that is formed by a part of the 30S subunit and the central part of the 50S subunit as indicated in figure 4.6. A cavity between the two subunits provides sufficient space to accommodate two tRNA molecules. It has been proposed that the TΨC corner of A-site tRNA is located between the central protuberance and the seat of the 50S subunit.

References

1. Cold Spring Harbor Symp. Quant. Biol. *34* (1969).

2. Nomura, N., Tissieres, A., and Lengyel, P., eds. (1974). Ribosomes (Cold Spring Harbor, N.Y.: Cold Spring Harbor Laboratory).

3. Weissbach, H., and Pestka, S., eds. (1977). Protein Biosynthesis (New York: Academic Press).

4. Nomura, M. (1973). Science *179*, 864.

5. Nomura, M., and Erdmann, V. A. (1970). Nature *228*, 144.

6. Nierhaus, K. H., and Dohme, F. (1974). Proc. Nat. Acad. Sci. USA *71*, 4713.

7. Gupta, S. L., Waterson, J., Sopori, M. L., Weissman, S. M., and Lengyel, P. (1971). Biochemistry *10*, 4410.

8. Thach, S. S., and Thach, R. E. (1971). Proc. Nat. Acad. Sci. USA *68*, 1791.

9. Ochoa, S. (1968). Naturwissenschaften *55*, 505.

10. Watson, J. D. (1964). Bull. Soc. Chim. Biol. *46*, 1399.

11. Steitz, J., and Jakes, K. (1975). Proc. Nat. Acad. Sci. USA *72*, 4734.

12. Clark, B. F. C., and Marcker, K. A. (1966). J. Mol. Biol. *17*, 394.

13. Vinuela, E., Salas, M., and Ochoa, S. (1967). Proc. Nat. Acad. Sci. USA *57*, 729.

14. Sanger, F., Air, G. M., Barrell, B. G., Brown, B. L., Coulson, A. R., Fiddes, J. C., Hutchinson, C. V., Slocombe, P. M., and Smith, M. (1977). Nature *265*, 687.

15. Crick, F. H. C. (1966). J. Mol. Biol. *19*, 548.

16. Garen, A. (1968). Science *160*, 149.

17. Beaudet, A. L., and Caskey, C. T. (1970). Nature *227*, 38.

18. Goodman, H. M., Abelson, J., Landy, A., Brenner, S., and Smith, J. D. (1968). Nature *217*, 1019.

19. Brownlee, G. G., Sanger, F., and Barrell, B. G. (1968). J. Mol. Biol. *34*, 379.

20. Miller, D. L., and Weissbach, H. (1977). In Protein Biosynthesis, H. Weissbach, and S. Pestka, eds. (New York: Academic Press), p. 352.

21. Maden, B. E. H., Traut, R. R., and Monro, R. E. (1968). J. Mol. Biol. *35*, 333.

22. Shine, J., and Dalgarno, L. (1974). Proc. Nat. Acad. Sci. USA *71*, 1342.

23. Stanley, W. M., Salas, M., Wahba, A. J., and Ochoa, S. (1966). Proc. Nat. Acad. Sci. USA *56*, 290.

24. Schreier, M., and Staehelin, T. (1973). Nature New. Biol. *242*, 35.

25. Jay, G., and Kämpfer, R. (1974). Proc. Nat. Acad. Sci. USA *71*, 3199.

26. Rich, A., and RajBhandary, U. L. (1976). Ann. Rev. Biochem. *45*, 805.

27. Schulman, L. H., Pelka, H., and Sundari, R. M. (1974). J. Biol. Chem. *249*, 7102.

28. Simsek, M., RajBhandary, U. L., Boisnard, M., and Pettissant, G. (1974). Nature *247*, 518.

29. Petersen, H. U., Danchin, A., and Grunberg-Manago, M. (1976). Biochemistry *15*, 1357.

30. Petersen, H. U., Danchin, A., and Grunberg-Manago, M. (1976). Biochemistry *15*, 1362.

31. Subramanian, A., and Davis, B. (1970). Nature *228*, 1273.

32. van Duin, J., Kurland, C. G., Dondon, J., Grunberg-Manago, M., Branlant, C., and Ebel, J. P. (1976). FEBS Letters *62*, 111.

33. Dondon, J., Godefroy-Colburn, T., Graffe, M., and Grunberg-Manago, M. (1974). FEBS Letters *45*, 82.

34. Pongs, O., Nierhaus, K. H., Erdmann, V. A., and Wittmann, H. G. (1974). FEBS Letters *40*, 528.

35. Brimacombe, R., Nierhaus, K. H., Garrett, R. A., and Wittmann, H. G. (1976). Prog. Nucl. Acid Res. Mol. Biol. *18*, 1.

36. Czernilofsky, A. P., Kurland, C. G., and Stöffler, G. (1975). FEBS Letters *58*, 281.

37. Noll, H., Noll, M., Hapke, B., and van Dieijen, G. (1973). In Regulation of Transcription and Translation in Eukaryotes, E. Bautz, ed. (New York: Springer-Verlag), p. 257.

38. Hershey, J. W. B. (1974). Abstract, EMBO Workshop Initiation Protein Synthesis Prokaryotic Eukaryotic Syst., 1974.

39. Shih, G. T., and Craven, R. T. (1973). J. Mol. Biol. *78*, 651.

40. Tischendorf, G. W., Zeichhardt, H., and Stöffler, G. (1975). Proc. Nat. Acad. Sci. USA *72*, 4820.

41. Lake, J. A., Rendergast, M., Kahan, L., and Nomura, M. (1974). Proc. Nat. Acad. Sci. USA *71*, 4688.

42. Nomura, M., Mizushima, S., Ozaki, M., Traub, P., and Lowry, C. V. (1969). Cold Spring Harbor Symp. Quant. Biol. *34*, 49.

43. Pongs, O., Lanka, E., Grunberg-Manago, M., and Stöffler, G. Unpublished experiments.

44. Girshovich, A. S., Bochkareva, E. S., Kamarov, V. A., and Ovchinnikov, Yu. A. (1974). FEBS Letters *42*, 213.

45. Lanka, E., Pongs, O., and Yagouchi, M. Unpublished experiments.

46. Pongs, O., Stöffler, G., and Lanka, E. (1975). J. Mol. Biol. *99*, 301.

47. Grunberg-Manago, M., Godefroy-Colburn, T., Wolfe, A. D., Dessen, P., Pantaloni, D., Springer, M., Graffe, M., Dondon, J., and Kay, A. (1973). In Regulation of Transcription and Translation in Eukaryotes, E. Bautz, ed. (New York: Springer-Verlag), p. 213.

48. Lelong, J. C., Gros, D., Gros, F., Bollen, A., Maschler, R., and Stöffler, G. (1974). Proc. Nat. Acad. Sci. USA *71*, 248.

49. Randall-Hazelbauer, L. L., and Kurland, C. G. (1972). Mol. Gen. Genet. *115*, 234.

50. Stöffler, G., Daya, L., Rak, K. H., and Garrett, R. A. (1971). J. Mol. Biol. *62*, 411.

51. Kay, A., Sander, G., and Grunberg-Manago, M. (1973). Biochem. Biophys. Res. Commun. *51*, 479.

52. Czernilofsky, A. P., Collatz, E. E., Stöffler, G., and Küchler, E. (1974). Proc. Nat. Acad. Sci. USA *71*, 230.

53. Sopori, M., Pellegrini, M., Lengyel, P., and Cantor, C. R. (1974). Biochemistry *15*, 5432.

54. Barta, A., Küchler, E., Branlant, C., Sriwada, J., Krol, A., and Ebel, J. P. (1975). FEBS Letters, *56*, 170.

55. Pestka, S. (1976). Prog. Nucl. Acid Res. Mol. Biol. *17*, 217.

56. Eisinger, J., and Gross, N. (1974). J. Mol. Biol. *88*, 165.

57. Grosjean, C., Söll, D., and Crothers, D. (1976). J. Mol. Biol. *102*, 499.

58. Thiebe, R., Harbers, K., and Zachau, H. G. (1972). Eur. J. Biochem. *26*, 144.

59. Pongs, O., and Reinwald, E. (1973). Biochem. Biophys. Res. Commun. *50*, 357.

60. Gorini, L. (1971). Nature New Biology *234*, 261.

61. Kurland, C. G., Rigler, R., Ehrenberg, M., and Blomberg, C. (1975). Proc. Nat. Acad. Sci. USA *72*, 4248.

62. Pongs, O., and Nierhaus, K. (1974). Mol. Gen. Genet. *131*, 215.

63. Ozaki, M., Mizushima, S., and Nomura, M. (1969). Nature *222*, 333.

64. Zimmermann, R. A., Garvin, R. D., and Gorini, L. (1971). Proc. Nat. Acad. Sci. USA *68*, 2263.

65. Pestka, S., Marshall, R., and Nirenberg, M. (1965). Proc. Nat. Acad. Sci. USA *53*, 639.

66. Gefter, M. L., and Russell, R. L. (1969). J. Mol. Biol. *39*, 145.

67. Nierhaus, D., and Nierhaus, K. H. (1973). Proc. Nat. Acad. Sci. USA *70*, 2224.

68. Pellegrini, M., and Cantor, C. R. (1977). In Molecular Mechanisms of Protein Biosynthesis, H. Weissbach, and S. Pestka, eds. (New York: Academic Press), p. 203.

69. Nierhaus, K. H., and Montejo, R. (1969). Proc. Nat. Acad. Sci. USA 70, 1931.

70. Highland, J. H., Howard, G. A., Ochsner, E., Stöffler, G., Hasenbank, R., and Gordon, J. (1975). J. Biol. Chem. 250, 1141.

71. Bodley, J. W., Lin, L., and Highland, J. H. (1970). Biochem. Biophys. Res. Commun. 41, 1406.

72. Erdmann, V. A. (1977). Prog. Nucl. Acid. Res. Mol. Biol. 18, 45.

73. Erdmann, V. A., Sprinzl, M., and Pongs, O. (1973). Biochem. Biophys. Res. Commun. 54, 942.

74. Richter, D., Erdmann, V. A., and Sprinzl, M. (1974). Proc. Nat. Acad. Sci. USA 71, 3226.

75. Zimmermann, R. A., Mackie, G. A., Muto, A., Garrett, R. A., Ungewieckel, E., Ehresmann, C., Stiegler, P., Ebel, J. P., and Fellner, P. (1975). Nucl. Acids Res. 2, 279.

76. Ofengand, J. Personal communication.

77. Eilat, D., Pellegrini, M., Oen, H., Lapidot, Y., and Cantor, C. R. (1974). J. Mol. Biol. 88, 831.

78. Van Duin, J., Van Knippenberg, P. H., Dieben, M., and Kurland, C. G. (1972). Mol. Gen. Genet. 116, 181.

79. Pongs, O., Reinwald, E., Lanka, E., and Stöffler, G. Unpublished experiment.

80. Riddle, D. L., and Carbon, J. (1973). Nature New Biol. 242, 230.

5 TRANSFER RNA–MEDIATED SUPPRESSION

A. M. Körner,
S. I. Feinstein,
and S. Altman

5.1 Introduction

Suppression can be defined as the restoration of a normal phenotype in a mutant organism by a mutation at a site separate from that responsible for the original mutation. It is the purpose of this chapter to show how mutations in genes coding for tRNA molecules can suppress mutations in genes coding for proteins. Such tRNA gene mutations allow the translation of mRNAs that contain codon alterations to yield complete and functional polypeptides. Besides tRNA-mediated suppression, there are other mechanisms of suppression such as intragenic frameshift suppression and ribosome-mediated suppression. We will not consider these here, but they have been reviewed elsewhere (1).

Transfer RNAs can mediate suppression of mutations that have been assigned to the following classes: nonsense mutations, missense mutations, and frameshift mutations. Each class will be discussed separately, initially with respect to procaryotic systems where the subject has been most extensively investigated. Transfer RNA–mediated suppression in eucaryotes has been unequivocally demonstrated only in yeast, and this will be discussed at the end of the chapter.

5.2 Nonsense Codon Suppression

Three codons, U-A-G, U-A-A, and U-G-A, specify polypeptide chain termination during mRNA translation. These nonsense codons normally occur at the end of translated portions of mRNAs. However, they can also be generated anywhere in an mRNA molecule as a result of a mutation at the site of a single base. An "internally" occurring nonsense codon leads to premature chain termination. The insertion of an amino acid into a nascent polypeptide chain at the site of a nonsense codon is defined as nonsense suppression. Translation of any of the three nonsense codons occurs when a normal tRNA species is altered to permit efficient base pairing with the nonsense codon and subsequent peptide chain elongation on the ribosome. Such an altered suppressor tRNA can prevent premature chain termination of any polypeptide whose mRNA contains similar nonsense mutations. Thus strains of *E. coli* containing nonsense suppressors can in principle be pleiotropic in their suppressing phenotype. These strains also exhibit a low level of suppression of the nonsense codons that normally occur at the end of translated portions of mRNA. The U-A-G and U-A-A codons are frequently referred to as amber and ochre codons, respectively, so tRNAs that suppress these codons are designated as amber and ochre suppressors, respectively.

Since the most common event generating a nonsense-suppressing tRNA is a single mutation leading to a change in the tRNA anticodon, tRNAs that can mutate to become suppressors are predominantly those whose anticodons are related by a single base change to C-U-A, U-U-A, and U-C-A, the sequences complementary to the codons U-A-G, U-A-A, and U-G-A. This relationship, ultimately confirmed, was suggested by the first studies of nonsense codons.

5.3 Elucidation of Nonsense Codons

The initial investigation into the nature of tRNA-mediated suppression followed the amino acid assignment of 61 of 64 possible nucleotide triplets in the genetic code. The preliminary evidence for the nature of the amber and ochre nonsense triplets derived from genetic experiments using the rII mutations of T4 bacteriophage. Amber and ochre mutants in the rII A gene were mutagenized with hydroxylamine, 2-aminopurine, and 5-bromouracil. From the known base changes in DNA induced by these mutagens, for example, G-C to A-T transitions by hydroxylamine, and from the interconversions induced by them between amber, ochre, and wild-type mutations in the A cistron, it was deduced that triplets for amber and ochre mutations are U-A-G and U-A-A, respectively (2). By independent reasoning U-A-G and U-A-A were shown to be nonsense codons following observations of the various amino acid substitutions resulting from reversions of a nonsense mutation in the *E. coli* alkaline phosphatase gene (3, 4). The codon assignments of the inserted amino acids were in each case related to U-A-G or U-A-A by a single base change. Subsequently it was found that the presence of mutations designated su_4^+ and su_5^+ could suppress both the premature termination of the polypeptide chain at the U-A-G codon and a second class of alkaline phosphatase mutants—those that gave premature termination at the site of a U-A-A codon (5).

In *E. coli* all the suppressors of premature termination at U-A-A codons that have been isolated so far suppress termination at U-A-G codons. This is consistent with the postulates of Crick's wobble hypothesis, which allow base pairing of the complementary triplet to U-A-A, namely U-U-A, with both U-A-A and U-A-G. This is because the U can wobble to pair with either G or A. The complementary triplet to U-A-G, namely C-U-A, cannot base pair with U-U-A, offering an explanation of the fact that amber suppressors can suppress termination only at the site of amber codons. Theoretically a suppressor tRNA that recognizes only ochre codons, with an I-U-A anticodon, is also possible in *E. coli* (6).

Ochre suppressors are generally less efficient than amber suppressors (7, 8). Even so they often have a harmful effect on the cellular growth rate, suggesting that very efficient ochre suppressors may be lethal to the cell.

A third class of suppressible mutations is associated with the codon U-G-A. A set of mutant strains of T4 having previously identified mutations in the rII region were strongly suppressed in strains of *E. coli* that could not suppress amber or ochre codons at the same sites. The nature of the nucleotide triplets at the site of the suppressible mutations was confirmed by their conversion to ochre mutants, but not to amber mutants, by ethyl methane sulfonate mutagenesis, which is specific for G-to-A base transitions in DNA (9).

5.4 Transfer RNA as the Nonsense Suppressor

Early mapping studies of suppressor phenotypes showed that they mapped as discrete genetic loci at various places on the *E. coli* genetic map. This result gave strong support to the notion that suppression is mediated by a single gene product and this was subsequently confirmed by both nucleotide sequence analysis and functional studies of suppressor tRNAs. Not all possible suppressors have been isolated; some of them are probably lethal. But some lethal suppressors have been isolated through the use of partial diploid (merodiploid) strains. A suppressor might be lethal in a haploid strain if its formation involves a mutational change in a tRNA species that can then no longer fulfill its original function of uniquely translating a particular codon. A suppressor might also be lethal if it inserts the incorrect amino acid into certain codons or if it is very efficient in translating naturally occurring termination signals. In addition the absence of a normal tRNA species could interfere with the regulation of some biosynthetic process within the cell. Other suppressor tRNAs may suppress so weakly that detection is impossible; or they may insert an amino acid, at the site of the nonsense codon, that generates an inactive protein, making selection by genetic techniques impossible. Nonsense suppressors, in particular, are listed in table 5.1 and illustrated in figure 5.1, which also includes missense suppressors. Not all suppressor tRNAs are clustered on the *E. coli* genetic map (figure 5.1 and table 5.2), nor do all families of isoaccepting species cluster. For example, the two isoaccepting species of $tRNA^{Tyr}$ are located far from each other, so there is no possibility that the transcription of the two genes could be under the same operator and/or promoter control.

Proof that nonsense suppression is tRNA-mediated originally came from studies of suppression in vitro. Messenger RNAs from derivatives of RNA bacteriophages R17 and f2 containing amber mutants in the coat protein genes were translated in vitro by protein-synthesizing systems derived from suppressor strains of *E. coli*. The factor present in suppressor strains, and absent from wild-type strains, which enabled complete translation was in the tRNA fraction. In the case of the amber suppressor strain su^+_1 it could be shown that a new species of $tRNA^{Ser}$ was responsible for the insertion of serine into the now complete polypeptide chain at the site of the nonsense codon. The normal anticodon of $tRNA^{Ser}_1$ is only one base change removed from an anticodon that can suppress the amber codon U-A-G (10, 11). While these experiments demonstrated the functional involvement of tRNA in nonsense codon suppression, the first direct proof that an anticodon change in a tRNA could allow a tRNA to become a suppressor came from studies of suppressors derived from $tRNA^{Tyr}_1$ of *E. coli*.

5.5 Amber and Ochre Suppressors Derived from $tRNA^{Tyr}_1$

The amber suppressor su^+_3 is located at 27 minutes on the *E. coli* chromosome, close to the attachment site of the lysogenic bacteriophage Φ80. A transducing derivative of Φ80 carrying the gene for the su^+_3 suppressor was isolated as the result of an abnormal excision event of a lysogenized Φ80

Table 5.1 Some nonsense suppressors in *E. coli*

Suppressor Locus	Codon Suppressed	Amino Acid Inserted	Method of Identification
Su-1	U-A-G	serine	Insertion of serine in translation of bacteriophage RNA.
Su-2	U-A-G	glutamine	Synthesis in vitro of T4 lysozyme from amber mutant.
Su-3	U-A-G	tyrosine	Incorporation of tyrosine into polypeptides
Su-4	U-A-A, U-A-G	tyrosine	Incorporation of tyrosine into polypeptides
Su-5	U-A-A, U-A-G	lysine	Incorporation of lysine into polypeptides
Su-6	U-A-G	leucine	Ribosome binding of leucyl-tRNA, stimulated by U-A-G
Su-7	U-A-G	glutamine	Insertion of glutamine into amber mutant of tryptophan synthetase
		tryptophan	Insertion of tryptophan into amber mutants of bacteriophage T4 head protein. Ratio of tryptophan to glutamine inserted in this system 1:9.
U-G-A-1	U-G-A	tryptophan	Derivative of Su-7, identified by genetic and biochemical methods.
U-G-A-2	U-G-A	tryptophan	As U-G-A-1.

Note: This table illustrates a variety of nonsense suppressors that have been demonstrated in *E. coli*, with the codons suppressed, the amino acid inserted, and the method of identification. Suppressor genes for which the data are less complete have been omitted. This table does not, therefore, represent the full range of suppressors. The nomenclature for amber and ochre suppressors is that of Garen (see refs. 3, 4).

phage from the chromosome of an *E. coli* su$_3^+$ strain. Subsequently the transducing phage Φ80 su$_3^+$ was used to infect *E. coli* and during the course of the infection the suppressor gene carried by the phage was amplified about 100-fold. The product of the Su-3 gene is also similarly amplified compared to its normal intracellular levels, and when[32] P is added during infection, this product can be isolated from infected cell extracts in radiochemically pure form. The nucleotide sequence analysis of the tRNA coded by the Su-3 gene reveals that it is tRNA$_1^{Tyr}$ with a single base change in the anticodon (figure 5.2), demonstrating unequivocally the role of tRNA in nonsense codon suppression. Confirmatory evidence of the role of the anticodon in suppression mediated by tRNA^{Tyr} came from studies of an ochre-

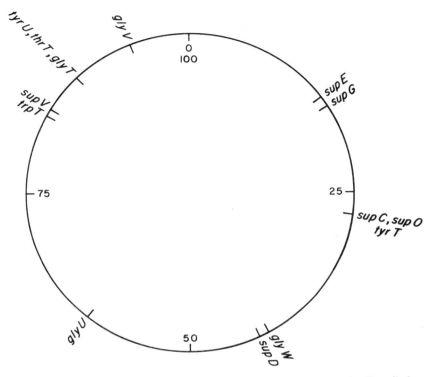

Figure 5.1 A map showing the positions of various tRNA genes on the *E. coli* chromosome (91). The map shows those suppressor genes with which particular tRNAs are associated. The positions of other tRNA genes are indicated even though these have been located by different methods, such as hybridization or nucleotide sequence analysis of tRNA precursor molecules. (See table 5.2.)

suppressing derivative of Φ80 su$_3^+$ (13, 14). The tRNA^{Tyr} encoded for by Φ 80su$_{oc}^+$ has a single nucleotide change in the first anticodon position when compared to tRNA^{Tyr}su$_3^+$—C to U—satisfying the wobble rules predicted by Crick (15).

Revertants of Φ80su$_3^+$ that have partially or completely lost suppressor activity have provided interesting information regarding the function of tRNA in suppression and in other intracellular processes.

5.6 Mutant Forms of tRNA^{Tyr}su$_3^+$

A major contribution to our understanding of the importance of the conformational integrity of tRNA in its biosynthesis and interaction with other

Table 5.2 Guide to the map of tRNA genes in figure 5.1

Gene Symbol	Alternate Symbol	Map Position (min)	Trait Affected
glyT	supA36	88	tRNA$_2^{Gly}$
glyU	suA36, sufD	61	tRNA$_1^{Gly}$
glyV	suA58, suA78	94	tRNA$_3^{Gly}$ (duplicate gene)
glyW	suA58, suA78	42	tRNA$_3^{Gly}$ (duplicate gene)
supC	Su-4	27	Ochre suppressor, inserts tyrosine
supD	Su-1	43	Amber suppressor, inserts serine
supE	Su-2	15	Amber suppressor, inserts glutamine
supG	Su-5	16	Ochre suppressor, inserts lysine
supO		27	Ochre suppressor (may be identical to supC)
supV	Su-8	84	Ochre suppressor
thrT		88	tRNA$_3^{Tyr}$
trpT	Su-7, supU	83	tRNA^{Trp}
tyrT	supF, Su-3	27	tRNA$_1^{Tyr}$
tyrU	supM	88	tRNA$_2^{Tyr}$

Note: Map positions taken from ref. 91.

macromolecules has come from the availability of mutant forms of the suppressing tRNA^{Tyr}su$_3^+$. These mutant tRNAs are derivatives of tRNA$_1^{Tyr}$ that contain one base change in the anticodon Q$_{35}$ (or G*) to C (where Q is a modified guanosine residue), and another base change elsewhere in the molecule. These double mutant forms of tRNA^{Tyr} were obtained following the selection of *E. coli* cells that were carrying or infected with the transducing bacteriophage Φ80 su$_3^+$ and showed reduced or absent suppression ability. Some of the nonsuppressing strains are of course revertants to the wild-type tRNA$_1^{Tyr}$. The strains of mutant transducing phage are referred to according to the position of the base change in the nucleotide sequence of tRNA^{Tyr}su$_3^+$; thus A15 designates a mutant in which the guanosine at position 15 has been altered to adenosine. One group of mutants (16) all differ from the immediate parent in that the nucleotide sequence of tRNA^{Tyr}su$_3^+$ contains a single G-to-A change at different locations: A15, A17, and A31. Functional analysis of A31 and A15 reveals that the alteration G to A at position 31 reduces the affinity of tRNA^{Tyr}su$_3^+$ for its homologous aminoacyl synthetase, whereas the same change at position 15 has no effect. In ribosome binding experiments tRNA^{Tyr}su$_3$ A15 bound to ribosomes in the presence of the U-A-G triplet to the same extent as tRNA^{Tyr}su$_3^+$, but the corresponding tRNA^{Tyr}su$_3$ A31 hardly bound at all. However, the A15 mutant suppresses the U-A-G codon weakly and anomalously in vitro and is therefore probably defective in some step in protein synthesis that occurs after the aminoacylated tRNA is bound to the ribosome. Recent studies of the solution conformation of tRNA^{Tyr}su$_3$ A15, performed by temperature-jump methods, have shown that the thermodynamics of folding and unfolding of the three-dimensional tRNA structure are significantly different from those of tRNA^{Tyr} (17).

In another study of mutant tyrosine tRNAs (18), temperature-sensitive suppressor mutants were examined. The nucleotide sequence of the mutant tRNA^{Tyr} from these phages revealed that they contained, in addition to the anticodon mutation, single base changes that prevented hydrogen bonding to form base pairs in the amino acid acceptor stem. When a second site revertant of the mutant A2 was obtained having suppressor activity in vivo identical to the wild-type su$_3^+$, it was found to have the mutations A2 and U80 (figure 5.2) thereby restoring base pairing in the stem. Similarly the suppressor function of su$_3$ A25 is restored in the double mutant A25Ull where base pairing in the dihydrouracil stem is restored. Other mutants and double mutants in suppressor activity have also been examined, and it is clear that the amount of tRNA synthesized by the mutant phage is directly related to the change in tRNA structure, since double mutants that have restored suppressor activity and structure produce nearly the same amounts of tRNA as su$_3^+$ strains (see table 3.2).

Another interesting phenomenon arising from the study of mutant forms of tRNA^{Tyr}su$_3^+$ is that some mutant forms mischarge; together with the result for A31 and A15, this phenomenon indicates important sites for the recognition by tyrosyl aminoacyl synthetase of tRNA^{Tyr} (19, 20). Mutants of Φ80su$_3^+$ that exhibit missuppression, the insertion of an amino acid other than tyrosine at the nonsense codon, have also been examined. They include A2, A1, A81, U81, and G82; each mutation occurs in the aminoacyl stem of the tRNA (see figure 5.2), indicating that it is a site of primary importance in the recognition of the tRNA substrate by the aminoacyl synthetase. These mutants can translate the U-A-G codon as glutamine in vivo (A81 has only been shown to insert a neutral amino acid). In aminoacylation experiments in vitro A2, U81, A81, and G82 can all be aminoacylated with glutamine in the presence of high concentrations of the synthetase, although with the exception of G82 they retain the ability to insert tyrosine as well as glutamine in vivo.

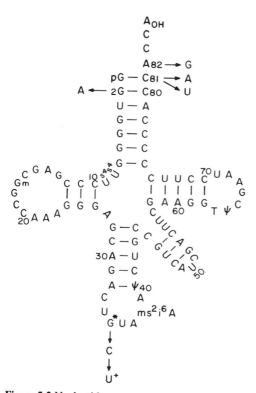

Figure 5.2 Nucleotide sequence of suppressor derivatives of *E. coli* tRNA$_1^{Tyr}$. The wild-type, nonsuppressing anticodon has a modified G (designated Q; see ch. 7) in the first, or wobble, position. The amber-suppressing derivative su$_3^+$ has a C in the first position, and the ochre-suppressing derivative su$_{oc}^+$ A2 is a double mutant containing a modified U (U$^+$) in the first anticodon position and a second mutation A2, which confers temperature sensitivity on the function of the suppressor tRNA. Also shown are several other secondary mutations of the amber-suppressing derivative: A2, A81, U81, and G82. All these confer temperature sensitivity on the suppressor function, and all also make the tRNA capable of mischarging with glutamine.

These data, resulting from single base changes in the tRNAs' nucleotide sequence, offer important insight into the recognition of tRNA by the homologous aminoacyl synthetase, which has only been made accessible through genetic manipulation of a suppressor tRNA.

5.7 T4 Suppressor tRNAs

In addition to the studies of mutant forms of tRNA^{Tyr} isolated from *E. coli*, studies of bacteriophage T4-encoded amber- and ochre-suppressing tRNAs have furthered our understanding of the relationship between the structure of tRNA and its interaction with macromolecules, particularly the biosynthetic processing enzymes. When bacteriophage T4 infects *E. coli*, eight new tRNAs coded for by the T4 genome are produced. Suppressor mutants have been derived from two of these eight tRNAs. Transfer RNA^{Ser}, whose anticodon is N-G-A (where N is an unidentified modified nucleotide), has been shown to mutate at low frequency to an amber suppressor psu$_1^+$ with the anticodon C-U-A (21). The two mutations in the anticodon required to make it an amber suppressor account for the low frequency with which this mutant is obtained. The nature of the suppressor was ascertained by analysis of the polypeptide made when an amber mutation of a T4 head protein is suppressed. The amino acid inserted at the site of the amber codon was serine. The suppressor identified as psu$_1^+$ is thought to be identical to the suppressor psu$_a^+$. An ochre suppressor allelic to psu$_a^+$, named psu$_b^+$, has also been identified (22). A second ochre suppressor, psu$_2^+$, has been derived from the T4 tRNA^{Gln} with the anticodon altered from N-U-G to N-U-A, where N is an unidentified derivative of uridine (23). Originally isolated by suppression of ochre mutations in the gene for the T4 lysozyme, the mutant was identified using secondary revertants with no suppressor activity. The loss of suppressor activity was accompanied by the disappearance of tRNA^{Gln} from infected cells; this disappearance was easily monitored by analyzing the tRNA species, made when *E. coli* was infected with T4, by polyacrylamide gel electrophoresis. Subsequent studies have shown that the secondary mutation causes defects in processing of the tRNA.

The low efficiency of psu$_2^+$ as a suppressor led to a search for mutations of psu$_2^+$ with increased suppression efficiency. A more efficient suppressor shows the phenotype expected of an amber suppressor, suppressing amber mutations exclusively and the expected anticodon change from N-U-A to C-U-A. Whereas N can wobble to pair with either A or G, allowing the original psu$_2^+$ to suppress both U-A-G and U-A-A mutations, C can pair only with G in this position allowing suppression only of the U-A-G mutation.

5.8 Suppression of the U-G-A Nonsense Triplet

The derivatives of only one tRNA in *E. coli* have been definitively shown to be involved in suppression of the U-G-A codon, namely tRNA^{Trp}. The first tRNA^{Trp} suppressor of U-G-A mutations that was isolated and for which the nucleotide sequence was obtained differs from the amber and ochre suppressors in that the base change in the tRNA responsible for the suppressor activity is not in the anticodon of the tRNA. Hirsh (24) determined the nucleotide sequences of the normal and suppressing species of tRNA^{Trp} from the corresponding strains of *E. coli*. Hirsh found that in the suppressing species of tRNA^{Trp} base 24 is altered from adenine to guanine (see figure 5.3), but that in both the suppressing and the normal tRNA^{Trp} the anticodon is C-C-A. According to the wobble hypothesis this anticodon should recognize only the tryptophan codon U-G-G, and not the nonsense codon U-G-A. The change at base 24 must contribute in some fashion to the stabilization of the interaction between the anticodon and U-G-A, allowing this unusual (C-A) base pairing to occur in the wobble position. The nonsuppressing form of tRNA^{Trp} can also translate the U-G-A codon as tryptophan in polypeptide synthesis in vitro, but it does this with a much lower efficiency than the suppressor. Similarly the suppressor tRNA can translate U-G-G in vitro. This corresponds to the observation in vivo that the suppressor mutation is not lethal, since the mutant tRNA^{Trp} can recognize both the tryptophan codon U-G-G and the nonsense codon U-G-A.

In other independent genetic experiments Soll isolated two different U-G-A-suppressing forms of tRNA^{Trp}, designated U-G-A-1 and U-G-A-2 (25). The first of these is presumably identical to the U-G-A suppressor characterized by Hirsh since it can be converted by mutation to an amber suppressor. This would happen if the anticodon of the U-G-A-suppressing tRNA is C-C-A, which could mutate in a single step to C-U-A (figure 5.3). The second U-G-A suppressor, U-G-A-2, is presumed to have the anticodon U-C-A, which is different from that of wild-type tRNA^{Trp}. The suppressing strain U-G-A-2 can be derived either through mutation of the ochre suppressor derived from su$_7^+$, in which the tRNA^{Trp} has the anticodon U-U-A, or through mutation of the Su-7 wild-type gene, in which the tRNA^{Trp} has the anticodon C-C-A.

The su$_7^+$ amber suppressor is lethal in haploid strains presumably because there is only one gene copy of tRNA^{Trp} and because the anticodon change from C-C-A to C-U-A, which allows this tRNA to recognize the amber codon, also prevents it from reading the tryptophan codon any longer. The C-to-U base change in the middle position of the anticodon has the unusual effect of changing the aminoacyl charging specificity of the tRNA, resulting in mischarging with glutamine instead of charging with tryptophan. According to the wobble hypothesis, however, the U-G-A suppressor with

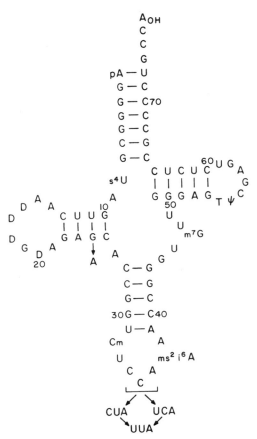

Figure 5.3 Nucleotide sequence of suppressor derivatives of *E. coli* tRNA^{Trp}. The wild-type, nonsuppressing anticodon has the sequence C-C-A. The suppressor characterized by Hirsh (24) has the wild-type anticodon but also a G-to-A change at position 24. Soll (25) has characterized amber-, U-G-A-, and ochre-suppressing derivatives of the tRNA that have changes in the anti-codon only, as shown in the diagram.

the anticodon change from C-C-A to U-C-A should still recognize the tryp-tophan codon (U-G-G). Therefore it is assumed that the change in the first anticodon position also results in mischarging and that the subsequent insertion of glutamine at normal tryptophan codons enhances the lethality of this tRNA species in the haploid state. A similar recessive lethal form of U-G-A suppressor has been isolated in *Salmonella typhimurium* by Miller and Roth (26).

5.9 Suppression of U-G-A in *Salmonella typhimurium* by supK Strains

A novel recessive suppressor of U-G-A mutations has been demonstrated in *S. typhimurium* (27). In wild-type strains, as is the case in *E. coli*, there is a low level of misreading of U-G-A mutations, but in mutant strains designated supK there is an increased level of suppression that appears to be correlated with a partial loss of tRNA methylase activity. The undermethylated tRNA or tRNAs responsible for the suppression of U-G-A codons have not been identified, nor has an analogous situation been revealed thus far in *E. coli*. Some strains of *Salmonella* that carry the supK mutation also suppress certain frameshift mutations. Total absence of the supK gene product is lethal, but merodiploid strains that contain both the wild-type (supK$^+$) and the suppressor supK genes display no suppressor activity. In the presence of the supK mutation growth of the cells is very slow, probably resulting from the loss of normal coding properties by the undermethylated tRNAs. Further information concerning the role of nucleotide modification in codon-anticodon recognition and how this relates to suppression has been derived from studies of *E. coli*.

5.10 A Role for Nucleotide Modification in Nonsense Codon Suppression

Transfer RNAs that contain the sequence U-U in the first two positions of the anticodon often have a modification of the aromatic ring of the first U. This is true of certain species of tRNA^{Glu} and tRNA^{Lys} in both *E. coli* and yeast, in which the modified base has been found to be a 2-thiouridine derivative (28–33), as well as for two ochre-suppressing tRNAs of know sequence, *E. coli* tRNA^{Tyr}su$^+_{oc}$ A2 (14) and T4 tRNA psu$^+_2$ (34). In the latter two cases the exact nature of the modified uracil derivative is not known, but any modification may very well have an effect on codon-anticodon interactions. For example, a 2-thiouridine derivative in the first anticodon position should in principle restrict wobble (15) so that the modified U pairs only with A rather than with A and G. In fact, tRNA^{Glu} species containing such a modification appear to pair better with A than with G (29, 35), but the specificity is not absolute. Thus considerations of codon-anticodon interactions based solely on canonical Watson-Crick hydrogen-bonding possibilities must be too restrictive. Nucleotide modification may also affect the ability of the three anticodon bases to stack and form stable base-paired sections of helices with the codon segment of mRNA (36, 37).

The exact nature of the influence of modification on codon-anticodon interaction may vary with the tRNA species, again underscoring the complex nature of this phenomenon. Sulfur-deficient *E. coli* tRNA$_2^{Glu}$ pairs more efficiently in vitro with G-A-G than with G-A-A, unlike the fully 2-thio-substituted parent species (38). However, yeast tRNA$_3^{Lys}$, chemically treated to remove the 2-thio modification of the uridine in the first anticodon position, still pairs better in vitro with A-A-A than with A-A-G, as does the untreated parent species (33).

An ochre-suppressing tRNA in which the wobble position U (or modified U) pairs better with A rather than G would more efficiently suppress ochre (U-A-A) than amber (U-A-G) codons. T4 tRNA psu$_2^+$ favors either ochre or amber codons, depending on the context of the nonsense mutations being examined (23, 39). *E. coli* tRNA^{Tyr}su$_{oc}^+$ A2, on the other hand, differs from most other ochre suppressors in favoring the amber codon at most (but not all) sites examined (37). It seems unlikely that this latter suppressor has a thiolated U in the first anticodon position. In fact, attempts to label this nucleotide with ^{35}S have proved futile (14).

A most convincing demonstration of the importance of nucleotide modification in nonsense codon recognition and the complexity of this effect arises from the isolation of an *E. coli* mutant strain that fails to modify the first position of the anticodon of T4 tRNA psu$_2^+$ (40). In this strain the efficiency of suppression of psu$_2^+$ and another ochre suppressor, psu$_b^+$, are often much reduced. Further, the extent of the reduction in psu$_2^+$ efficiency varies not only with the context of the nonsense codon (in the T4 lysozyme gene), but also with the nature of the codon—amber or ochre.

5.11 Effects of mRNA Context on Nonsense Codon Suppression

The efficiency of nonsense suppression by a particular nonsense-suppressing tRNA can vary greatly, depending on the position of the nonsense codon in a particular mRNA. Variations of this kind have been observed in certain genes at sites where the nature of the amino acid being inserted is not thought to affect protein function (41, 42, 43). The use of homotopic amber and ochre mutations (a homotopic amber and ochre pair are defined as amber and ochre mutations occurring at the same codon) allows a measurement of the ratio of suppression of an ochre codon to its homotopic amber codon by any particular ochre-suppressing tRNA. This ratio measured at various sites in several *E. coli* and T4 genes can vary over a tenfold range with a particular suppressor tRNA, clearly showing the effects of mRNA context on nonsense codon suppression (23, 37, 39, 44). Used in this manner, the homotopic ratios also eliminate the problem of differential effects of the inserted amino acid on protein function.

The effect of mRNA context on nonsense codon suppression could come about through interactions (1) between the suppressing tRNA and nucleotides adjacent to the nonsense codon in the mRNA, (2) between the suppressing tRNA and an adjacent tRNA on the ribosome, and (3) between release factors and certain sequences adjacent to the nonsense codon in the mRNA. Different suppressing tRNAs can manifest different homotopic amber–ochre ratios at sites where the inserted amino acid is unimportant for protein function (37). Thus part of the tRNA structure other than the primary anticodon sequence must be involved in the codon recognition process. These effects may be due in part to nucleotide modifications. In addition attempts have been made to study the effect of particular mRNA contexts on nonsense codon suppression in genes where the nucleotide sequence adjacent to nonsense codons are known (37, 44–46). Some sequences, for example C-A following the nonsense codon, appear to have systematic effects on all suppressors tested, but no general rules relating nonsense codon suppression efficiency and mRNA context have yet emerged. Detailed knowledge of the structure of the tRNA anticodon loop while on the ribosome and its relation to the segment of mRNA being decoded will probably be required before generalizations can be made. If mRNA context affects the efficiency of translation of naturally occurring codons, as seems likely from studies of nonsense codon suppression, it may be one means by which in vivo translation efficiency and utilization of iso-accepting tRNA species are modulated. An apparent case of such modulation has been observed in the rescue by a minor tRNA species of normal protein synthesis in extracts of interferon-treated eucaryotic cells (47).

5.12 Suppression of Naturally Occurring Nonsense Codons: The Mechanism of Chain Termination

When the nonsense codons U-A-A, U-A-G, and U-G-A occur in the middle of messages being translated, premature termination of the incomplete polypeptide occurs. These codons, however, are the same as those that signal accurate termination at the end of mRNA translation. In the presence of suppressors of the termination codons there is significant and deleterious read-through of the termination signals, resulting in the biosynthesis of longer polypeptides than are normally produced.

The phenomenon of read-through of natural termination signals has been investigated in several systems. The activity of both β-galactosidase and transacetylase in strains of *E. coli*–carrying ochre suppressors is found to be severely reduced when compared to the activity of these enzymes in wild-type cells. This may be due to the production of anomalous enzyme molecules following the suppression of the naturally occurring U-A-A termination signals (48). A similar effect has been observed following the infection by MS2 bacteriophage of *E. coli* carrying an amber suppressor. This phage normally codes for three proteins: coat protein, A protein, and replicase; but in amber-suppressing hosts a fourth protein is produced, shown by gel electrophoresis to be about 30 amino acids longer than the normal A protein and not incorporated into mature virus particles. This result suggests that the termination signal in the mRNA for the A protein is U-A-G (49); this has been subsequently confirmed by nucleotide sequence analysis (92). An investigation of translation of bacteriophage mRNAs in vitro by extracts of *E. coli* amber suppressors showed that for R17, MS2, and f2 a longer-than-normal synthetase protein is produced resulting from the read-through of U-A-G termination codons by the amber suppressors (50).

Translation of the Qβ bacteriophage coat protein in a U-G-A-suppressing host has provided evidence that U-G-A can act as a natural terminator. Accompanying the synthesis of the Qβ coat protein is normally a small amount of inefficiently terminated longer protein, 200 amino acids longer than the major translation product. In a U-G-A-suppressing host this longer protein is present in larger amounts. Peptide analysis of this longer protein shows that it contains tryptophan at the site of the unrecognized termination codon and that translation has continued beyond it (51).

The mechanism of peptide chain termination requires not only the presence of the terminator or nonsense codons, but also the soluble release factors RF1 and RF2, which interact with U-A-A and U-A-G or with U-A-A and U-G-A, respectively. This interaction on the ribosome is stimulated by a third factor RF3 which itself interacts with GDP and GTP (for a review see ref. 52).

In general the growth rate of *E. coli* cells containing amber suppressors is unaffected by the presence of the amber suppressors; but ochre suppressors, in spite of their comparatively low efficiency of suppression, do seem to have a deleterious effect on cell growth and enzyme production. This may be because ochre codons are the most widely used natural termination signal or are used in genes for which effective termination is an absolute requirement for efficient growth. Alternatively ochre suppressors are the only type of suppressor that can translate both ochre and amber codons, thereby causing read-through of any genes that employ tandem amber-ochre termination signals.

The context—that is, the nature of the surrounding codons—of naturally occurring termination signals varies considerably in the different mRNAs that have been examined from *E. coli* and various bacteriphages. No particular sequence of nucleotides seems to be required on either side of the natural terminators, nor is there such a sequence on either side of nonsense codons arising from mutation. In this respect inferences regarding the mechanism of chain termination derived from studies of internal nonsense codon suppression should be relevant to the mechanism of natural chain termination.

5.13 Missense Suppression

The presence of a single base mutation in the mRNA for a particular protein can give rise to a protein that has a mutant phenotype because an incorrect amino acid can be inserted at the site of the altered codon. Insertion of the correct amino acid, in spite of the mutant codon, is termed missense suppression. This requires that a tRNA that can be aminoacylated with the correct amino acid be able to translate the codon for the incorrect amino acid. All the available data on missense suppression in *E. coli* concern species of $tRNA^{Gly}$ that can insert glycine at the sites of codons specifying other amino acids. There are three isoaccepting species of $tRNA^{Gly}$, each of which normally responds to one or two of the codons G-G-G, G-G-U, G-G-C, and G-G-A. They are coded for by four genetic loci glyT, glyU, glyV, and glyW (figure 5.1 and table 5.3). The study of the $tRNA^{Gly}$ missense suppressors has been of particular interest because of the involvement of the whole family of isoaccepting tRNAs and the way in which the species of $tRNA^{Gly}$ at the four different loci have been mutated in their anitcodons to suppress a variety of missense mutations. The data on missense suppression derive from extensive analysis of different amino acid insertions into the tryptophan synthetase A gene (53).

5.14 Missense Suppression in the Tryptophan Synthetase A Gene

Examination of a variety of mutant forms of the tryptophan synthetase A (TSA) protein has revealed a set of mutations in the codons originally specifying glycine residues that result in the insertion of other amino acids at

Table 5.3 Missense suppressors of *E. coli* derived from tRNA^{Gly} genes

Locus	tRNA	Codon Recognized	Wild-Type Anticodon	Suppressor Anticodon	Suppressor
glyT	tRNA$_2^{Gly}$	G-G-A/G	U-C-C	N-C-U	glyTsu36$^+$ (HA)
glyU	tRNA$_1^{Gly}$	G-G-G	C-C-C		glyUsu36$^+$
				C-U-C	supT (G-A-G)
				U-U-C	supT (G-A-A)
glyV	tRNA$_3^{Gly}$	G-G-U/C	G-C-C	U-C-C	glyVins
				U-U-C	glyVins (G-A-A)
glyW	tRNA$_3^{Gly}$	G-G-U/C	G-C-C	G-C-A	glyWsu78

Note: References given in text.

these sites. From the nature of the amino acids inserted and a knowledge of the possible codons for these amino acids, it is possible to deduce the nature of the original glycine codons. Specific missense suppressor strains that cause the correct insertion of glycine in the polypeptide specified by the mutant mRNAs have been isolated for particular sites in the TSA gene. For mutants in the TSA gene designated A36, A58, and A78 extragenic suppressor classes have been obtained; these are named su36$^+$, su58$^+$, and su78$^+$, respectively.

5.15 glyTsu36$^+$ and glyUsu36$^+$

Arginine is inserted into tryptophan synthetase A in place of glycine at a single specific site in the mutant strain A36. The missense suppressor su36$^+$ inserts glycine at the arginine codon A-G-A. When purified su36$^+$ tRNA is used in a cell-free protein-synthesizing system with poly rAG as template, glycine is occasionally inserted in place of arginine in the newly formed polypeptide (54, 55). Transfer RNA from nonsuppressing strains does not allow insertion of glycine. Similarly, with tRNA from su78$^+$, glycine is sometimes inserted in place of cysteine in response to the U-G-U codon when poly rUG is used as template (normally only cysteine and valine are incorporated; see ref. 56). A second suppressor of the A36 mutation, with similar activity in the assay in vitro, is located at a different place on the *E. coli* chromosome (57). As a result there are two suppressors of A36, designated glyTsu36$^+$ and glyUsu36$^+$.

Viable suppressor strains carrying the glyTsu36$^+$ suppressor are heteroploid for the suppressor locus. Haploid glyTsu36$^+$ strains grow very poorly and require the insertion of wild-type glyT genes for normal growth. The glyUsu36$^+$ gene on the other hand has no deleterious effects on cell growth. Chromatographic separation and gene mapping have shown that the glyU locus is the structural gene for tRNA$_1^{Gly}$ that in the presence of ribosomes binds the triplet G-G-G. The glyT locus is the structural gene for tRNA$_2^{Gly}$ that binds the triplets G-G-A and G-G-G (57, 58). Clearly the mutation that gives rise to the altered coding properties of tRNA$_2^{Gly}$ in glyTsu36$^+$ strains eliminates the presence of a tRNA that can translate G-G-A, while alteration of tRNA$_1^{Gly}$ in glyUsu36$^+$ still leaves tRNA$_2^{Gly}$ capable of translating the G-G-G codon. In both su36$^+$ suppressor strains the capacity of the mutant tRNA to interact with glycyl-tRNA synthetase is reduced by several orders of magnitude. This can explain, at least in part, the low efficiency of the missense suppression.

There are two types of glyT suppressors. One is glyTsu36$^+$ (HA), easily induced by hydroxylamine and semilethal when haploid. The other is a rare, spontaneously occurring mutant form glyTsu$_{159}$, whose effect on cell growth is not severe. It has been suggested that the latter mutation allows ambiguous reading of both G-G-A and A-G-A codons (57). Transfer RNA^{Gly} from glyTsu36$^+$ (HA) has been studied in detail and the nucleotide sequence of the anticodon compared to that of the wild-type tRNA$_2^{Gly}$ (codon recognized: G-G-A/G). The wild-type anticodon is U-C-C or N-C-C, and the mutant anticodon is U-C-U or N-C-U (N is an unidentified derivative of U). The base adjacent to the anticodon on the 3'-side is changed from A to a derivative of threonyl carbamoyl adenosine (59). The new anticodon sequence is complementary to A-G-A, the codon recognized in vivo and in vitro. The appearance of this hypermodified nucleotide adjacent to the anticodon is the result of the newly available recognition site for the enzymes responsible for its biosynthesis. Threonyl carbamoyl adenosine and its derivatives are only found in tRNA from *E. coli* and other organisms when there is a uridine residue in the third position of the anticodon, as has become the case in the missense suppressor glyTsu36$^+$ (HA). The presence of this modification in the missense suppressor tRNA may enhance the interaction between the anticodon and the missense codon and the ribosome binding of the complex.

Investigation of an independently isolated missense suppressor mutation supT, which suppresses G-A-G, reveals that it also gives rise to mutant forms of tRNA$_1^{Gly}$ (60). Neither the supT mutation nor a derivative of it suppresses the A36 mutation, but recombination experiments show that supT and glyU are very closely linked and probably allelic. The anticodon of the supT missense suppressor is C-U-C, corresponding to the codon G-A-G for glutamic acid. This suppressor can be converted to yet another suppressor, and the probable anticodon U-U-C, specific for G-A-A, is also a glutamic acid codon.

5.16 glyVsu58$^+$ and glyVins

Two genetic loci specify the gene for tRNA$_3^{Gly}$ that inserts glycine at the sites of G-G-U and G-G-C triplets. At one of these loci, glyV, there must be several copies of the gene for the tRNA, and a single mutation alters only from one-quarter to one-third of the total glyV product. The second locus specifying tRNA$_3^{Gly}$ is glyW. An *E. coli* precursor tRNA containing three copies of the tRNA$_3^{Gly}$ sequence has been isolated by Schedl et al. (61). Thus one or both of the sites glyV and glyW contain multiple gene copies that are transcribed together.

The A58 mutation in TSA is suppressed in glyVsu58 where a mutation in the anticodon, from G-C-U to tRNA$_3^{Gly}$, allows it to translate the aspartic acid triplet A-G-U/C as glycine. A second mutant form of tRNA$_3^{Gly}$, specified by glyVins, is a pseudorevertant of glyTsu$_{159}$. In the glyVins mutant the anticodon of tRNA$_3^{Gly}$ is changed from G-C-C to U-C-C so that the tRNA can read either G-G-A or G-G-G and can thus compensate for the deficiency due to the glyTsu$_{159}$ mutation (62).

In a search for missense suppressors of a TSA mutant A46, which would insert glutamic acid in place of glycine as the result of a G-G-A to G-A-A change in the codon, mutant strains containing glyVins were mutagenized with hydroxylamine. A suppressor of A46 was obtained, a mutant of tRNA$_3^{Gly}$ that could translate G-A-A as glycine. Most probably the new anticodon is U-U-C, derived by a single-step mutation from glyVins tRNA$_3^{Gly}$ whose anticodon is U-C-C (63).

5.17 glyWsu78$^+$

The missense suppressor su78$^+$, which inserts glycine at the cysteine codon U-G-U/C, is also the result of a mutation of tRNA$_3^{Gly}$. However, this mutation maps not at glyV (at 86 minutes on the *E. coli* chromosome) but at glyW (at 37 minutes), and it alters only about one-tenth of all tRNA$_3^{Gly}$ molecules. The tRNA specified by glyWsu78$^+$ has anticodon G-C-A instead of G-C-C as found in the wild-type, and the base adjacent to the 3'-end of the anticodon is now methylthio-isopentenyl adenosine rather than adenosine (64). The alterations in the anticodon result in a 200-fold decrease in the speed (V_{max}) of the aminoacylation by glycyl tRNA synthetase.

In summary there are four loci for the three isoaccepting species of tRNA^{Gly}: glyT, glyU, glyV, and glyW. Alterations in the anticodons of the three isoacceptors allow the tRNAs specified by these loci to become missense suppressors, inserting glycine at the sites of codons for other amino acids. The changes in the anticodons are often accompanied by other changes affecting the tRNAs, such as biosynthesis of new modified nucleotides adjacent to the anticodon, and decreased affinity for the homologous aminoacyl synthetase, demonstrating the importance of the structural integrity of a tRNA molecule in its interaction with other macromolecules.

5.18 Suppression of Frameshift Mutations

The accurate translation of mRNA requires that the tRNA molecules read the codons for the correct amino acids in phase, triplet after triplet. The addition or deletion of nucleotides from the mRNA resulting from the insertion or removal of nucleotide pairs from the DNA alters the reading frame of the triplets unless complete triplets or multiples thereof are added or deleted. Incorrect amino acids are then inserted following this frameshift until a nonsense codon is reached and translation is terminated. Neither the tRNA nor the ribosome has any mechanism for recognizing that the message is being read in an incorrect phase. It appears that most spontaneous frameshift mutations are base-pair deletions (-1 frameshifts), but those induced synthetically by intercalating agents or selected by testing for reverse mutations are usually base insertions ($+1$ frameshifts). Frameshift mutations can be suppressed intragenically by a second insertion or deletion of a nucleotide close to the site of the original mutation, to restore the correct phase of translation. Frameshift mutations can also be suppressed by secondary mutation external to the gene containing the original mutation. All these frameshift suppressors have been found in S. *typhimurium*.

The first extragenic suppressor of a frameshift mutation that was found was the U-G-A suppressor associated with the supK mutation in *Salmonella*. This suppressor, resulting from a mutation in a tRNA methylase, allows suppression of both frameshifts and the U-G-A nonsense mutation, but the mechanism by which this comes about is unclear (65).

The first suppressor of a frameshift mutation that could be definitively demonstrated to involve tRNA was a suppressor of a frameshift mutation in the histidine operon, in the histidine dehydrogenase gene of S. *typhimurium*

(66). A comparison of the wild-type protein and the amino acid sequence of several revertants of the frameshift mutant showed that the frameshift was due to the existence of an extra G·C base pair in the gene (67).

Examination of a variety of frameshift mutations in the histidine operon and their extragenic suppressors revealed 48 separate suppressor mutations. These mutations map as six groups on the *Salmonella* chromosome, but they fall into two groups on the basis of their suppressor ability (68). The six genetic groups, designated sufA through sufF (suf = suppressor frame-shift) are distributed throughout the *Salmonella* genetic map. The suppressors sufA, sufB, and sufC all correct the phase of frameshift mutations in which there is an extra C within a sequence of several other C residues; sufD, sufE, and sufF correct the phase of reading when there is an extra G within a sequence of several other G residues. The sufB mutation that suppresses the frameshift mutation in the histidine dehydrogenase mutant, D3018, allows the production of wild-type protein, translating the four-base codon C-C-C-U as a proline codon (69). Chromatographic analysis of the tRNA^{Pro} produced by sufA (which also suppresses this mutation) and sufB shows differences in the elution profiles for these mutants as compared to the wild type, supporting the hypothesis that in both cases an altered tRNA^{Pro} species recognizes the suppressible proline codon containing an extra C residue.

The suppressing activity of both sufD and sufF results from altered species of tRNA^{Gly}. The suppressor frameshift sufD maps on the *Salmonella* chromosome at a position corresponding to the glyU gene in *E. coli* that codes for tRNA$^{Gly}_{GGG}$. The complete nucleotide sequence of the tRNA^{Gly} species responsible for the suppression activity of sufD, obtained by Riddle and Carbon, reveals that the anticodon that interacts with G-G-G-G is C-C-C-C—most unusually, a four-base anticodon rather than a triplet (70). Although sufD and sufF suppress mutations of a similar type, sufD is a dominant mutation and sufF is recessive. sufF appears to code for an altered tRNA^{Gly} corresponding to *E. coli* tRNA$^{Gly}_2$, but it is probable that the altered chromatographic profile is due to a deficiency in a tRNA modification enzyme that allows some misreading of the frameshift mutation. This may be compared to the low level of frameshift suppression by the nonsense suppressor supK, also the result of inadequate nucleotide modification. The existence of two modes of frameshift suppression demonstrates a structural and functional flexibility of tRNA, the details of which we have yet to understand.

Although all the frameshift suppressors so far described act at sites where there are several G residues or several C residues in sequence in the mRNA, this apparent specificity is probably due to the mechanism of selection of the frameshift mutations to be suppressed. The intercalating compound ICR-191, an acridinelike compound, was used to induce most of the frameshift mutations in bacteria, and it is specific for runs of G·C base pairs. There is no theoretical reason to suppose that frameshifts and the tRNAs that suppress them could not be found for other sequences, runs of A, or runs of U. The mechanism for tRNA-mediated suppression of frameshift mutations is unclear, except in the case of the four-base anticodon of tRNA^{Gly} from sufD. It is not known how a tRNA with three bases in the anticodon could cause the ribosome to advance four bases along the message at the site of a frameshift mutation, nor how alteration of the modification of a tRNA could facilitate this. Frameshift mutations are suppressible by streptomycin-induced ribosomal ambiguity at a low level, indicating that suppression can be mediated by the ribosome as well as by tRNA. (For a review of frameshift mutation and suppression see ref. 71.)

5.19 Transfer RNA–Mediated Suppression in Yeast

The only eucaryotic systems in which tRNA-mediated nonsense suppression has been unequivocally demonstrated are in the yeasts *Saccharomyces cerevisiae* and *Schizosaccharomyces pombe*. We shall discuss primarily results obtained from work with *S. cerevisiae* although there is a large body of information pertaining to nonsense suppression in the fission yeast *S. pombe* (72). No basic differences in suppression mechanisms have been observed, but the biochemistry of this latter system has not been so extensively investigated.

The nature of suppression in yeast, so fas as it has been investigated, differs from nonsense suppression in procaryotes primarily in the higher redundancy of the genes for tRNA^{Tyr} that can be mutated to suppressor alleles and in the existence of exclusively ochre suppressors (rather than, as in the case in *E. coli*, ochre suppressors that can also suppress amber mutations). It is estimated that there are 320 to 400 genes specifying tRNAs in *S. cerevisiae* that would allow for the redundancy of many tRNA species (73). Suppressors of amber and ochre mutations have been identified and shown to be tRNAs, but no tRNA suppressor for U-G-A has been directly demonstrated, although there is evidence for the existence of suppressible U-G-A mutations (74). Since yeast may be manipulated either as a haploid or a diploid organism, with or without extrachromosomal, cytoplasmically inheritable plasmids, it has been possible to show the existence of suppressor genes that are lethal to the haploid organism and of genes that modify suppressor activity. Recent work on yeast tRNA suppressors has shown how amber and ochre codons in procaryotic mRNA can be suppressed during translation in vitro by tRNA from yeast suppressor strains. This work shows particular promise since it affords a method for the assay of suppressor tRNAs from other eucaryotic system (85, 86).

The initial identification of yeast tRNA suppressors resulted from the isolation of mutants that could simultaneously revert the phenotype of as many as five separate mutations. This type of mutation was originally called a super-suppressor, but this term has fallen from use (75, 76). Amber and ochre mutations were identified and located by amino acid replacement in intragenic revertants of the yeast protein iso-1-cytochrome c, whose amino acid sequence has been completely determined. Eight independent suppressor mutations were found that could suppress the ochre mutations and insert tyrosine at the site of U-A-A nonsense codons. However, the suppressors of the ochre mutations could not suppress any of the amber mutations (77). These suppressors are designated SUP2-1, SUP3-1, SUP4-1, SUP5-1, SUP6-1, SUP7-1, SUP8-1, and SUP11-1 and have been classified as class 1 set 1. Since each of the suppressors inserts tyrosine but maps at a separate location on the yeast chromosome, each probably derives from a different redundant gene for an isoacceptor of tRNA^{Tyr}. Subsequently allelic mutations were obtained that could suppress amber mutations (but not ochre), and these were named SUP6-2 and SUP7-2 (78). These amber-suppressing mutants cause the insertion of tyrosine into amber mutants of iso-1-cytochrome c, which leads to the conclusion that the gene products of the allelic amber and ochre suppressors are differently altered forms of the same tRNA^{Tyr}. The specificity of the ochre suppressor that can suppress U-A-A codons but not U-A-G codons is presumed to result from the presence of inosine in the first position of the anticodon. Inosine can form base pairs with uridine, cytosine, and adenosine but not with guanosine. In *E. coli* where inosine has not been found in the anticodon of ochre suppressors, the possibility for base pairing with guanosine exists, allowing ochre suppressors to suppress amber mutations.

The alteration of an isoaccepting species of tRNA^{Tyr} in yeast-suppressing strains was shown by the use of reverse phase chromatography in which tyrosyl-tRNA^{Tyr} from wild types and suppressors was compared. In each ochre-suppressing strain tested there was an extra early eluting peak of tyrosyl-tRNA, and in a strain with an allelic amber suppressor there was a new late eluting species. In each case a peak representing a species of tyrosyl-tRNA normally present in the wild type was missing (79). Unequivocal evidence of an altered species of tRNA^{Tyr} in an amber suppressor has been obtained by Piper and co-workers who compared the nucleotide sequence of a minor species of tRNA^{Tyr} from SUP5-a, an amber suppressor allelic to the ochre suppressor SUP5-1, with that from the nonsuppressor strain. They found that the original wild-type anticodon GΨA had been changed to CΨA so that the new species of tRNA^{Tyr} could now translate the codon U-A-G. In this case at least, it is clear that the tRNA-mediated suppression in an eucaryote, as in *E. coli*, works through classical hydrogen-bonding patterns in codon-anticodon interaction (80).

Allelic amber suppressors have been found for each of the eight independent ochre suppressors isolated by Gilmore (81). In all cases the suppressor activity is dominant, and tyrosine is inserted at the site of the nonsense codon. The picture is complicated, however, by the presence of an extra chromosomal element psi (82). This is an element of genetic material that is cytoplasmically inherited but whose nature and mode of action is not understood. Amber suppressors are not affected by the presence of psi, but the efficiency of suppression of ochre suppressors is increased by the presence of psi. If the efficiency of suppression is too greatly enhanced, then the ochre suppressor becomes lethal to the cell. Recently an ochre suppressor SUQ5-1 that inserts serine at the U-A-A codon has been identified (83). In the absence of the psi factor the suppressor activity is too weak to be detectable, but psi somehow enhances the activity by an unknown mechanism.

The redundancy of the genes for tRNA^{Tyr} may be the reason that suppressors, being altered forms of single isoacceptors, have been found corresponding only to this tRNA in haploid cells. Alterations in other tRNA species might remove certain indispensable unique species from the tRNA pool and therefore be lethal to the cell. This has been shown to be the case by Brandriss et al., who have isolated amber suppressors in diploid strains that are lethal in haploid strains. One of these suppressor strains, SUP-RL1, causes serine to be inserted at the site of an amber codon in iso-1-cytochrome c (84).

The role of yeast tRNA in suppression has been most clearly demonstrated in systems that synthesize protein in vitro from mRNA containing nonsense codons. It has not yet proved possible to synthesize proteins in vitro in a system derived exclusively from yeast, but procaryotic mRNA may be translated into protein in systems that contain mammalian ribosomes and initiation factors and yeast tRNA (85, 86). Effective suppression of amber codons in the Qβ coat protein and synthetase genes has been shown in such systems to depend on the presence of tRNA from suppressing strains of yeast. In the presence of tRNA from an ochre-suppressing strain, an elongated form of Qβ synthetase is made, indicating that U-A-A is the normal termination codon and that it is suppressed by a suppressor of U-A-A from yeast.

5.20 Antisuppressors in Yeast

The accuracy of translation of mRNA in *E. coli* can be affected by mutations in the ribosomal proteins that influence the fidelity of codon-anticodon interactions. In *S. cerevisiae* mutant strains in which this interaction is affected have also been isolated, and they have been investigated through the antisuppression activity that they manifest. A highly specific and restricted pattern of suppression of mutations in the structural gene for L-histidinol phosphatase has been interpreted to mean that the sites of the suppressible mutations must contain not nonsense but rather specific missense mutations (87). Three independent suppressors were isolated for the mutations in question: SUP-H1 and SUP-H3, which are closely linked, and SUP-H2. However, in the presence of another mutation, the antisuppressor, no suppression of the original mutations could be detected. The antisuppressor mutation is designated Sin1-1, for *s*uppressor *int*eracting. An investigation of the corresponding Sin1 mutation in *S. pombe* found that this gene is probably the structural gene for a protein since nine independent mutations of this gene map within a length of chromosome corresponding to the length of a structural gene for an average-size protein (72). The protein might be a ribosomal protein or a tRNA-modifying enzyme, for example. Subsequently 14 unlinked *sin* genes have been identified, all of which can be mutated to recessive antisuppressor alleles and whose characteristics are again compatible with mutations in these types of proteins (88). A second sin mutation, Sin2-1, has been reported in *S. cerevisiae*. It could also be a mutation in a ribosomal protein, since it alters the susceptibility of different strains to cycloheximide, a translational inhibitor (89).

We have already discussed the effect of the extrachromosomal element psi in enhancing the efficiency of ochre suppression in *S. cerevisiae*. A class of antisuppressor mutants differing from those already mentioned has been isolated by McCready and Cox (90), and these can reverse the lethal phenotype in strains that contain both an ochre suppressor and the psi factor. Since the mechanism of enhancement of suppression by psi is unknown, it is not possible to deduce at what step in the biosynthesis of the suppressor tRNA or in translation the antisuppressor might act.

Antisuppressors could be the result of mutations in any of the following: a ribosomal protein affecting tRNA-mRNA interaction, a termination factor, an aminoacyl-tRNA synthetase, a tRNA modification enzyme affecting the anticodon, an enzyme affecting the maturation of the suppressor tRNA, or a factor controlling total tRNA synthesis (90).

5.21 Conclusion

The phenomenon of tRNA-mediated suppression has led to a variety of studies of both genetic and biochemical importance. Initially, recognition of the suppressor phenotype facilitates mapping of tRNA genes (in cases in which the suppression is tRNA-mediated) and gives information concerning the possible modes of transcriptional control of those genes. An understanding of the biochemistry of tRNA-mediated suppression and the determination of the nucleotide sequence of the tRNAs responsible for the suppression confirmed the classical dogma concerning the importance of Watson-Crick base pairing in the process of translation. More recent investigations show that many factors can influence the precise efficiency of hydrogen bond formation between anticodon and codon. Among these are nucleotide modifications of the first anticodon base, mRNA context, and as yet undefined features of tRNA structure outside the anticodon. Last, the recognition and study of revertants of suppressor tRNAs, having lost part or all of their suppressor activity, yields information concerning the effects of mutations in various segments of tRNAs, which alter the interaction of precursor tRNA with enzymes involved in processing during tRNA biosynthesis and the interaction of the mature molecule with aminoacyl synthetases. The ability to select mutants of tRNAs through recognition of their suppressor phenotypes will remain a critical feature in the continued elucidation of tRNA structure-function relationships.

Acknowledgment

The work performed in our laboratory was supported by USPHS grant GM-19422 to S. Altman.

References

1. Hartman, P. E., and Roth, J. R. (1973). Advances in Genetics vol. 17, E. W., Caspari, ed. (New York: Academic Press), pp. 1–105.

2. Brenner, S., Stretton, A. O. W., and Kaplan, S. (1965). Nature 206, 994–998.

3. Weigert, M. G., and Garen, A. (1965). Nature 206, 992–994.

4. Weigert, M. G., Lanka, E., and Garen, A. (1967). J. Mol. Biol. 23, 401–404.

5. Gallucci, E., and Garen, A. (1966). J. Mol. Biol. 15, 193–200.

6. Bock, R. M. (1967). J. Theoret. Biol. 16, 438–439.

7. Brenner, S., and Beckwith, J. R. (1965). J. Mol. Biol. 13, 629–637.

8. Ohlsson, B. M., Strigini, P. F., and Beckwith, J. R. (1968). J. Mol. Biol. 36, 209–218.

9. Sambrook, J. F., Fan, D. P., and Brenner, S. (1967). Nature 214, 452–453.

10. Engelhardt, D. L., Webster, R. E., Wilhelm, R. C. and Zinder, N. D. (1965). Proc. Nat. Acad. Sci. USA 54, 1791–1797.

11. Capecchi, M. R., and Gussin, G. N. (1965). Science 149, 417–422.

12. Goodman, H. M., Abelson, J., Landy, A., Brenner, S. and Smith, J. D. (1968). Nature 217, 1019–1024.

13. Altman, S., Brenner, S., and Smith, J. D. (1971). J. Mol. Biol. 56, 195–197.

14. Altman, S. (1976). Nucl. Acids Res. 3, 441–448.

15. Crick, F. H. C. (1966). J. Mol. Biol. 19, 548–555.

16. Abelson, J. N., Gefter, M. L., Barnett, L., Landy, A., Russell, R. L., and Smith, J. D. (1970). J. Mol. Biol. 47, 15–28.

17. Leon, V., Altman, S., and Crothers, D. M. (1977). J. Mol. Biol. 113, 253–265.

18. Smith, J. D., Barnett, L., Brenner, S., and Russell, R. L. (1970). J. Mol. Biol. 54, 1–14.

19. Hooper, M. L., Russell, R. L., and Smith, J. D. (1972). FEBS Letters 22, 149–155.

20. Celis, J. E., Hooper, M. L., and Smith, J. D. (1973). Nature New Biol. 244, 261–264.

21. McClain, W. H., Guthrie, C., and Barrell, B. G. (1973). J. Mol. Biol. *81*, 157–171.

22. Wilson, J. H., and Kells, S. (1972). J. Mol. Biol. *69*, 39–56.

23. Comer, M. M., Guthrie, C., and McClain, W. H. (1974). J. Mol. Biol. *90*, 665–676.

24. Hirsh, D. (1971). J. Mol. Biol. *58*, 439–458.

25. Soll, L. (1974). J. Mol. Biol. *86*, 233–243.

26. Miller, C. G., and Roth, J. R. (1971). J. Mol. Biol. *59*, 63–75.

27. Reeves, R. H., and Roth, J. R. (1975). J. Bacteriol. *124*, 332–340.

28. Ohashi, Z., Saneyoshi, H., Harada, F., Hara, H., and Nishimura, S. (1970). Biochem. Biophys. Res. Commun. *40*, 866–872.

29. Ohashi, Z., Harada, F., and Nishimura, S. (1972). *FEBS* Letters *20*, 239–241.

30. Yoshida, M., Takeishi, K., and Ukita, T. (1970). Biochem. Biophys. Res. Commun. *5*, 852–857.

31. Yoshida, M., Takeishi, K., and Ukita, T. (1971). Biochim. Biophys. Acta *228*, 153–166.

32. Chakraburtty, K., Steinschneider, A., Case, R. V., and Mehler, A. H. (1975). Nucl. Acids Res. *2*, 2069–2075.

33. Sen, G. C., and Ghosh, H. P. (1976). Nucl. Acids Res. *3*, 523–535.

34. Seidman, J. G., Comer, M. M., and McClain, W. H. (1974). J. Mol. Biol. *90*, 677–689.

35. Sekiya, T., Takeishi, K., and Ukita, T. (1969). Biochim. Biophys. Acta *182*, 411–426.

36. Grosjean, H., Söll, D. G., and Crothers, D. M. (1976). J. Mol. Biol. *103*, 499–519.

37. Feinstein, S. I., and Altman, S. (1977). J. Mol. Biol. *112*, 453–470.

38. Agris, P. F., Söll, D., and Seno, T. (1973). Biochemistry *12*, 4331–4336.

39. Comer, M. M., Foss, K., and McClain, W. H. (1975). J. Mol. Biol. *99*, 283–293.

40. Colby, D. S., Schedl, P., and Guthrie, C. (1976). Cell *9*, 449–463.

41. Salser, W. (1969). Mol. Gen. Genet. *105*, 125–130.

42. Salser, W., Fluck, M. M., and Epstein, R. (1970). Cold Spring Harbor Symp. Quant. Biol *34*, 513–520.

43. Fluck, M. M., Salser, W., and Epstein, R. H. (1977). Mol. Gen. Genet. *151*, 137–149.

44. Feinstein, S. I., and Altman, S. (1977). Genetics (in press).

45. Yahata, H., Ocada, Y., and Tsugita, A. (1970). Mol. Gen. Genet. *106*, 208–212.

46. Akaboshi, E., Inouye, M., and Tsugita, A. (1976). Mol. Gen. Genet. *149*, 1–4.

47. Zilberstein, A., Dudock, B., Berissi, H., and Revel, M. (1976). J. Mol. Biol. *108*, 43–54.

48. Kantor, G. J., Person, S., and Anderson, F. A. (1969). Nature *223*, 535–537.

49. Remaut, E., and Fiers, W. (1972). J. Mol. Biol. *71*, 243–261.

50. Atkins, J. F., and Gesteland, R. F. (1975). Mol. Gen. Genet. *139*, 19–31.

51. Weiner, A. M., and Weber, K. (1973). J. Mol. Biol. *80*, 837–855.

52. Tate, W. P., and Caskey, C. T. (1974). Mol. Cell Biochem. *5*, 115–126.

53. Berg, P. (1972). Harvey Lectures, Series 67 (New York: Academic Press), pp. 247–272.

54. Carbon, J., Berg, P., and Yanofsky, C. (1966). Cold Spring Harbor Symp. Quant. Biol. *31*, 487–495.

55. Carbon, J., Berg, P., and Yanofsky, C. (1966). Proc. Nat. Acad. Sci. USA *56*, 764–771.

56. Gupta, M., and Khorana, H. G. (1966). Proc. Nat. Acad. Sci. USA *56*, 772–779.

57. Hill, C. W., and Squires, C., and Carbon, J. (1970). J. Mol. Biol. *52*, 557–569.

58. Carbon, J., and Squires, C. (1971). Cancer Res. *31*, 663–666.

59. Roberts, J. W., and Carbon, J. (1974). Nature *250*, 412–414.

60. Hill, C. W., Combriato, G., and Dolph, W. (1974). J. Bacteriol. *117*, 351–359.

61. Schedl, P., Primakoff, P., and Roberts, J. (1974). Brookhaven Symp. Biol. *26*, 53–76.

62. Squires, C., and Carbon, J. (1971). Nature New Biol. *233*, 274–277.

63. Murgola, E. J., and Yanofsky, C. (1974). J. Bacteriol. *117*, 439–443.

64. Carbon, J., and Fleck, E. W. (1974). J. Mol. Biol. *85*, 371–389.

65. Riyasaty, S., and Atkins, J. F. (1968). J. Mol. Biol. *34*, 541–557.

66. Yourno, J., and Tanemura, S. (1970). Nature *225*, 422–426.

67. Yourno, J., and Heath, S. (1969). J. Bacteriol. *100*, 460–468.

68. Riddle, D. L., and Roth, J. R. (1972). J. Mol. Biol. *66*, 483–493.

69. Riddle, D. L., and Roth, J. R. (1972). J. Mol. Biol. *66*, 495–506.

70. Riddle, D. L., and Carbon, J. (1973). Nature New Biol. *242*, 230–234.

71. Roth, J. R. (1974). Ann. Rev. Genet. *8*, 319–346.

72. Hawthorne, D. C., and Leupold, U. (1974). Current Topics in Microbiological Immunology *64*, 1–47.

73. Schweitzer, E., Mackechnie, C., and Halvorson, H. O. (1968). J. Mol. Biol. *40*, 261–277.

74. Hawthorne, D. C. (1976). Biochimie *58*, 179–182.

75. Hawthorne, D. C., and Mortimer, R. K. (1963). Genetics *48*, 617–620.

76. Gilmore, R. A. (1967). Genetics *56*, 641–658.

77. Gilmore, R. A., Stewart, J. W., and Sherman, F. (1971). J. Mol. Biol. *61*, 157–173.

78. Sherman, F., Liebman, S. W., Stewart, J. W., and Jackson, M. (1973). J. Mol. Biol. *78*, 157–168.

79. Bruenn, J., and Jacobson, B. (1972). Biochim. Biophys. Acta *287*, 68–76.

80. Piper, P. W., Wasserstein, M., Engbaek, F., Kaltoft, K., Celis, J. E., Zeuthen, J., Liebman, S., and Sherman, F. (1976). Nature *262*, 757–761.

81. Liebman, S. W., Sherman, F., and Stewart, J. W. (1976). Genetics *82*, 251–272.

82. Cox, B. S. (1971). Heredity *26*, 211–232.

83. Liebman, S. W., Stewart, J. W., and Sherman, F. (1975). J. Mol. Biol. *94*, 595–610.

84. Brandriss, M. C., Stewart, J. W., Sherman, F., and Botstein, D. (1976). J. Mol. Biol. *102*, 467–476.

85. Capecchi, M. R., Hughes, S. H., and Wahl, G. M. (1975). Cell *6*, 269–277.

86. Gesteland, R. F., Wolfner, M., Grisafi, P., Fink, G., Botstein, D., and Roth, J. R. (1976). Cell *7*, 381–390.

87. Gorman, J. A., and Gorman, J. (1971). Genetics *67*, 337–352.

88. Thuriaux, P., Minet, M., Hofer, F., and Leupold, U. (1975). Mol. Gen. Genet. *142*, 251–261.

89. Gorman, J. A., and Gorman, J. (1971). Genetics Suppl. *68*, 524.

90. McCready, S. J., and Cox, B. S. (1973). Mol. Gen. Genet. *124*, 305–320.

91. Bachmann, B. J., Low, K. B., and Taylor, A. L. (1976). Bacteriol. Rev. *40*, 116–169.

92. Contreras, R., Ysebaert, M., Min Jou, W., and Fiers, W. (1973). Nature New Biol. *241*, 99–101.

6 OTHER ROLES OF tRNA

R. LaRossa
and D. Söll

6.1 Introduction

Aside from the major role of tRNA in ribosomal protein synthesis, many other processes have been discovered in which tRNA or aminoacyl-tRNA is implicated. Unfortunately very few of these processes are understood in detail, and often the mechanism of a particular reaction or its significance has remained elusive. It is clear, however, that in some cases tRNA and aminoacyl-tRNA act as regulators of particular metabolic processes. These multiple roles of tRNA, of aminoacyl-tRNA, and of the aminoacyl-tRNA synthetases lend a new dimension to the study of these macromolecules. There is a renaissance in the search for and investigation of these other functions of tRNA, but there may be a tendency to ascribe regulatory roles to still poorly understood tRNA functions. Although the regulatory role of tRNA is in some cases well documented (for example, the biosynthesis of histidine), additional complexity arises from the fact that tRNA may be only one part of a regulatory chain. The actual regulatory signal molecule may not be tRNA but may correlate its levels with tRNA. For instance, the level of aminoacyl-tRNA, amino acid concentration, and the amounts of magic spot compounds are interrelated in the bacterial stringent response. Transfer RNA acts as a signal for magic spot formation; however, does magic spot trigger a change in tRNA levels?

Since a detailed discussion of all the other known roles of tRNA is impossible in a short review, we will limit ourselves to areas in which significant advances have been made during the past few years. Excellent reviews of a general nature (1, 2) or those dealing with particular areas (3–8) provide the detailed background information not given here. This chapter will by nature be heterogeneous as these other tRNA reactions are so diverse.

6.2 Experimental Approaches

Because of the diversity of tRNA involvement in cellular functions, there is no straightforward strategy for exploring old or finding new roles for tRNA. Some of the most significant findings come from biochemical studies of viral RNAs (for example 9) or from detailed genetic and biochemical analysis of amino acid biosynthetic pathways.

One type of in vivo experiment has afforded much general information. This experiment utilizes temperature-sensitive strains containing a thermolabile aminoacyl-tRNA synthetase. If growth of the strain is conducted under semipermissive conditions, then the concentration of the cognate aminoacyl-tRNA in vivo should be much reduced compared to that found in wild-type cells or in cells grown at permissive temperatures, although in most cases this has not been verified. The steady-state levels of specific cellular components can then be determined. Their amounts reflect the influence of the in vivo concentration of a particular aminoacyl-tRNA in regulating their levels. This type of experiment, for instance, showed the involvement of aminoacyl-tRNA or of the aminoacyl-tRNA synthetase in the regulation of certain amino acid operons (for example 10) in the transport of branched-chain amino acids (11), in the synthesis of the magic spot compounds (p)ppGpp (6), and in cell division (12). Similar experiments could also be done with mutants giving rise to altered tRNA modification. These mutants, however, are much more difficult to isolate; ironically the most useful ones so far (13) have been obtained by studying the regulation of amino acid biosynthesis.

One in vitro experiment is providing specific information on the role of particular compounds in regulation: the coupled transcription-translation system (14). Such systems lead to the formation of certain easily measurable gene products. The starting material for such experiments, the operon DNA, is available as the DNA of specific transducing phages or most recently through recombinant DNA technology.

6.3 Genetics of Aminoacyl-tRNA Synthetases

Since mutant aminoacyl-tRNA synthetases have played an important part in uncovering tRNA involvement in other processes, we will describe briefly the present status of genetic characterization of these enzymes.

Although aminoacyl-tRNA synthetases represent a class of indispensable enzymes in all cells (15), many mutants have been obtained in *Escherichia coli, Salmonella typhimurium, Bacillus subtilis,* yeast, and in mammalian cell lines (table 6.1). They can be divided into three major classes which often suggest the experimental procedure by which they were obtained. Most prevalent are conditional lethal mutants harboring thermolabile aminoacyl-tRNA synthetases. These confer thermosensitivity to the organism at the nonpermissive temperature probably because a particular aminoacyl-tRNA is not formed. Biochemical studies with these mutants are often hampered by the lability of the mutant enzymes in vitro, which makes rigorous characterization of the enzymes impossible. When the biochemical properties of the aminoacyl-tRNA synthetases from such mutants were investigated, a decreased affinity for their substrates was often found to accompany the thermolability.

A second class of aminoacyl-tRNA synthetase mutants is the analog-resistant type. An aminoacyl-tRNA synthetase sometimes activates and even transfers onto tRNA a close analog of the cognate amino acid. If this "wrong" aminoacyl-tRNA leads to proteins that are nonfunctional in vivo, or if the analog severely inhibits the enzymatic function of the aminoacyl-tRNA synthetase, then the organism may not grow. Many known mutations in the structural genes of aminoacyl-tRNA synthetases confer a resistance to amino acid analogs. Usually the mutant enzyme has a much decreased affinity for the analog; thus the mutant strain is not affected by the presence of the analog in the medium.

A third class of aminoacyl-tRNA synthetase mutants that can be clearly characterized comprises mutations that confer auxotrophy for the particular amino acid to the organism. These strains contain an altered aminoacyl-tRNA synthetase with a lowered affinity for the cognate amino acid. Thus the mutant strain grows only if the required amino acid is exogenously supplied, since the intracellular amino acid concentration is too low for efficient aminoacyl-tRNA formation.

Genetic evidence to date is the main support for the idea that bacteria possess only one aminoacyl-tRNA synthetase for each amino acid. A possible exception is *E. coli* glutamyl-tRNA synthetase, where three dif-

Table 6.1 Aminoacyl-tRNA synthetase mutants

tRNA Synthetase	Organism	Type of Mutant	Enzyme Defect[a]	Genetic Location	Reference
Alanyl-	Chinese hamster	Auxotrophi	—	—	136
	Escherichia coli	Temperature-sensitive	tRNA	Near recA	133–135, 187
	Mouse cells	Temperature-sensitive	—	—	137
Arginyl-	Chinese hamster	Temperature-sensitive	—	—	141
	E. coli	Analog-resistant	ATP, tRNA	Between pheS and his	138–140
Asparaginyl-	Chinese hamster	Temperature-sensitive	—	—	142
	E. coli	Temperature-sensitive	—	At min 21[b]	166
	Hamster	Temperature-sensitive	—	—	143
Glutaminyl-	Chinese hamster	Temperature-sensitive	—	—	141
	E. coli	Temperature-sensitive	—	Near lip	127
Glutamyl-	E. coli	Temperature-sensitive	—	Near dsdA	144
		Streptomycin-dependent	Amino acid	Near xyl, near his	145, 146
Glycyl-	E. coli	Auxotrophic	Amino acid	Near xyl	147, 148
		Temperature-sensitive	—	Near xyl	149
Histidyl-	Chinese hamster	Temperature-sensitive	—	—	141
	E. coli	Bradytrophic	Amino acid	—	150
	Salmonella typhimurium	Analog-resistant	Amino acid	Near strB	26

Table 6.1 (continued)

tRNA Synthetase	Organism	Type of Mutant	Enzyme Defect [a]	Genetic Location	Reference
Isoleucyl-	E. coli	Auxotrophic	ATP, amino acid	Between thr and pyrA	151, 152
		Analog-resistant			153
		Temperature-sensitive			154
	S. typhimurium	Auxotrophic		Near pyrA	155
	Saccharomyces cerevisiae	Temperature-sensitive		On Chromosome 2	156, 157
Leucyl-	Chinese hamster	Temperature-sensitive			158
	E. coli	Temperature-sensitive		Near lip	70
	S. typhimurium	Analog-resistant	Amino acid	Near gal	159, 160
	Neurospora crassa	Temperature-sensitive, auxotrophic	Amino acid		161
Lysyl-	Bacillus subtilis	Temperature-sensitive	Amino acid, ATP	Between purA and sul	162, 163
	E. coli	Analog-resistant			164
Methionyl-	Chinese hamster	Temperature-sensitive			141
	E. coli	Auxotrophic Analog-resistant	Amino acid tRNA	At min 47[b]	165, 167
	S. typhimurium	Auxotrophic	Amino acid	At min 67[c]	168, 169
	S. cerevisiae	Temperature-sensitive	Amino acid		157, 170
Phenylalanyl-	E. coli	Analog-resistant	Amino acid		171
		Temperature-sensitive		Near pps	144, 172
Seryl-	E. coli	Temperature-sensitive Serine hydroxamate-resistant		Near serC	18, 70
					173

			Amino acid[d], ATP[d]	At min 37.7[a]	129, 189
Threonyl-	E. coli	Borrelidin-resistant			
		Auxotrophic	Amino acid	Near trp	21
Tryptophanyl	B. subtilis	Temperature-sensitive, analog-resistant	—	Between argC and metA	174
	E. coli	Auxotrophic	Amino acid	—	175, 188
		Temperature-sensitive	—	Between pubA and aroB	144
	N. crassa	Auxotrophic	—	—	
Tyrosyl-	E. coli	—	Amino acid	At min 32[a]	177, 178
		Analog-resistant	—	—	144
	S. typhimurium	Auxotrophic	—	—	179
Valyl-	E. coli	Temperature-sensitive	ATP, amino acid	Near pyrB	172, 180, 181

[a] The affinity of the enzyme for the specified substance in this column is lowered unless otherwise indicated.
[b] On the genetic map (190).
[c] On the genetic map (191).
[d] Affinity increased.

ferent genetic loci have been found to affect the activity of this enzyme. This may be related to its composition of nonidentical subunits (16), although there is no agreement on this (17). In *E. coli*, for which 17 of the 20 aminoacyl-tRNA synthetases have been genetically identified and mapped (table 6.1), the map locations (figure 6.1) indicate that there is no clustering of aminoacyl-tRNA synthetase genes in the *E. coli* genome. Although much less is known about locations of tRNA genes (see chapter 5), it appears that the aminoacyl-tRNA synthetase genes are also separated from the genes for the cognate tRNAs, the tRNA-modifying enzymes, and from the genes responsible for the biosynthesis of the cognate amino acids. A possible exception is the gene (serS) for *E. coli* seryl-tRNA synthetase mapping very close to the *serC* gene that specifies phosphohydroxypyruvate transaminase, an enzyme in serine biosynthesis. However, no coupling of the expression of these two genes has been found yet (18).

The situation in eucaryotic cells is more complex since multiple amino-acyl-tRNA synthetases for the same amino acid have been found in the cytoplasm, while different enzymes are contained in organelles, such as mitochondria, chloroplasts, and possibly nuclei (15). The isolation of mutants of aminoacyl-tRNA synthetases in eucaryotic organisms may help clarify the origin of the aminoacyl-tRNA synthetases found in the various organelles. An example of this is seen in *Neurospora crassa* for which a mitochondrial leucyl-tRNA synthetase is coded for by a nuclear gene (19).

6.4 Involvement of tRNA in Amino Acid Biosynthesis

The involvement of aminoacyl-tRNA in the regulating operons for the biosynthesis of several amino acids has been demonstrated in procaryotic (3) and eucaryotic (20) organisms. No uniform picture of such regulatory mechanisms has yet emerged. Although the general features of the repres-sion-derepression phenomenon in the different biosynthetic pathways are similar, the detailed regulation of each could follow different mechanisms. In view of the variety of metabolic products and reaction sequences, it may be surprising to find only one mechanism. Direct evidence (3) has been obtained for a regulatory role of tRNA in the biosynthesis of histidine; tryptophan; the branched-chain amino acids leucine, isoleucine, and valine; arginine, methionine; threonine (21); and glutamine. However, only in a few cases, such as the histidine operon in *S. typhimurium* and the trypto-phan operon in *E. coli*, is there enough detailed information to pinpoint

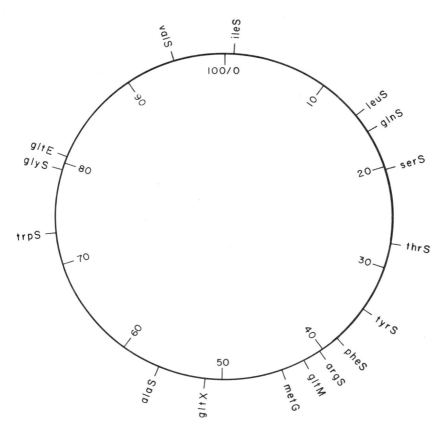

Figure 6.1 Location of aminoacyl-tRNA synthetase genes on the genetic map of *E. coli* (190).

where tRNA acts in these processes. Two points of interaction have been demonstrated so far: (i) tRNA may interact with the first enzyme of the biosynthetic pathway and thus exert a regulatory role over this enzyme and accordingly over the whole reaction sequence; (ii) aminoacyl-tRNA (or the synthetase) may interact with a regulatory sequence in the operator region (the attenuator site) causing transcription termination before the first structural gene of the operon. In addition the degree of nucleotide modification of tRNA is important; this could affect either regulatory role. Most of this information is covered in a recent review (3); we would like to discuss in detail only the histidine and the tryptophan operons and more recent information not covered in that review.

6.5 Histidine Biosynthesis

The histidine operon comprises the structural genes for nine enzymes, which catalyze the formation of histidine, and the operator and promoter control regions. The histidine regulatory genes were identified by selecting *S. typhimurium* mutants resistant to the histidine analog 1, 2, 4 triazole-3-alanine under conditions where derepression of the *his* operon is necessary for growth. This analog is incorporated into protein and causes repression of the *his* operon (22). Genetic analysis of such revertants revealed six loci that cause derepression of the histidine operon (23, 24).

The first regulatory locus *hisO* represents a *cis*-dominant operator-promoter region (23, 25).

A second regulatory locus *hisS* codes for the histidyl-tRNA synthetase (26). Temperature-sensitive, auxotrophic, and analog-resistant *hisS* mutants have been isolated. The manner in which many of these mutations affect *his* operon expression has not been elucidated, although the properties and kinetic parameters of histidyl-tRNA synthetase from many of these strains has been determined.

Another regulatory locus *hisR* encodes at least tandem structural genes for tRNA^{His}. In *hisR* strains a 25 to 50 percent reduction in histidine acceptor RNA is found (27, 28) with minor variations in other amino acid acceptance levels (28).

The fourth regulatory gene *hisT* was shown to code for an altered tRNA-modifying enzyme, pseudouridine synthetase I. This enzyme converts certain uridine residues in the tRNA anticodon region into pseudouridine. This was established by nucleotide sequence analysis of tRNA^{His} from wild-type (29) and *hisT* strains (30) and confirmed by similar analysis of tRNA^{Leu} (31). This change in nucleotide modification appears to alter the elution profile of several aminoacyl-tRNA species (Tyr-tRNA, His-tRNA, Ile-tRNA, Leu-tRNA; see refs. 30–32). In all these cases, pseudouridine is found in the anticodon of wild-type tRNAs (32). The pleiotropic effect of the *hisT* mutations also includes the derepression of the leucine and isoleucine-valine operons (31–33) and the resistance to many amino acid analogs (32).

Two other regulatory loci, *hisU* and *hisW*, are pleiotropic and have been implicated in the processing of tRNA species. (i) The acceptor activity of tRNA isolated from *hisU* and *hisW* strains is lower compared to that of tRNA isolated from the parental *his*$^+$ strain for many amino acids (28). (ii)

In *hisU* and *hisW* strains the level of tRNA^{His} has been reduced 20 to 30 percent (34). (iii) Gene dosage experiments reveal that *hisW* probably does not code for tRNA^{His}. However, the derepressed phenotype of a *hisR*$^+$ *hisW*$^-$ strain is suppressed in a *hisR*$^+$/*hisR*$^+$ *hisW*$^-$ meridiploid (35). Recent evidence indicates that hisU indeed codes for a nuclease activity similar to *E. coli* RNase P and involved in processing of precursor tRNAs (36).

The levels of His-tRNA in vivo have been measured in strains containing representative mutations of one of the six regulatory genes (34). An inverse relation between His-tRNA levels and histidine operon expression was found for mutants involved in production (*hisU, hisW, hisR*) or amino-acylation (*hisS*) or tRNA^{His}. This relation (as predicted by the operon theory) did not hold for *hisO* mutants. Surprisingly this relationship was not true for *hisT* mutants. These results indicate that histidyl-tRNAs having pseudouridine residues in the anticodon region are a signal for repressing the histidine operon.

The molecular details of how this molecule causes repression are unknown. With what if any macromolecule does this repression signal interact? To answer this question investigators utilized in vitro transcriptional systems (37, 38), coupled in vitro transcription-translation systems (39) and physiological-genetic approaches (40). Mutants in *hisG* (for a review see ref. 41), *hisO* (37, 39, 40), and *hisS* (42) were used in these studies.

The original procedure (23) for the selection of histidine regulatory mutants excluded histidine auxotrophs; thus certain putative regulatory mutants blocking His-tRNA formation—including those mutations in the structural genes for both the feedback-sensitive first enzyme of the pathway, N-1-(5′-phosphoribosyl) ATP:pyrophosphate phosphoribosyl-transferase (*hisG*), and histidyl-tRNA synthetase (*hisS*)—were excluded by the selection protocol.

It was found that when the first enzyme of the pathway was altered either by mutation or by the addition of histidine analogs, the kinetics of repression were altered (41). *HisG* mutants that are partially constitutive for *his* operon expression have been isolated (42). These *hisG*$^-$ alleles are trans-recessive to wild-type *hisG* (43). Thus the enzyme exerts its effect as a diffusable gene product. The *hisG* protein has an affinity for tRNA; its activity is inhibited by His-tRNA more than by any other tRNA tested (44). Binding of the *hisG* protein to *hisO* DNA has also been demonstrated (45). Thus the *hisG* enzyme fits some requirements for a classical repressor with His-tRNA serving as the co-repressor. Strains containing extensive internal deletions of *hisG*, however, can be derepressed physiologically (by growth on histidinol) or genetically (for example, in combination with *hisT⁻*) suggesting at least two regulatory mechanisms, one independent of *hisG*, for histidine operon expression (46).

Paradoxes also appear in descriptions of *hisS* mutations. The most noteworthy of these show up in the affinity for ATP, tRNA, and histidine of mutant histidyl-tRNA synthetases. In two *hisS* mutants the enzyme has clearly altered affinities. In other *hisS⁻* enzymes slight, if any, alterations of affinity for substrates were observed. Only in the two *hisS* strains with the well-characterized enzyme defects were His-tRNA levels correlated with expression of the *his* operon (34). There may exist *hisS* mutants in which histidyl-tRNA synthetase plays a role in stimulating histidine operon expression. In fact, strains with elevated histidyl-tRNA synthetase levels display enhanced expression of histidine biosynthetic enzymes (42). This implies that histidyl-tRNA synthetase is a positive regulator of the *his* operon.

Over 100 regulatory mutations in the *hisO* (operator-promoter) region have been isolated and mapped (25). These mutants, selected by many methods, were originally classified as either operatorlike (having high *his* enzyme levels) or promoterlike (having low *his* enzyme levels). In vivo levels of histidine biosynthetic enzymes and histidine mRNA from the *hisO* strains were measured in a genetic background where histidyl-tRNA synthetase levels are elevated (*strB*), where the *his* operon is derepressed by unlinked regulatory mutations (*hisT* or *hisR*), or where a second *hisO* mutation is present (37, 40). *HisO* mutations (such as *hisO1812*) that elevate enzyme levels in response to elevated histidyl-tRNA synthetase levels but not in response to *hisT* or *hisR* mutations have been found. This again indicates that the aminoacyl-tRNA synthetase positively regulates *his* operon expression rather than exerting its effect by complexing the purported negative co-repressor His-tRNA. In addition the enzyme appears to exert an effect at a site near the RNA polymerase binding site rather than at the downstream site where His-tRNA exerts its negative control. Thus histidyl-tRNA synthetase may be a positive control element that binds at or near the RNA polymerase binding site and travels with RNA polymerase until it exerts its effect at a transcriptional barrier.

Measurement of *his* mRNA levels of promoterlike mutants in *hisT⁻*, *hisO1242*, and wild-type backgrounds, led to classification of the promoterlike mutations into three types (37). The set of mutants that in combination with $hisT^-$ retain their promoterlike character but whose phenotype is cis-recessive to that of $hisO1242$ leads to the argument that some promoter-like mutations decrease the binding of a positive factor whose need is obviated by the $hisO1242$ mutation.

In a minimal transcription system it was shown that $hisO^+$ DNA contained a transcriptional barrier, the attenuator site, leading to production of small amounts of his mRNA in vitro. This site was absent in the $hisO1242$ deletion from which large amounts of his mRNA could be transcribed (37). Further evidence that the major controlling mode of the his operon is a positive one came from transcription-translation experiments (39). With various amounts of DNA isolated from $\Phi80dhisO1242$, histidine enzyme levels do not reach a limiting value (in the range tested), while with $hisO^+$ DNA a plateau level of his enzyme synthesis, is observed. This indicates that some component necessary for his synthesis that acts in a positive fashion is limiting. Further, using DNA from a $\Phi80$ phage whose $hisO$ region is putatively defective in the binding of a positive factor (a promoter-like mutation) lowers maximal his enzyme synthesis. These experiments mirror in vivo observations and again define the attenuator site at which the positive factor exerts its effect, even though the binding site of the positive factor is thought to be near that for RNA polymerase. An unexpected finding in these experiments was that a coupling of transcription and translation is needed for expression of $hisO^+$ DNA, while transcription and translation may occur sequentially in the expression of $hisO1242$ DNA in vitro. This implies that translation may be necessary for expression of the his operon.

These observations brought about the present view of histidine operon expression (39). In the major repression-derepression phenomenon His-tRNA binds to a positive antitermination factor. When complexed with His-tRNA this factor is thought to be inactivated; thus transcription terminates at the attenuation site. Whether histidyl-tRNA synthetase is this antitermination factor remains to be answered.

Transfer RNA plays at least one other role in the regulation of his operon. Uncharged tRNA in the ribosomal A-site ("idling ribosomes") is known to trigger magic spot production. (p)ppGpp stimulates transcription and translation of the histidine operon in vitro and in vivo (48); evidence has been presented (48) that this may be a general phenomenon for many amino acid biosynthetic operons in vivo (the alarmone hypothesis).

6.6 Tryptophan Biosynthesis

The enzymes of the *trp* operon, a cluster of five genes, catalyze the specific reactions that lead from the common branch point of aromatic amino acid biosynthesis, chorismate, to tryptophan. A number of possible *trp* regulatory mutants have been identified by isolation of mutants resistant to the tryptophan analogs 5-methyl-tryptophan or 6-fluoro-tryptophan (5-methyl-tryptophan is not attached to tRNA by tryptophanyl-tRNA synthetase in vitro yet it causes repression).

TrpO mutations are defined as *cis*-dominant, transrecessive mutations linked to the *trp* structural genes (49, 50).

The product of the *trpR* gene (51) is the tryptophan repressor (52). When complexed with tryptophan it binds to the *trpO* DNA in vitro, preventing interaction of RNA polymerase with the *trp* promoter (see figure 6.2); thus transcription is blocked in a manner analogous to that found for the *lac* and *gal* operons (53). Conversely RNA polymerase already bound to DNA is insensitive to the addition of the *trp* repressor-tryptophan complex (53). Such in vitro repression is not found with DNA from *trpO* mutants (52).

Mtr mutants, which are also 5-methyltryptophan–resistant, do not have normal *trp* enzyme levels when grown in casamino acid–supplemented media; further these mutants are not completely repressed by the addition of tryptophan (54). Although these mutants have been mapped, the biochemical basis of their phenotype is unknown.

Another set of 5-methyltryptophan–resistant strains, the *trpT* mutants, contain constitutive tryptophan enzyme levels (49). The *trpT* gene product and map location have not yet been identified.

Some mutations in *trpS*, the structural gene for tryptophanyl-tRNA synthetase, also effect *trp* operon expression. One *trpS* mutant (selected as a tryptophan auxotroph in a *trp*$^+$/*trp*$^+$ merodiploid) can neither completely repress nor completely derepress the *trp* operon (55). Reversion of a *trpS*$^-$ auxotroph was accomplished by a second mutation in *mtr*; the *trpS mtr* double mutant has repressed *trp* enzyme levels (56). This implies some interplay between the gene product of two loci, rather than that *mtr* causes an elevated tryptophan pool which might allow aminoacylation of tRNA^{Trp} sufficient for growth by the defective tryptophanyl-tRNA synthetase.

Roles for tRNA^{Trp} and for tryptophanyl-tRNA synthetase in the repression of the *trp* operon seemed possible, although a repressor-operator system independent of Trp-tRNA levels had been established by in vitro and in vivo experiments. Thus if Trp-tRNA and/or tryptophanyl-tRNA synthe-

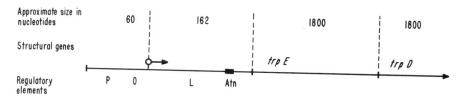

Figure 6.2 The beginning of the tryptophan operon in *E. coli*. The arrow indicates the site of mRNA initiation, and the dark block indicates the site of transcription termination. P promoter; O operator; L leader; Atn attenuator.

tase had a role in regulation of the *trp* operon, a second regulatory site within the *trp* operon had to be postulated.

Evidence for such a second site was first presented in 1973 (57). The mapping of internal deletion mutations in the *trp* operon revealed some mutations that terminated operator-proximal to all *trpE* point mutations (the first structural gene of the operon). These mutants responded to the *trpR* allele (indicating that *trpO* region is not altered) and caused altered levels of *trp* gene products. This proved that a second mechanism works in concert with the repressor-operator prevention of transcription initiation to control *trp* operon expression. These deletions define a site that could be analogous to the region deleted by the *hisO1242* mutation. Examination of the sequence of trp-mRNA indicated that a leader sequence of 162 nucleotides precedes the initiation codon of *trpE,* the first structural gene of the operon (38). This suggests that there exists a third *cis*-dominant regulatory site within the leader region (*trpL*) before *trpE*, which was designated the *trp* attenuator. *Trp* mRNA synthesized in vivo (59–61) and in vitro (62, 63) has been shown to terminate at the *trp* attenuator. Thus as in the *his* operon there exists a barrier to transcription that is promoter-proximal to the first structural gene.

Trp-tRNA has been shown to act at the *trp* attenuator (61). When tryptophan auxotrophs are grown in media containing tryptophan, 90 percent of the RNA polymerase molecules that reach the *trp* attenuator terminate (59). During tryptophan starvation, but not during isoleucine starvation (imposed by valine addition), such termination at the attenuator is decreased eight- to tenfold (60, 61). This implies that attenuation is relieved specifically by tryptophan starvation. Utilization of *trpS⁻* strains containing an altered tryptophanyl-tRNA synthetase and addition of various tryptophan analogs have demonstrated that Trp-tRNA is most likely involved in terminating transcription at the attenuator (61). This is supported by the report that some *trpS⁻* mutants possess elevated tryptophan biosynthetic enzyme levels in both *trpR⁺* and *trpR⁻* strains (60, 64). Thus either tryptophanyl-tRNA synthetase or Trp-tRNA regulates *trp* operon expression independent of the *trpO-trpR* mediated control. Further, in the *trpX* mutation the level of *trp* biosynthetic enzymes is elevated. The *trpX* mutation causes an alteration of a single modified nucleotide in tRNA^Trp (the isopentenyl modification of isopentenyladenosine is lacking); this does not interfere with its aminoacylation. Thus tRNA^Trp most probably has a regulatory role (182).

A factor that acts positively to stimulate in vitro synthesis of *trp* enzymes has been described (65). It has been shown that this positive factor antagonizes termination at the attenuator. Polarity suppressors bearing altered rho factors (rho is a protein that causes RNA polymerase to terminate transcription) increase expression of the wild-type *trp* operon by decreasing termination at the attenuator (66, 67).

Thus the tryptophan operon is regulated by a repression system involving the tryptophan repressor and free tryptophan and by an attenuation system involving a metabolic product of tryptophan (probably Trp-tRNA), rho factor, and an unidentified antitermination factor. During amino acid deprivation *trp* mRNA levels are higher in *trpR⁻ relA⁺* strains than in *trpR⁻ relA⁻* strains (61). Thus magic spot compounds are also assumed to stimulate *trp* as well as *his* expression, supporting the alarmone hypothesis (48).

6.7 Isoleucine, Valine, and Leucine Biosynthesis

The biosynthesis of isoleucine, valine, and leucine in *E. coli* and *S. typhimurium* is regulated in a very complex fashion; leucine, isoleucine, and valine are all required for maximal repression of both the *leu* and *ilv* genes. The experimental evidence establishing a role for aminoacyl-tRNA in the regulation of the *leu* and *ilv* operons is sparser that that accumulated for histidine or tryptophan biosynthesis. One complication may be that multiple isoaccepting species of tRNA^Leu, tRNA^Ile, and tRNA^Val exist, while only a single species of both tRNA^His and tRNA^Trp exists; thus if only one tRNA species is involved in regulation, then studies of total tRNA or aminoacyl-tRNA levels may not properly reflect the situation.

Growth of either an *ileS* (68), *valS* (69), or *leuS* (70) mutant under semipermissive conditions results in derepression of the *leu* and *ilv* gene clusters. This implies that either the aminoacyl-tRNA synthetase or the cognate aminoacyl-tRNA functions in a regulatory manner. Studies with valine analogs show that Val-tRNA, rather than free valine, mediates repression (71, 72). a-Aminobutyrate can be activated but not attached to tRNA; a-amino-β-chlorobutyrate can be both activated and attached. Addition of a-aminobutyrate to cultures did not change enzyme levels, while addition of a-amino-β-chlorobutyrate to cultures caused repression. A recent unrelated study of leucyl-tRNA synthetase overproducing strains demonstrated an inverse relation between Leu-tRNA levels and expression of the *leu* and *ilv* genes (73). This implicates at least one tRNA^{Leu} isoaccepting species in this process.

The most direct proof of tRNA affecting *leu* and *ilv* enzyme levels comes from the study of the pleiotropic response of *hisT* mutants (30–33). Such mutants show altered isoaccepting patterns for tRNA^{Leu} and tRNA^{Ile}. The isoacceptor pattern of tRNA^{Val}, however, is constant. This is predicted because tRNA^{Val} lacks pseudouridine in its anticodon loop. Together with the results from studies of aminoacyl-tRNA synthetase mutants, these studies imply that the tRNA^{Leu} and/or tRNA^{Ile} species with pseudouridine in their anticodon regions may function in the regulatory mechanisms.

How does such regulation take place? As observed in the histidine operon, it was found several years ago that one species of Leu-tRNA, specifically Leu-tRNA$_4$, binds well to immature threonine deaminase, the first enzyme of the pathway and the *ilvA* gene product (4, 74). This suggested this enzyme's regulatory properties on the whole pathway. However, more recently it was shown that strains with *ilvA* deletions (183) or with mu insertions in *ilvA* (184) can still function in the repression phenomenon, indicating at least one other regulatory mechanism.

Evidence is increasing that the branched-chain aminoacyl-tRNAs have a role in transcriptional control. Little *ilv* mRNA is transcribed from $\Phi 80 \lambda$ d*ilv* DNA in vitro. This is reminiscent of the effect caused by the attenuator region found in $\Phi 80$d*hisO*$^+$ DNA (37). When the rates of *ilv* mRNA formation in wild-type *S. typhimurium* were compared to those in the *hisT* mutant and in isoleucyl- and leucyl-tRNA synthetase mutants, three- to eightfold higher rates were found in the latter three strains (75). Similar studies of *E. coli ilv* mRNA have been performed (76). In addition *rho*$^-$ mutants have lower *ilv* mRNA and enzyme levels than isogenic *rho*$^+$ strains (77), indicating that the transcription terminator factor rho is involved in the regulation of the *ilv* cluster. Thus the *ilv* cluster shows regulatory characteristics of an attenuation mechanism. Demonstration of a specific role for aminoacyl-tRNA in regulating the *ilv* and *leu* gene clusters awaits the definition of the lesions caused by many of the *ilv* and *leu* regulatory mutations (3) and development of in vitro transcription and coupled transcription-translation systems.

6.8 Glutamine Biosynthesis

The biosynthesis of glutamine is regulated by a very complex mechanism (78, 79). There are two observations that tRNA may be involved in this process: (i) When a temperature-sensitive *E. coli* strain carrying a thermolabile glutamyl-tRNA synthetase was grown at semipermissive temperature, a derepression of glutamate synthase and of glutamine synthetase was observed while the level of glutamate dehydrogenase remained unchanged (80). In addition an inverse correlation between the extent of in vivo charged Glu-tRNA and the level of glutamate synthase was observed (80), although the determination of charged tRNA was carried out by an inaccurate method. (81). (ii) Recently derepressed levels of glutamine synthetase were found in mutants having elevated levels of tRNA^{Gln} or of glutaminyl-tRNA synthetase (82, 83). It is too early to say whether this phenomenon is expressed throught the in vivo level of Gln-tRNA or whether it is a direct effect on glutamine synthetase, since the determination of the activities of glutamate synthase, of glutamate dehydrogenase, and of the concentration of some metabolites crucial for their regulation has not been completed.

6.9 Amino Acid Biosynthesis in Eucaryotic Cells

In the past few years a significant number of temperature-sensitive aminoacyl-tRNA synthetase mutants in mammalian cell lines (CHO cells) have been obtained (see table 6.1). These mutants now allow the classical experiment of growing such strains in semipermissive temperature and observation of the levels of amino acid biosynthetic enzymes. Recently such an experiment was done. (20). The growth of a CHO cell line harboring a thermosensitive asparaginyl-tRNA synthetase under semipermissive conditions also resulted in derepression of asparagine synthetase activity. Removal of asparagine from wild-type Chinese hamster ovary cells likewise resulted in a derepression of asparagine synthetase. Asparaginyl-tRNA was implicated in this process.

6.10 Involvement of tRNA in Amino Acid Transport

Leucine, isoleucine, and valine are taken up in *E. coli* by multiple transport systems (for a review see refs. 84, 85). This is mediated by and depends on the binding proteins for these amino acids (86). It was shown earlier that the branched-chain amino acid transport system and the biosynthetic enzymes for these amino acids in *E. coli* or *S. typhimurium* are not regulated together in a *cis*-dominant type mechanism, although both systems appear to share some common regulatory components (87). When an *E. coli* strain containing a deletion of the leucine operon and a temperature-sensitive leucyl-tRNA synthetase mutation was grown at semipermissive temperature, a large increase in the high-affinity transport of branched-chain amino acids was observed while the transport of unrelated amino acids remained unchanged. Concomitantly the leucine-isoleucine-valine binding protein activity increased (11). These studies indicate that Leu-tRNA or leucyl-tRNA synthetase plays a role in the repression of branched-chain amino acid transport, as similar studies using *ileS⁻* and *valS⁻* mutants do not cause derepression (88). Recently it has been reported that a mutation in *rho* causes elevated levels of branched-chain amino acid transport components (89). These two factors suggest that an attenuation mechanism may regulate the synthesis of these proteins.

6.11 Transfer RNA and the Stringent Response

When enteric bacteria are subjected to amino acid deprivation (or other "mistreatment"), cells adjust their metabolism by ceasing to make ribosomal RNA. This is known as the stringent response, which is governed by the *relA* gene (see 6, for example). A mutation in *relA* causes relaxed control (90), allowing the continued production of stable RNA under amino acid limitation conditions (91). The magic spot compounds ppGpp and pppGpp play a central role in this response (92). They are synthesized by a ribosome-associated protein (coded for by the *relA* gene; ref 93) in the presence of mRNA and deacylated tRNA bound at the ribosomal A-site (94). These phosphorylated nucleotides appear to cause a major pleiotropic alteration of cellular metabolism (6) including shutdown of stable RNA synthesis, although the synthesis of some tRNA species may not be controlled by these nucleotides (95). Thus tRNA in conjunction with other components of the protein-synthetic apparatus plays a major regulatory role. Not only does the protein-synthesizing apparatus act during synthesis, but it also signals the cessation of such synthesis. The requirement for tRNA in magic spot formation has been further studied. In vitro experiments with enzymatically modified tRNAs have shown the absolute necessity of a free 3'-hydroxyl group of the terminal adenosine of the tRNA molecule (96). The tetranucleotide T-Ψ-C-G can substitute for tRNA in this reaction (97).

In vivo experiments have shown that the lack of formylation of Met-tRNA$_f^{Met}$ is not sufficient to cause magic spot formation (98). Experiments with $E.\ coli$ mutants with increased tRNA^{Gln} levels have shown that the role of magic spot formation does not depend on the absolute concentration of uncharged tRNA; rather the ratio of aminoacyl-tRNA to tRNA is important (82).

ppGpp appears to act as a diffusable agent, differentially controlling the synthesis of many gene products (derepressing amino acid–biosynthetic operons (48, 61) and repressing rRNA operons (91)) perhaps through interactions with promoters and/or RNA polymerase.

6.12 Inhibitory Effect of tRNA on the Activity of Enzymes

The phenomenon that tRNA binds to enzymes (although not being a substrate for those) and reduces their activity deserves brief discussion in view of its possible importance in regulation. The binding of tRNA or aminoacyl-tRNA to the first enzyme in the biosynthesis of the cognate amino acids may represent an important regulatory tool. For instance it was shown that the His-tRNA binds to the feedback-sensitive but not to the feedback-resistant G-enzyme, the first enzyme in histidine biosynthesis (99). However, the tRNA specificity in such a reaction needs to be carefully examined. It was established several years ago that Tyr-tRNA or Phe-tRNA inhibits the activity of the feedback-sensitive DAHP synthetase, an enzyme of tyrosine and phenylalanine biosynthesis of $S.\ cerevisae$ (100). While this result is correct, recent work has shown that this inhibition is not specific, since unrelated tRNAs exhibit the same inhibitory effect. The inhibition appears to be caused by the removal of a metal ion, required for enzymatic activity through chelation with the added tRNA (101). Thus caution is advised in assessing reports of tRNA inhibition of various enzymatic activities. In this context it is pertinent to mention that the inhibition of tryptophan pyrrolase from the vermilion mutant of $Drosophila$ vy tRNA^{Tyr} (102)

was artifactual (103), and the suppression mechanism is still unknown (104).

6.13 Aminoacyl-tRNA–Protein Transferases

The transfer of aminoacyl groups from aminoacyl-tRNA into peptide linkages with amino-terminal amino acids of proteins has been demonstrated in procaryotic and eucaryotic cell-free systems (8). In *E. coli* this reaction is carried out by a soluble enzyme, leucyl, phenylalanyl-tRNA–protein transferase (105). This enzyme catalyzes the transfer of leucine, phenylalanine, and methionine from their respective aminoacyl-tRNAs into acceptor proteins (8, 106, 107). Although the physiological significance of this enzyme is still poorly understood, the characterization of a poorly growing *E. coli* mutant lacking the aminoacyl transferase suggests a regulatory role in proline metabolism at the level of proline oxidase (108). A multiplicity of effects is associated with this mutation (109).

In cell-free systems from mammalian cells a transfer of arginine form arginyl-tRNA to certain acceptor proteins was demonstrated. Such proteins include erythrocyte ghost membrane proteins (110) and chromatin proteins (111); this may suggest a regulatory role for this enzyme.

6.14 Viruses and tRNA

A puzzling phenomenon attracting much attention is that the RNA of many plant viruses can be aminoacylated with crude or pure aminoacyl-tRNA synthetases (see chapter 2). These studies have demonstrated that large RNA molecules can be aminoacylated at the 3′-end and have led to the establishment of the nucleotide sequence of the 112-nucleotide-long 3-terminal fragment of TYMV RNA (185, 186); this fragment can be valylated. While these studies have interesting implications for the problem of recognition of tRNA by aminoacyl-tRNA synthetases, there is still no explanation of why this phenomenon occurs at all. In the case of mammalian RNA viruses aminoacylation of the RNA contained in the virus genome has also been observed (112, 113). Moreover it has been well established that specific tRNA species noncovalently bound to the viral RNA in the virions of RNA tumor viruses serve as the obligatory primer for the reverse transcriptase (9, 114, 115). It is particularly interesting that these are not virus-specific tRNAs; rather they are selected tRNAs from the host (116).

6.15 Regulation of Biosynthesis of Aminoacyl-tRNA Synthetases and of tRNA

Since tRNA, aminoacyl-tRNA synthetases, and their product, aminoacyl-tRNA, are crucial for many cellular processes, an understanding of the factors controlling the levels of these macromolecules is desirable. In the bacterial cell an apparent stoichiometric relationship between the amount of tRNA and the cognate aminoacyl-tRNA synthetase has been reported (117–120); thus it is possible that their regulation may be coupled. In addition models in which aminoacyl-tRNA acts as an autoregulatory compound (4) affecting aminoacyl-tRNA synthetase or tRNA synthesis may be imagined (121). Despite much work (for an authoritative review see ref. 5) little is known about the synthesis and degradation of these molecules in enteric bacteria. For other organisms our knowledge is much less advanced. Since studies of mutants have been successful in uncovering regulatory mechanisms (such as histidine biosynthesis), it is desirable to isolate regulatory mutants affecting tRNA and aminoacyl-tRNA synthetase levels. Such mutants may provide answers to the following questions: (i) Is there coordinate control of all aminoacyl-tRNA synthetases or of a subset thereof? Alternatively are some aminoacyl-tRNA synthetases regulated independently? (ii) Is there coordinate regulation of all tRNA species? Alternatively is each isoacceptor family (say, all tRNA^{Leu} species) or each tRNA species (tRNA$_1^{Leu}$) individually controlled? (iii) Is the regulation of an aminoacyl-tRNA synthetase coupled to that of its cognate isoacceptor tRNA family? (iv) Is the regulation of amino acid biosynthesis linked to that of the cognate tRNA or aminoacyl-tRNA synthetase? A brief summary of our present knowledge of tRNA and aminoacyl-tRNA synthetase biosynthesis in *E. coli* is given here, since attempts to study this question have utilized this organism.

The biosynthesis of tRNA is a complicated process (see chapter 3) involving transcription of tRNA genes into precursor molecules larger than mature tRNA. The additional nucleotides are then removed by specific nucleases. Concomitant with this trimming process enzymatic nucleotide modifications yield the many modified nucleosides found in tRNA. To date there is little knowledge about the regulation of tRNA biosynthesis. It is known that the synthesis of tRNA and that of ribosomal RNA in bacteria depend on the growth rate and are regulated by similar mechanisms (95, 122–124). Because of the complexity of tRNA biosynthesis, regulation at many points can be envisaged. Thus one may expect a variety of regulatory mutants. None, however, has been isolated to date.

Aminoacyl-tRNA synthetases were once thought to be constitutive enzymes (125); their levels were thought to be unaffected by amino acid supply or growth rate. However, some imaginative work has shown that aminoacyl-tRNA synthetase formation depends on the amino acid concentration of the medium and on the growth rate of the cells (for review see ref. 5). Two controls may act on synthetase formation, one responsive to the demands of protein synthesis, the other to amino acid restriction (5). Further complexity arises from the inactivation and resynthesis of aminoacyl-tRNA synthetases (126). Moreover present evidence indicates that not all aminoacyl-tRNA synthetases are regulated in a uniform manner. When the cognate amino acid is limited, a repression-derepression phenomenon has been observed for some enzymes; for others such limitation does not lead to derepression (5). From such studies no clear picture has yet emerged.

Mutations affecting the levels of one or more aminoacyl-tRNA synthetases may provide more insight into the regulation of the levels of these macromolecules. To date few such mutants have been isolated. Recently temperature-resistant revertants of strains bearing defective aminoacyl-tRNA synthetases have been selected in the hope of isolating strains with elevated levels of specific tRNAs or aminoacyl-tRNA synthetases. These studies provided two mutants with elevated levels of tRNA. A strain bearing a defective thermolabile glutaminyl-tRNA synthetase, which is protected from thermal inactivation in vitro by the presence of tRNA, has been described (127). Temperature-resistant revertants of this strain, having elevated levels of tRNA^{Gln} resulting from a gene duplication, have been isolated (82). Only one tRNA species, tRNA$_1^{Gln}$, was overproduced. Similarly, stable mutants having elevated tRNA^{Gln} levels have been isolated from the same strain (128). Their biochemical characterization is not yet complete, but in one case a concomitant elevation of tRNA^{Gln} and tRNA^{Met} is seen. Whether this results from from the increased transcription of a common precursor remains to be seen. These may be the desired tRNA regulatory mutants. Thus use of this method with strains bearing specific mutant aminoacyl-tRNA synthetases having an altered interaction with tRNA may allow isolation of large numbers of the desired mutants.

Our knowledge of the genetic regulation of aminoacyl-tRNA synthetase levels is slightly more advanced.

$StrB^-$ mutants of S. typhimurium have elevated levels of histidyl-tRNA synthetase (42). The manner in which this effect is mediated is not known, although the levels of other aminoacyl-tRNA synthetases are not elevated.

The antibiotic borrelidin inhibits threonyl-tRNA synthetase of both procaryotic and eucaryotic organisms. Among borrelidin-resistant mutants have been found strains having either an altered enzyme or elevated levels of threonyl-tRNA synthetase (129). Unfortunately the mutations causing these phenotypes have not been mapped, making it difficult to determine the control mechanism(s) involved.

Amino acid analog–resistant mutants of *E. coli* having elevated amino-acyl-tRNA synthetase levels have been isolated using the serine analog, serine hydroxymate. Among serine hydroxymate–resistant mutants were found strains that contain elevated seryl-tRNA synthetase levels (130). Such mutations map close to the structural gene for seryl-tRNA synthetase (*serS*). Many other amino acid analogs interact with aminoacyl-tRNA synthetases in vitro. Mutants resistant to these analogs due to increased levels of the cognate aminoacyl-tRNA synthetase have not yet been isolated.

Among temperature-resistant revertants of strains having thermolabile aminoacyl-tRNA synthetases are many strains having elevated levels of the thermolabile enzyme. In this manner operatorlike mutation affecting seryl-(131) and leucyl- (132) tRNA synthetase levels have been identified. Analysis of such mutants has revealed a protein modulating the level of leucyl-tRNA synthetase in an unknown fashion (132). This approach has also identified a mutation that affects valyl-tRNA synthetase levels and maps close to the structural gene for valyl-tRNA synthetase. Other mutations causing elevated seryl-, leucyl-, and glutaminyl-tRNA synthetase levels have been isolated (128).

These mutants have been utilized to partially answer some of the questions raised at the beginning of this section and to question some of the currently accepted concepts. None of the aminoacyl-tRNA synthetase regulatory mutants that have been examined contain elevated levels of the cognate tRNA or elevated levels of other aminoacyl-tRNA synthetases. This demonstrates that there is no obligatory stoichiometric relationship between the level of an aminoacyl-tRNA synthetase and its cognate tRNA isoacceptor family.

6.16 Conclusion

This chapter attempted to present a brief picture of our knowledge of the roles of tRNA. It appears that tRNA involvement in metabolic reactions is widespread; presumably we have yet to discover many reactions. The unraveling of the many steps of tRNA biosynthesis—coupled with the production of mutants in many of these processes—may shed new light on tRNA involvement in the cell. The study of tRNA-modifying mutants in particular may be enlightening as is already clear from the work on *hisT* and *trpX* mutations.

With the renewed interest in regulatory roles of tRNA one can hope for advances that will clarify the role of tRNA found in viruses or identify and define specific tRNA species characteristic for certain tissues, certain organs, or for certain developmental stages of organisms. This should give us a better understanding of the delicate balance between aminoacyl-tRNA and general anabolic and catabolic reactions in the cell.

Acknowledgments

The authors thank M. Baer, R. Bryan, A Cheung, G. Theall, S. Morgan, and R. Wetzel for many discussions. Studies in the authors' laboratory were supported by grants from the National Institutes of Health.

References

1. Littauer, U. Z., and Inouye, H. (1973). Ann. Rev. Biochem. *42*, 439–470.

2. Ofengand, J. (1977). In Molecular Mechanisms of Protein Biosynthesis, H. Weissbach, and S. Pestka, eds, (New York: Academic Press), pp. 7–79.

3. Brenchley, J. E., and Williams, L. S. (1975). Ann. Rev. Microbiol. *29*, 251–274.

4. Calhoun, D. H., and Hatfield, G. W. (1975). Ann. Rev. Microbiol. *29*, 275–299.

5. Neidhardt, F. C., Parker, J., and McKeever, W. G. (1975). Ann. Rev. Microbiol. *29*, 215–250.

6. Cashel, M., and Gallant, J. (1974). In Ribosomes, M. Nomura, A. Tissières, and P. Lengyel, eds. (Cold Spring Harbor, New York: Cold Spring Harbor Laboratory), pp. 733–746.

7. Bertrand, K., Korn, L., Lee, F., Platt, T., Squires, C. L., Squires, C., and Yanofsky, C. (1975). Science *189*, 22–26.

8. Soffer, R. L. (1974). Adv. Enzymol. *40*, 91–139.

9. Faras, A. J., Taylor, J. M., Levinson, W. E., Goodman, H. M., and Bishop, J. M. (1973). J. Mol. Biol. *79*, 163–183.

10. Neidhardt, F. C. (1966). Bacteriol. Revs. *30*, 701–719.

11. Quay, S. C., Kline, E. L., and Oxender, D. L. (1975). Proc. Nat. Acad. Sci. USA *72*, 3921–3924.

12. Unger, M. W. (1977). J. Bacteriol. *130*, 11–19.

13. Singer, C. E., Smith, G. R., Cortese, R., and Ames, B. N. (1972). Nature New Biol. *238*, 72–74.

14. Zubay, G., Chambers, D. A., and Cheong, L. C. (1970). In The Lactose Operon, J. R. Beckwith, and D. Zipser, eds. (Cold Spring Harbor, New York: Cold Spring Harbor Laboratory), pp. 375–391.

15. Söll, D., and Schimmel, P. (1974). In The Enzymes, vol. 10, P. Boyer, ed. (New York: Academic Press), pp. 489–538.

16. Lapointe, J., and Söll, D. (1972). J. Biol. Chem. *247*, 4982–4985.

17. Willick, G. E., and Kay, C. M. (1976). Biochemistry *15*, 4347–4352.

18. Clarke, S. J., Low, B., and Konigsberg, W. H. (1973). J. Bacteriol. *113*, 1091–1095.

19. Beauchamp, P. M., Horn, E. W., and Gross, S. R. (1977). Proc. Nat. Acad. Sci. USA *74*, 1172–1176.

20. Arfin, S. M., Simpson, D. R., Chiang, C. S., Andrulis, I. L., and Hatfield, G. W. (1977). Proc. Nat. Acad. Sci. USA *74*, 2367–2369.

21. Johnson, E. J., Cohen, G. N., and Saint-Girons, I. (1977). J. Bacteriol. *129*, 66–70.

22. Levin, A. P., and Hartman, P. E. (1963). J. Bacteriol. *86*, 820–828.

23. Roth, J. R., Anton, D. N., and Hartman, P. E. (1966). J. Mol. Biol. *22*, 305–323.

24. Anton, D. N. (1968). J. Mol. Biol. *33*, 533–546.

25. Ely, B., Frankhauser, D. B., and Hartman, P. E. (1974). Genetics *78*, 607–631.

26. Roth, J. R., and Ames, B. N. (1966). J. Mol. Biol. *22*, 325–334.

27. Silbert, D. F., Fink, G. R., and Ames, B. N. (1966). J. Mol. Biol. *22*, 335–347.

28. Brenner, M., and Ames, B. N. (1972). J. Biol. Chem. *247*, 1080–1088.

29. Singer, C. E., and Smith, G. R. (1972). J. Biol. Chem. *247*, 2989–3000.

30. Singer, C. E., Smith, G. R., Cortese, R., and Ames, B. N. (1972). Nature New Biol. *238*, 72–74.

31. Allaudeen, H. S., Yang, S. K., and Söll, D. (1972). FEBS Letters *28*, 205–208.

32. Cortese, R., Landsberg, R. A., von der Haar, R. A., Umbarger, H. E., and Ames, B. N. (1974). Proc. Nat. Acad. Sci. USA *71*, 1857–1861.

33. Rizzino, A. A., Bresalier, R. S., and Freundlich, M. (1974). J. Bacteriol. *117*, 449–455.

34. Lewis, J. A., and Ames, B. N. (1972). J. Mol. Biol. *66*, 131–142.

35. Martin, R. G., Bagdasarian, M., Ames, B. N., and Roth, J. (1969). Fed. Eur. Biochem. Soc. Symp. *19*, 1.

36. Bossi, L., and Cortese, R. (1978). J. Bact. (in press).

37. Kasai, T. (1974). Nature *249*, 523–527.

38. Blasi, F., Bruni, C. B., Avitable, A., Deeley, R. G., Goldberger, R. F., and Meyers, M. M. (1973). Proc. Nat. Acad. Sci. USA *70*, 2692–2696.

39. Artz, S. W., and Broach, J. R. (1975). Proc. Nat. Acad. Sci. USA *72*, 3453–3457.

40. Ely, B. (1974). Genetics *78*, 593–606.

41. Goldberger, R. F. (1974). Science *183*, 810–816.

42. Wyche, J., Ely, B., Cebula, T., Snead, M., and Hartman, P. (1974). J. Bacteriol. *117*, 708–716.

43. Rothman-Denes, L., and Martin, R. G. (1971). J. Bacteriol. *106*, 227–237.

44. Kovach, J. S., Phang, J. M., Blasi, F., Barton, R. W., Ballesteros-Olmo, A., and Goldberger, R. F. (1970). J. Bacteriol. *104*, 787–792.

45. Meyers, M., Blasi, F., Bruni, C. B., Deely, R. G., Kovach, J. S., Leventhal, M., Mullinix, K. P., Vogel, T., and Goldberger, R. F. (1975). Nucl. Acids Res. *2*, 2021–2036.

46. Scott, J. F., Roth, J. R., and Artz, S. W. (1975). Proc. Nat. Acad. Sci. USA *72*, 5021–5025.

47. DeLorenzo, F., Straus, D. S., and Ames, B. N. (1972). J. Biol. Chem. *247*, 2302–2307.

48. Stephens, J. C., Artz, S. W., and Ames, B. N. (1975). Proc. Nat. Acad. Sci. USA *72*, 4389–4393.

49. Balbinder, E., Callahan III, R., McCann, P. P., Cordaro, J. C., Weber, A. R., Smith, A. M., and Angelosanto, F. (1970). Genetics *66*, 31–53.

50. Hiraga, S. (1969). J. Mol. Biol. *39*, 159–179.

51. Cohen, G., and Jacob, F. (1959). C. R. Acad. Sci. *248*, 3490–3492.

52. Rose, J. K., Squires, C. L., Yanofsky, C., Yang, H.-L., and Zubay, G. (1973). Nature New Biol. *245*, 133–137.

53. Squires, C. L., Lee, F. D., and Yanofsky, C. (1975). J. Mol. Biol. *92*, 93–111.

54. Hiraga, S., Ito, K., Matsuyama, T., Ozaki, H., and Yura, T. (1968). J. Bacteriol. *96*, 1880–1881.

55. Kano, Y., Matsushiro, A., and Shimura, Y. (1968). Mol. Gen. Genetics *102*, 15–26.

56. Ito, K., Hiraga, S., and Yura, T. (1969). Genetics *61*, 521–538.

57. Jackson, E. N., and Yanofsky, C. (1973). J. Mol. Biol. *76*, 89–101.

58. Squires, C., Lee, F., Bertrand, K., Squires, C. L., Bronson, M. J., and Yanofsky, C. (1976). J. Mol. Biol. *103*, 351–381.

59. Bertrand, K., Squires, C., and Yanofsky, C. (1976). J. Mol. Biol. *103*, 319–337.

60. Bertrand, K., and Yanofsky, C. (1976). J. Mol. Biol. *103*, 339–349.

61. Morse, D. E., and Morse, A. N. C. (1976). J. Mol. Biol. *103*, 209–226.

62. Pannekoek, H., Brammar, W. J., and Pouwels, P. H. (1975). Mol. Gen. Genetics *136*, 199–214.

63. Lee, F., Squires, C. L., Squires, C., and Yanofsky (1976). J. Mol. Biol. *103*, 383–393.

64. Ito, K., Hiraga, S., and Yura, T. (1969). J. Bacteriol. *99*, 279–286.

65. Pouwels, P. H., and van Rotterdam, J. (1975). Mol. Gen. Genetics *136*, 215–226.

66. Korn, L. J., and Yanofsky, C. (1976). J. Mol. Biol. *103*, 395–409.

67. Korn, L. J., and Yanofsky, C. (1976). J. Mol. Biol. *106*, 231–241.

68. Szentirmai, A., Szentirmai, M., and Umbarger, H. E. (1968). J. Bacteriol. *95*, 1672–1679.

69. Eidlic, L., and Neidhardt, F. C. (1965). Proc. Nat. Acad. Sci. USA *53*, 539–543.

70. Low, B., Gates, F., Goldstein, T., and Söll, D. (1971). J. Bacteriol. *108*, 742–750.

71. Freundlich, M. (1967). Science *157*, 823–824.

72. Williams, L. S., and Freundlich, M. (1969). Biochim. Biophys. Acta *186*, 305–316.

73. LaRossa, R., Mao, J., Low, B., and Söll, D. (1977). J. Mol. Biol. *117*.

74. Hatfield, G. W., and Burns, R. O. (1970). Proc. Nat. Acad. Sci. USA *66*, 1027–1035.

75. Childs, G., Sonnenberg, F., and Freundlich, M. (1977). Mol. Gen. Genetics *151*, 121–126.

76. Childs, G. J., and Freundlich, M. (1975). Mol. Gen. Genetics *138*, 257–268.

77. Wasmuth, J. J., and Umbarger, H. E. (1973). J. Bacteriol. *116*, 548–561.

78. Wohlheuter, R. M., Schutt, M., and Holzer, H. (1973). In The Enzymes of Glutamine Metabolism, S. Prusiner, and E. Stadtman, eds. (New York: Academic Press), pp. 45–64.

79. Ginsberg, A., and Stadtman, E. (1973). In The Enzymes of Glutamine Metabolism, S. Prusiner, and E. Stadtman, eds. (New York: Academic Press), pp. 9–44.

80. Lapointe, J., Delcuve, G., and Duplain, L. (1975). J. Bacteriol. *123*, 843–850.

81. LaRossa, R., Morgan, S., McCutchan, T., and Söll, D. (1978). In preparation.

82. Morgan, S., Körner, A., Low, B., and Söll, D. (1977). J. Mol. Biol. *117*.

83. Morgan, S., Cheung, A., Körner, A., and Söll, D. (1977). Fed. Proc. *36*, 659.

84. Oxender, D. L. (1972). In Metabolic Pathways, vol. 6., L. E. Hokin, ed. (New York: Academic Press), pp. 133–185.

85. Halpern, Y. S. (1974). Ann. Rev. Genet. *8*, 103–133.

86. Penrose, W. R., Nichoalds, G. E., Piperno, J. R., and Oxender, D. L. (1968). J. Biol. Chem. *243*, 5921–5928.

87. Quay, S. C., Oxender, D. L., Tsuyumu, S., and Umbarger, H. (1975). J. Bacteriol. *122*, 994–1000.

88. Quay, S. C., and Oxender, D. L. (1976). J. Bacteriol. *127*, 1225–1238.

89. Quay, S. C., and Oxender, D. L. (1977). J. Bacteriol. *130*, 1024–1029.

90. Alfoldi, L., Stent, G. S., and Clowes, R. C. (1962). J. Mol. Biol. *5*, 348–355.

91. Borek, E., Rochenbach, J., and Ryan, A. (1956). J. Bacteriol. *71*, 318–323.

92. Cashel, M. (1969). J. Biol. Chem. *244*, 3133–3141.

93. Block, R., and Haseltine, W. A. (1974). In Ribosomes, N. Nomura, A. Tissières, and P. Lengyel, eds. (Cold Spring Harbor, New York: Cold Spring Harbor Laboratory), pp. 747–761.

94. Haseltine, W. A., and Block, R. (1973). Proc. Nat. Acad. Sci. USA *70*, 1564–1568.

95. Ikemura, T., and Dahlberg, J. E. (1973). J. Biol. Chem. *248*, 5033–5041.

96. Sprinzl, M., and Richter, D. (1976). Eur. J. Biochem. *71*, 171–176.

97. Richter, D., Erdmann, V. A., and Sprinzl, M. (1974). Proc. Nat. Acad. Sci. USA *72*, 3226–3229.

98. Arnold, H.-H., and Ogilvie, A. (1977). Biochem. Biophys. Res. Commun. *74*, 343–349.

99. Smith, O., Meyers, M., Vogel, T., Deeley, R. G., and Goldberger, R. J. (1974). Nucl. Acids Res. *1*, 881–888.

100. Meuris, P. (1973). Mol. Gen. Genetics *121*, 207–218.

101. Bell, J. B., Gelugne, J. P., and Jacobson, K. B. (1976). Biochim. Biophys. Acta *435*, 21–29.

102. Twardzik, D. R., Grell, E. H., and Jacobson, K. B. (1971). J. Mol. Biol. *57*, 231–245.

103. Mischke, D., Kloetzel, P., and Schwochau, M. (1975). Nature *255*, 79–80.

104. Wosnick, M. A., and White, B. N. (1977). Nucl. Acids Res. *4*, 3919–3930.

105. Leibowitz, M. J., and Soffer, R. L. (1971). Proc. Nat. Acad. Sci. USA *68*, 1866–1869.

106. Horiuishi, H., Hashizume, S., Seguchi, M., and Takahashi, K. (1975). Biochem. Biophys. Res. Commun. *67*, 1136–1143.

107. Scarpulla, R. C., Deutch, C. E., and Soffer, R. L. (1976). Biochem. Biophys. Res. Commun. *71*, 584–589.

108. Soffer, R. L., and Savage, M. (1974). Proc. Nat. Acad. Sci. USA *71*, 1004–1007.

109. Deutch, C. E., Scarpulla, R. C., Sonnerblick, E. B., and Soffer, R. L. (1977). J. Bacteriol. *129*, 544–546.

110. Kaji, H., and Rao, P. (1976). FEBS Letters *66*, 194–197.

111. Kaji, H. (1976). Biochemistry *15*, 5121–5125.

112. Rosenthal, J. J., and Zamecnik, P. (1973). Proc. Nat. Acad. Sci. USA *70*, 865–869.

113. Wang, S., Kothari, R. M., Taylor, M., and Hung, P. (1973). Nature New Biol. *242*, 133–135.

114. Harada, F., Sawyer, R. C., and Dahlberg, J. E. (1975). J. Biol. Chem. *250*, 3487–3497.

115. Dahlberg, J. E., Sawyer, R. C., Taylor, J. M., Faras, A. J., Levinson, W. E., Goodman, H. M., and Bishop, J. M. (1974). J. Virol. *13*, 1126–1133.

116. Dahlberg, J. E., Harada, F., and Sawyer, R. C. (1974). Cold Spring Harbor Symp. Quant. Biol. *39*, 925–932.

117. Calendar, R., and Berg, P. (1966). Biochemistry *5*, 1681–1690.

118. Jacobson, K. B. (1969). J. Cellular Physiology *74*, suppl. 1, pp. 99–100.

119. DeLorenzo, F., and Ames, B. N. (1970). J. Biol. Chem. *245*, 1710–1716.

120. Joseph, D. R., and Muench, K. M. (1971). J. Biol. Chem. *246*, 7602–7609.

121. Ehresmann, B., Imbault, P., and Weil, J. H. (1974). Biochimie *56*, 1351–1355.

122. Rosset, R., Julien, J., and Monier, R. (1966). J. Mol. Biol. *18*, 308–320.

123. Dennis, P. P. (1972). J. Biol. Chem. *247*, 2842–2845.

124. Skjold, A. C., Juarez, H., and Hedgcoth, C. (1973). J. Bacteriol. *115*, 177–187.

125. Boman, H. G., Boman, I. A., and Maas, W. K. (1961). Biological Structure and Function, T. W. Goodwin and O. Lindburg, eds. (New York, London: Academic Press), pp. 297–308.

126. Williams, L. S., and Neidhardt, F. C. (1969). J. Mol. Biol. *43*, 529–550.

127. Körner, A., Magee, B. B., Linka, B., Low, K. B., Adelberg, E. A., and Söll, D. (1974). J. Bacteriol. *120*, 154–158.

128. Morgan, S., LaRossa, R., Cheung, A., Low, B., and Söll, D. (1978). Archives de Biologia y Medicina Experimentales (in press).

129. Paetz, W., and Nass, G. (1973). Eur. J. Biochem. *35*, 331–337.

130. Pizer, L. I., McKitrick, J., and Tosa, T. (1972). Biochem. Biophys. Res. Commun. *49*, 1351–1357.

131. Clarke, S. J., Low, B., and Konigsberg, W. (1973). J. Bacteriol. *113*, 1096–2003.

132. LaRossa, R., Vögeli, G., Low, B., and Söll, D. (1977). J. Mol. Biol. *117*.

133. Yaniv, M., and Gros, F. (1967). Genet. Elem. Prop. Funct. Symp. 3rd FEBS Meeting, 1966, pp. 157–178.

134. Lazar, M., Yaniv, M., and Gros, F. (1968). C.R. Acad. Sci. (Paris) *266*, 531–534.

135. Buckel, P., Lubitz, W., and Böck, A. (1971). J. Bacteriol. *108*, 1008–1016.

136. Hankinson, O. (1976). Somatic Cell Genet. *2*, 497–507.

137. Sato, K. (1975). Nature *257*, 813–815.

138. Cooper, P. H., Hirshfield, I. N., and Maas, W. K. (1969). Mol. Gen. Genet. *104*, 383–390.

139. Hirschfield, I. N., and Bloemers, H. P. (1969). J. Biol. Chem. *244*, 2911–2916.

140. Williams, L. S. (1973). J. Bacteriol. *113*, 1419–1432.

141. Thompson, L. H., Lofgren, D. J., and Adair, G. M. (1977). Cell *11*, 157–168.

142. Thompson, L. H., Stanners, C. P., and Siminovitch, L. (1975). Somatic Cell Genet. *1*, 187–208.

143. Wasmuth, J. J., and Caskey, C. T. (1976). Cell *9*, 655–662.

144. Russel, R. R. B., and Pittard, A. J. (1971). J. Bacteriol. *108*, 790–798.

145. Murgola, E. J., and Adelberg, E. A. (1970). J. Bacteriol. *103*, 20–26.

146. Murgola, E. J., and Adelberg, E. A. (1970). J. Bacteriol. *103*. 178–183.

147. Böck, A., and Neidhardt, F. C. (1966). Z. Verebungsl. *98*, 187–201.

148. Folk, W. R., and Berg, P. (1970). J. Bacteriol. *102*, 204–212.

149. Roback, E. R., Friesen, J. D., and Fiil, N. P. (1973). Can. J. Microbiol. *19*, 425–426.

150. Nass, G. (1967). Mol. Gen. Genet. *100*, 216–224.

151. Iaccarino, M., and Berg, P. (1971). J. Bacteriol. *105*, 527–537.

152. Treiber, G., and Iaccarino, M. (1971). J. Bacteriol. *107*, 828–832.

153. Coker, M., and Umbarger, H. E. (1970). Bacteriol. Proc. 135.

154. McGinnis, E., and Williams, L. S. (1974). Abstract, Ann. Meeting Amer. Soc. Microbiol., p. 172.

155. Blatt, J. M., and Umbarger, H. E. (1972). Biochem. Genetics *6*, 99–118.

156. Hartwell, L. L., and McLaughlin, C. S. (1968). Proc. Nat. Acad. Sci. USA *59*, 422–428.

157. McLaughlin, C. S., and Hartwell, L. H. (1969). Genetics *61*, 557–566.

158. Thompson, L. H., Harkins, J. L., and Stanners, C. P. (1973). Proc. Nat. Acad. Sci. USA *70*, 3094–3098.

159. Alexander, R. R., Calvo, J. M., and Freundlich, M. (1971). J. Bacteriol. *106*, 213–220.

160. Mikulka, T. W., Stieglitz, B. I., and Calvo, J. M. (1972). J. Bacteriol. *109*, 584–593.

161. Gross, S. R., McCoy, M. T., and Gilmore, E. B. (1968). Proc. Nat. Acad. Sci. USA *61*, 253–260.

162. Racine, F. M., and Steinberg, W. (1974). J. Bacteriol. *120*, 372–383.

163. Racine, F. M., and Steinberg, W. (1974). J. Bacteriol. *120*, 384–389.

164. Hirshfield, I. N., Tomford, J. W., and Zamecnik, P. C. (1972). Biochim. Biophys. Acta *259*, 344–356.

165. Somerville, C. R., and Ahmed, A. (1977). J. Mol. Biol. *111*, 77–82.

166. Yamamoto, M., Nomura, M., Ohsawa, H., and Maruo, B. (1977). J. Bact. *132*, 127–131.

167. Archibold, E. R., and Williams, L. S. (1973). J. Bacteriol. *114*, 1007–1013.

168. Sanderson, K. E., and Demerec, M. (1965). Genetics *51*, 897–913.

169. Gross, T. S., and Rowbury, R. J. (1969). Biochim. Biophys. Acta *184*, 233–236.

170. DeRobichon-Szulmajster, H. (1975). Unpublished.

171. Fangman, W. L., and Neidhardt, F. C. (1964). J. Biol. Chem. *239*, 1839–1843.

172. Eidlic, L., and Neidhardt, F. C. (1965). J. Bacteriol. *89*, 706–711.

173. Tosa, T., and Pizer, L. I. (1971). J. Bacteriol. *106*, 972–982.

174. Steinberg, W., and Anagnostopoulos, C. (1971). J. Bacteriol. *105*, 6–19.

175. Doolittle, W. F., and Yanofsky, C. (1968). J. Bacteriol. *95*, 1283–1294.

176. Nazario, M., Kinsey, J. A., and Ahmad, M. (1971). J. Bacteriol. *105*, 121–126.

177. Schlessinger, S., and Nester, E. W. (1969). J. Bacteriol. *100*, 167–175.

178. Buonocore, V., Harris, M. H., and Schlesinger, S. (1973). J. Biol. Chem. *247*, 4843–4849.

179. Heinonen, J., Artz, S. W., and Zalkin, H. (1972). J. Bacteriol. *112*, 1254–1263.

180. Tingle, M. A., and Neidhardt, F. C. (1969). J. Bacteriol. *98*, 837–839.

181. Yaniv, M., and Gros, F. (1969). J. Mol. Biol. *44*, 31–45.

182. Yanofsky, C., and Soll, L. (1977). J. Mol. Biol. *113*, 663–677; Yanofsky, C., personal communication.

183. Kline, E. L., Brown, C. S., Coleman, W. G., Jr., and Umbarger, H. E. (1974). Biochem. Biophys. Res. Comm. *57*, 1144–1151.

184. Smith, J. M., Smolin, D. E., and Umbarger, H. E. (1976). Mol. Gen. Genet. *148*, 111–124.

185. Briand, J-P., Jonard, G., Guilley, H., Richards, K., and Hirth, L. (1977). Eur. J. Biochem. *72*, 453–463.

186. Silberklang, M., Prochiantz, A., Haenni, A-L., and RajBhandary, U. L. (1977). Eur. J. Biochem. *72*, 465–478.

187. Theall, G., Low, K. B., and Söll, D. (1977). Mol. Gen. Genet. *156*, 221–227.

188. Kano, Y., Matsushiro, A., and Shimura, Y. (1968). Mol. Gen. Genet *102*, 15–26.

189. Hennecke, H., Böck, A., Thomale, J., and Nass, G. (1977). J. Bact. *131*, 943–950.

190. Bachmann, B. J., Low, K. B., and Taylor, A. L. (1976). Bacteriol. Rev. *40*, 116–167.

191. Sanderson, K. W., and Demerec, M. (1965). Genetics *51*, 897–913.

7 MODIFIED NUCLEOSIDES AND ISOACCEPTING tRNA

S. Nishimura

7.1 Introduction

A typical characteristic of tRNA is that it contains a variety of modified nucleosides. These modified nucleosides differ in two ways from those in other nucleic acids. One is that they are present in high proportion: for example, 17 of the total 87 nucleosides in $tRNA_1^{Ser}$ from rat liver are modified (1). The other is that they show a wide range of structural variation. More than 50 modified nucleosides have been isolated and characterized (see refs. 2, 3); some of these are modified only by methylation of the base or at the 2′ hydroxyl of the ribose moiety, while others are so-called hypermodified nucleosides with more complex modifications. The structures of the modified nucleosides, other than 2′-O-methylated nucleosides, found in tRNA are shown in figure 7.1.

7.2 Structural Elucidation of Modified Nucleosides

Inosine, pseudouridine, 4-thiouridine, and a number of methylated nucleosides were isolated and characterized more than a decade ago. Many of these modified nucleosides are present in most tRNA molecules, and so their detection and isolation in relatively large quantities from unfractionated tRNA is straightforward. In contrast hypermodified nucleosides are usually located in the first position in the anticodon or next to the anticodon, in only one or several species of tRNA from a given source. Thus it is difficult to detect them in unfractionated tRNA, and this is probably one reason why hypermodified nucleosides were not isolated in earlier work. On the other hand they are fairly easy to detect in pure species of tRNA, because they constitute at least 1 percent of the total nucleoside content. When the presence of a new modified nucleoside in a specific tRNA has been established, its isolation from unfractionated tRNA is facilitated by some knowledge of its chemical properties. In some instances, such as that of dihydrouridine from yeast tRNA^{Ala}, relatively abundant modified nucleosides have been characterized during primary sequence determination (4). It is often preferable to increase the modified nucleoside content by first separating oligonucleotides that contain the component of interest. Knowledge of the exact location of the modified nucleoside in a specific tRNA molecule is possibly the only way to establish with certainty that the nucleoside is a legitimate component of tRNA and not an impurity carried through the isolation procedure. When unfractionated tRNA is used as a source, this problem can be solved by isolating the corresponding mono- or oligonucleotide before converting it to the nucleoside on which most of the structural work is to be carried out.

For detection of modified nucleosides in particular tRNA, the tRNA is first hydrolyzed by RNase T$_2$ and the resulting nucleotide mixture is then subjected to two-dimensional thin-layer or two-dimensional paper chromatography (2). If this procedure is used, about 0.1 mg of purified tRNA is sufficient for detecting a modified component present at a level of one residue per tRNA molecule. Enzymatic hydrolysis of tRNA by RNase T$_2$ (5, 6), snake venom phosphodiesterase (7), or nuclease P$_1$ (8) is preferable to alkaline hydrolysis, because some modified nucleosides are labile under alkaline conditions. In general 0.2A$_{260}$ unit (approximately 10μg) of a modified nucleoside is sufficient for examining its UV and mass spectra and for measuring its chromatographic and electrophoretic mobilities. A known modified nucleoside can be characterized simply by comparing data from these measurements with data for authentic samples. As a minimum criterion for positive identification the UV spectrum should be recorded. If possible, chromatographic mobility should be measured using several different solvent systems. Caution should be exercised in relying only on the elution position on high-pressure liquid chromatography or only on the electrophoretic mobility of the ^{32}P-labeled nucleotide. For example, 5-methoxy-uridine, recently isolated from *Bacillus subtilis* tRNA, exhibits the same chromatographic behavior as uridine (9). If the structure of the modified nucleoside has not previously been established, several column-chromato-

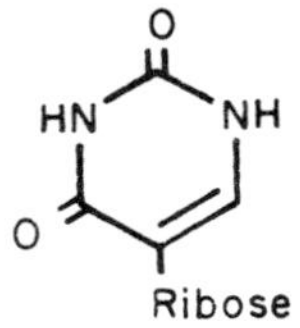
Pseudouridine (Ψ)

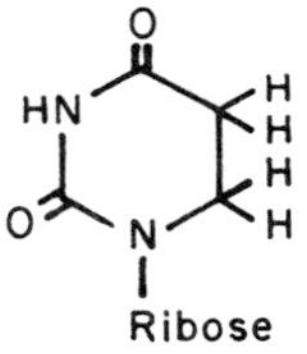
Dihydrouridine (D or hU)

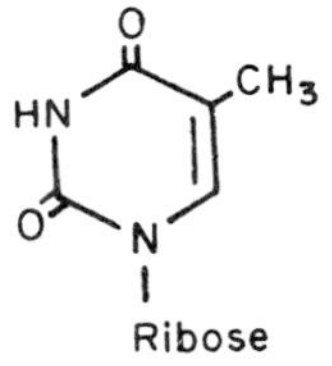
Ribothymidine (T)

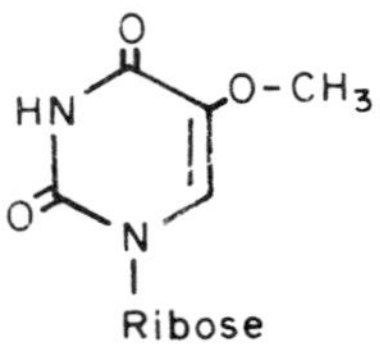
5-Methoxyuridine (mo^5U)

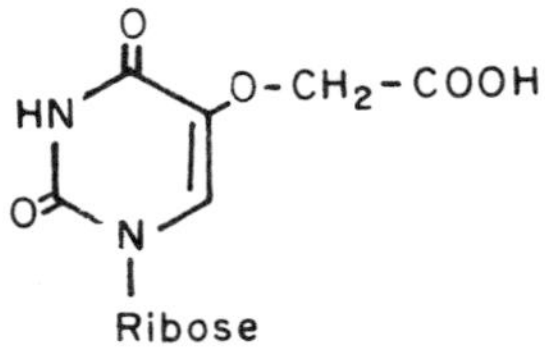
Uridin-5-oxyacetic acid (V)

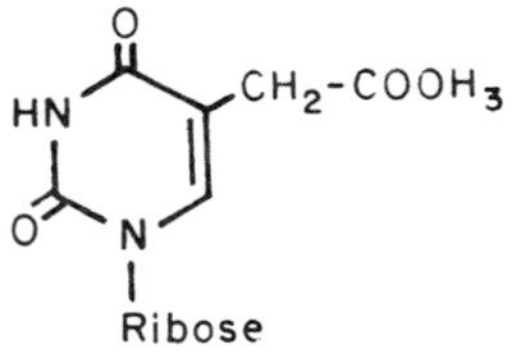
5-(Methoxycarbonylmethyl) uridine (mcm^5U)

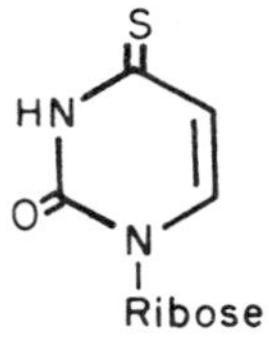
4-Thiouridine (s^4U)

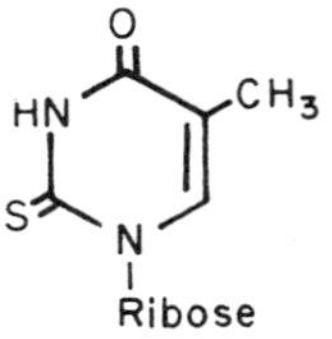
5-Methyl-2-thiouridine (m^5s^2U)

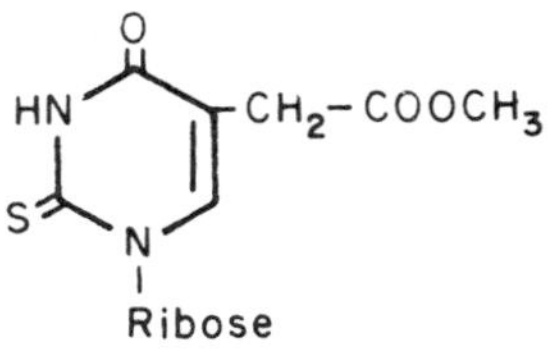
5-(Methoxycarbonylmethyl)-2-thiouridine (mcm^5s^2U)

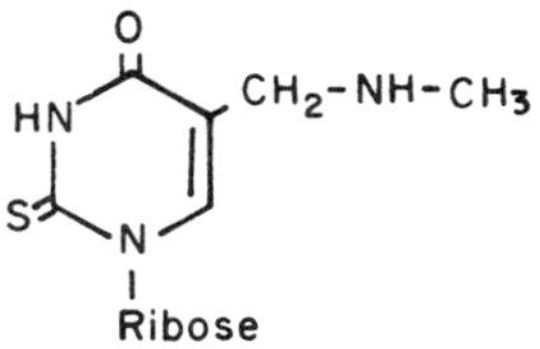
5-Methylaminomethyl-2-thiouridine (mnm^5s^2U)

Figure 7.1 The structures of the modified nucleosides, other than 2′-*O*-methylated nucleosides, found in the sequence of tRNA. In the case of base Y, Yt, and peroxy Y, the ribosyl group is presumably attached to N-9. References for structural characterization of the modified nucleosides are cited in review articles (2, 3).

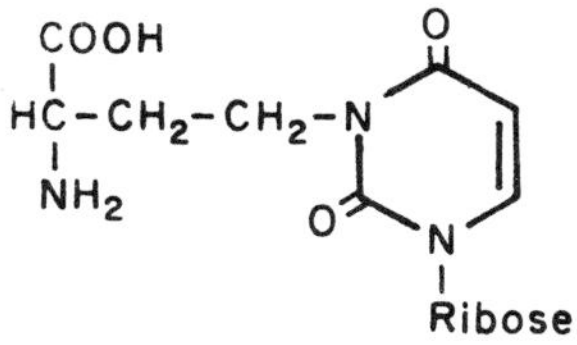

3-(3-Amino-3-
carboxypropyl)
uridine (acp³U)

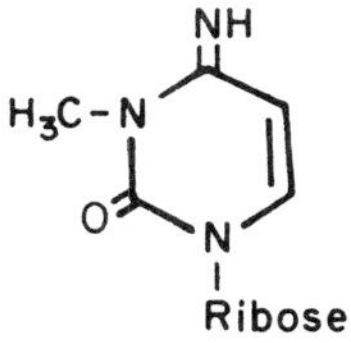

3-Methylcytidine
(m³C)

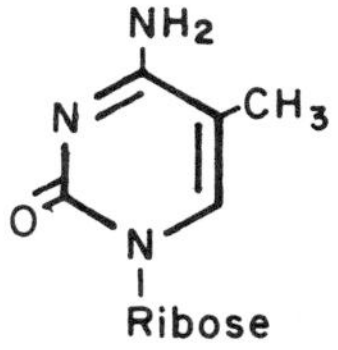

5-Methylcytidine
(m⁵C)

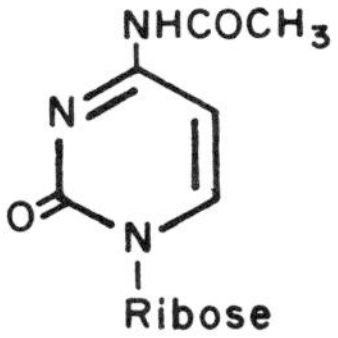

N⁴-Acetylcytidine
(ac⁴C)

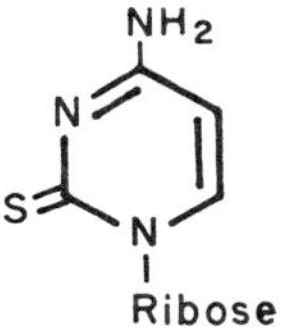

2-Thiocytidine (s²C)

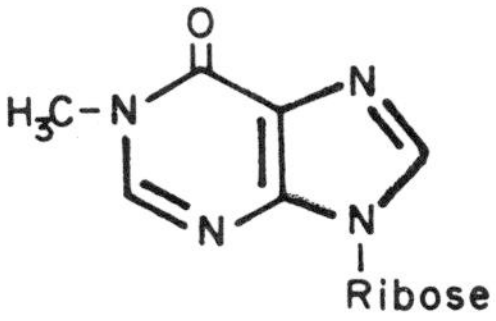

1-Methylinosine
(m¹I)

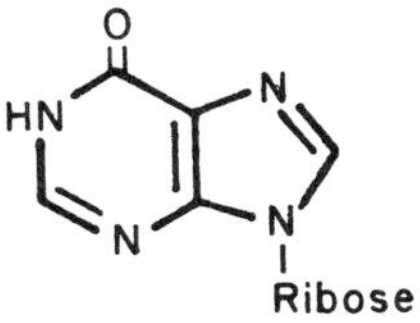

Inosine (I)

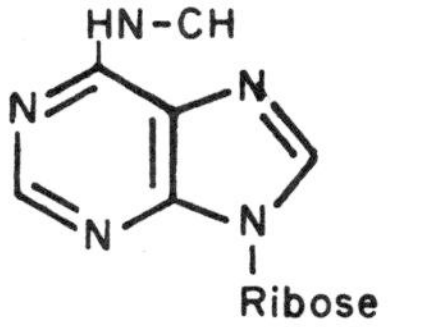

N⁶-Methyladenosine
(m⁶A)

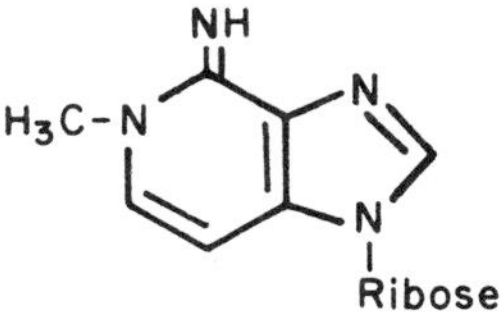

1-Methyladenosine
(m¹A)

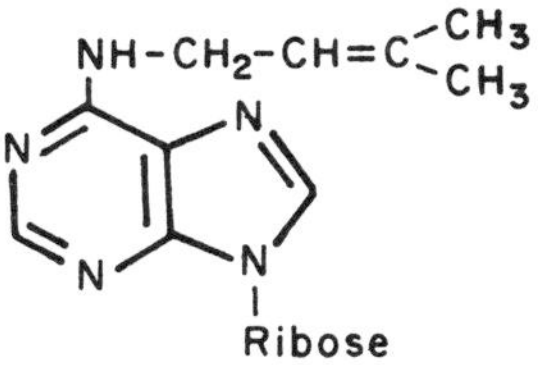

N⁶-Isopentenyl-
adenosine (i⁶A)

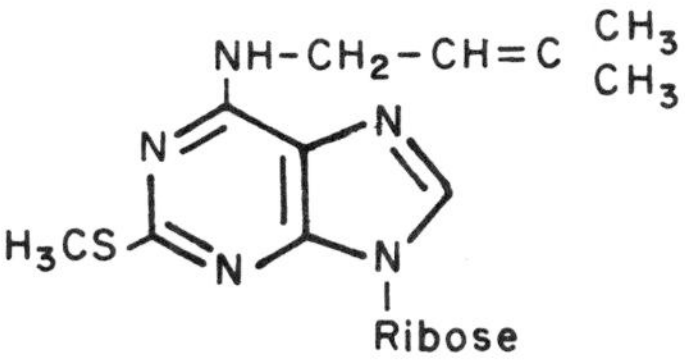

2-Methylthio-N⁶-
isopentenyladenosine
(ms²i⁶A)

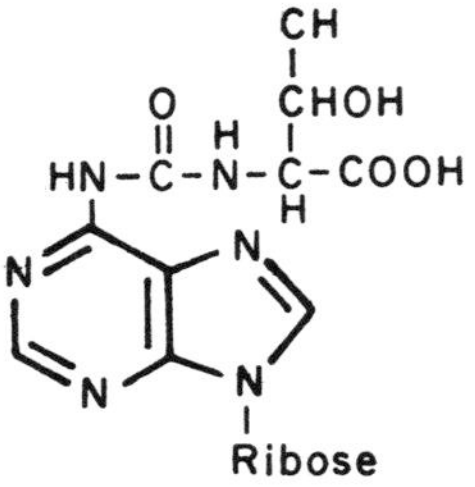

N-[N-(9-β-D-Ribo-
furanosylpurin-6-yl)
carbamoyl]threonine
(t^{6}A)

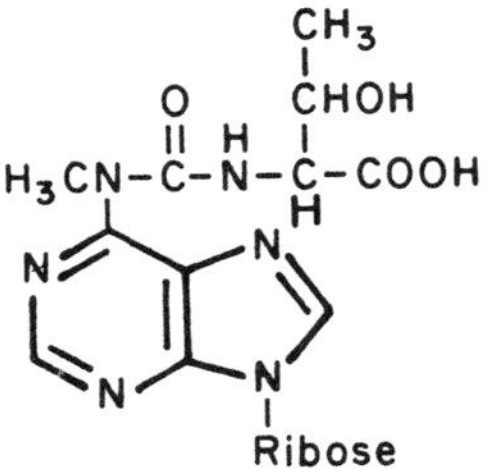

N-[N-(9-β-D-Ribo-
furanosylpurin-6-yl)
N-methylcarbamoyl]
threonine (mt^{6}A)

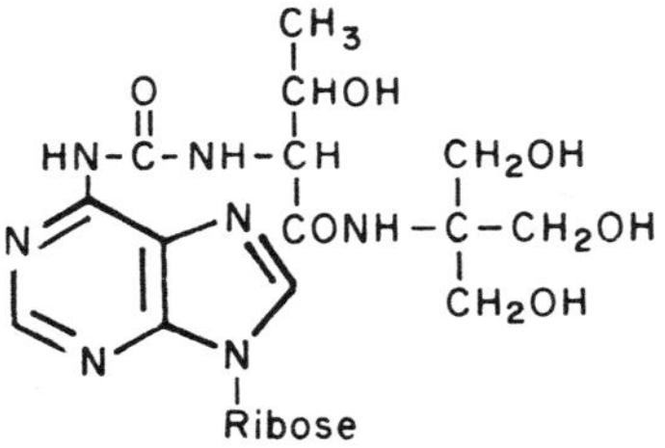

N-[N-[(9-β-D-Ribo-
furanosylpurin-6-yl)
carbamoyl]threonyl]
2-amido-2-hydroxy-
methylpropane-1,3-
diol[161]

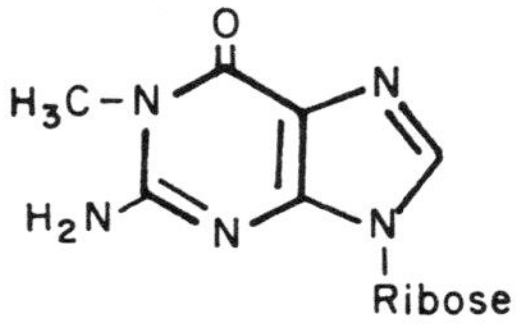

1-Methylguanosine
(m^{1}G)

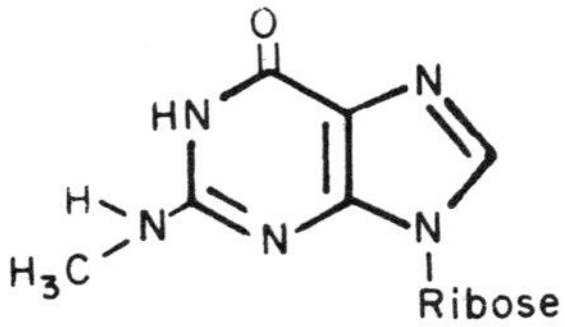

N^2-Methylguanosine
(m^{2}G)

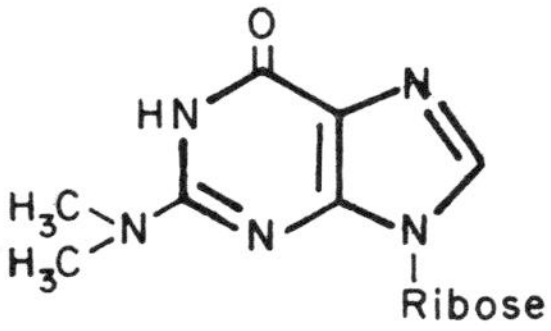

N^2,N^2-Dimethyl-
guanosine (m$_2^2$G)

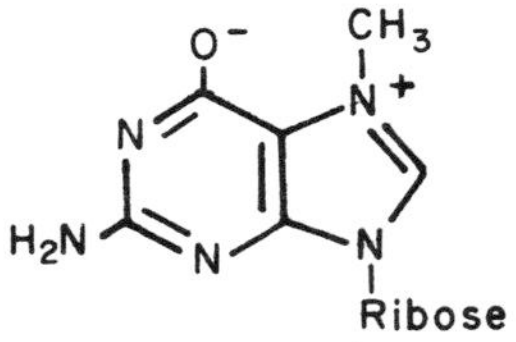

7-Methylguanosine
(m^{7}G)

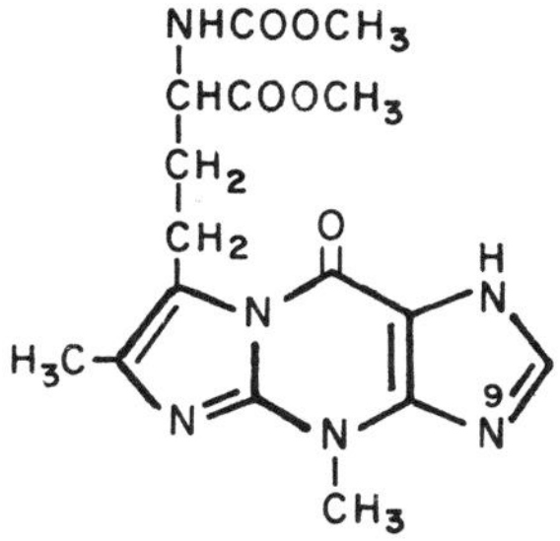

Base "Y" (yW)

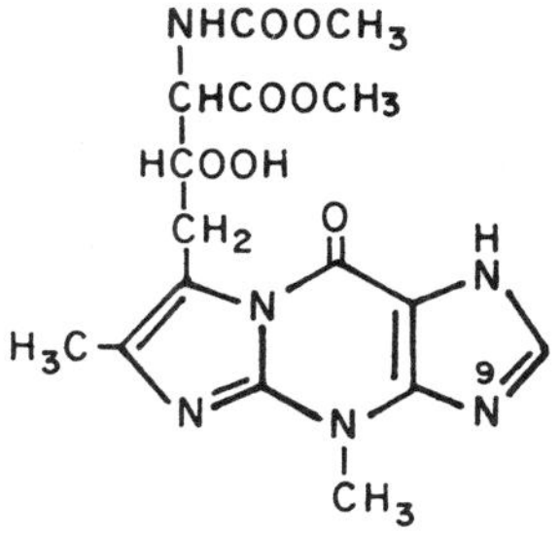

Base "peroxy Y"
(oyW)

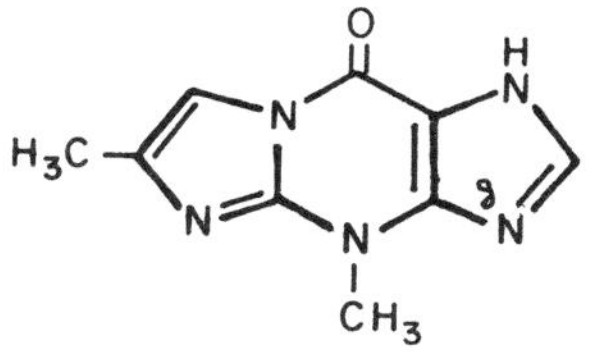

Base "Yt" (W)

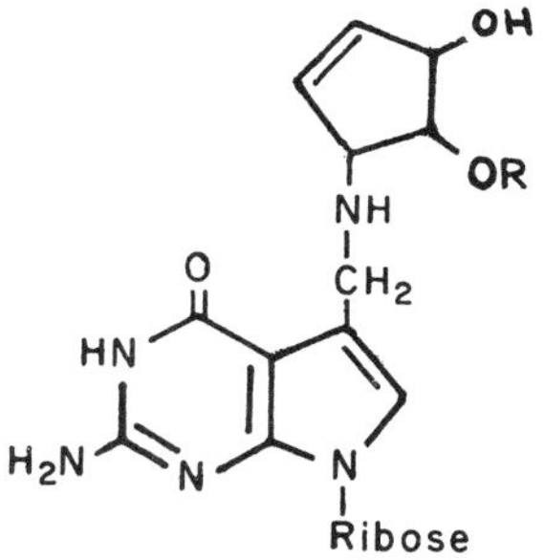

R = H: Q (queuosine or Quo[162])
R = β-D-mannosyl: manQ[163]
R = β-D-galactosyl: galQ[163]

graphic procedures can be used to isolate the compound in large quantity; among the most useful are those involving DEAE-Sephadex A-25 (10) and Dowex 1 (7), as well as paper chromatography (6). Limitation in the quantity of material available poses the greatest problem in structural characterization of a new modified nucleoside. The isolation of 1 to 10 mg of a new nucleoside is usually a laborious process, and physiochemical techniques, such as FT-NMR (Fourier Transform–Nuclear Magnetic Resonance) and mass spectrometry have proved especially valuable when the quantity of sample available is limited; FT-NMR spectra can be recorded with as little as 50 to 100 μg of material in favorable circumstances (11), while mass spectrometry generally requires 2 to 10 μg of material, or even a magnitude less in fortunate cases (6, 10). The greatest limitation to the use of mass spectrometry results from the high polarity of many modified nucleosides, necessitating their inital conversion to suitable chemical derivatives (12). The constraint of volatility can often be overcome by field desorption (13) or the innovative technique of plasma desorption mass spectrometry (10).

Extensive efforts to isolate new nucleosides have primarily been directed to tRNA from yeast and *E. coli*. However, examination of various other sources, in particular gram-positive bacteria, mammalian tissues, and plant cells, should also be fruitful. Many of the tRNAs whose primary structures have been determined have been found to contain unknown nucleosides (14, 15). These unknown compounds are generally located in the anticodon region, and so it is of added interest to establish their structures in order to understand the mechanism of codon-anticodon interactions. The main practical difficulty in this regard is the isolation of sufficient material for mass spectrometry and for NMR. The extent of this problem can be illustrated by the fact that $tRNA_{minor}^{Ile}$ from *E. coli*, which recognizes only the codon A-U-A, contains an unknown modified nucleoside in the first position of the anticodon (16), but $tRNA_{minor}^{Ile}$ constitutes less than 5 percent of the total $tRNA^{Ile}$.

7.3 Function of Modified Nucleosides

In the cloverleaf model of tRNA modified nucleosides are located in specific positions, depending on their structure, rather than at random in the polynucleotide chain. The structural complexities of many modified nucleosides, their locations in specific sites in the tRNA molecule, and their ubiquitous presence in a wide variety of organisms encourage us to believe that they play an important role in tRNA function, although direct proof of this is presently lacking. Here we present several lines of strong suggestive evidence for certain functions of modified nucleosides, together with contradictory evidence.

7.3.1 Modified Nucleosides Located Adjacent to the Anticodon

Striking consistencies are observed between the presence of particular modified nucleosides adjacent to the anticodon and codon recognition of corresponding tRNAs (2). Transfer RNAs that recognize codons starting with U almost always contain hydrophobic modified nucleosides, such as N^6-isopentenyladenosine (i^6A) or its derivatives. On the other hand most tRNAs that recognize codons starting with A contain hydrophilic modified nucleosides, such as N-[9-(β -D-ribofuranosyl)purin-6-ylcarbamoyl]threonine (t^6A) or its derivatives. No similar consistency appears in tRNAs that recognize a codon starting with C or G. In these tRNAs rather simple modified purine nucleosides, such as 6-methyladenosine (m^6A), 2-methyladenosine (m^2A), 1-methylguanosine (m^1G), 1-methylinosine (m^1I), or unmodified adenosine occupy the position next to the anticodon. The presences of m^2A and m^6A seem to be specific for tRNA from microorganisms. These modified nucleosides probably facilitate the formation of strictly correct base pairing between the base in the third position of the anticodon and the base in the first code word. Since the A·U pair has less energy than the G·C pair, hypermodified nucleosides may be necessary for either stabilization of the A·U pair or enhancement of fidelity of the base pair in the case of tRNAs that recognize A or U in the first position of codon sequences. To support this idea, note that $E.$ $coli$ $tRNA_f^{Met}$, which contains adenosine instead of t^6A, recognizes G-U-G and U-U-G as well as A-U-G for initiator codon, while $tRNA^{Met}$, which contains t^6A, apparently recognizes only A-U-G as an internal codon (17). Exceptions to this regularity are also known. It has recently been shown that rat liver $tRNA^{Tyr}$ contains t^6A instead of i^6A(18). Moreover $B.$ $subtilis$ $tRNA^{Met}$ contains m^6A instead of t^6A (19). It is also known that yeast $tRNA_3^{Leu}$ with anticodon m^5C-A-A and $Mycoplasma$ sp. (kid) $tRNA^{Phe}$ contain 1-methylguanosine (m^1G) and that yeast $tRNA^{Phe}$, mammalian tissue, and $Bacillus$ $stearothermophilus$ contains Y-base (wyosine) or its derivatives instead of i^6A. Further, $tRNA^{Trp}$ from brewer's yeast and from mammalian cells contains adenosine and m^1G, respectively, although $tRNA^{Trp}$ should contain i^6A in that position according to the rule (14, 15).

Several workers have shown that the presence of modified nucleosides next to the anticodon is essential for the amino acid transfer function of tRNAs. *E. coli* suppressor tRNA^{Tyr}, which contains unmodified adenosine adjacent to the anticodon, is inactive with respect to amino acid transfer or binding to ribosomes, although its aminoacylation activity is normal (20). Similar results were obtained with *E. coli* tRNA^{Ile} isolated from cells of a mutant (thr$^-$, rel$^-$) grown with a suboptimal concentration of threonine and presumably lacking t^{6}A (21). Binding of the complementary oligonucleotide to the anticodon is enhanced by the presence of the modified nucleoside next to the anticodon (22). Chemically modified tRNAs, in which only modified nucleosides adjacent to the anticodon are affected, are also inactive in both template-dependent binding of tRNA to ribosomes and amino acid transfer (23). However, in tRNA from *Lactobacillus acidophilus,* the absence of an isopentenyl group has no effect on the amino acid transfer function (23, 24). Transfer RNA^{Phe} from *Mycoplasma* sp. (kid) lacking hypermodified nucleosides functions normally in a cell-free protein-synthesizing system derived from *E. coli* (25). Removal of Y-base from yeast tRNA^{Phe} has no effect on its formation of a ternary complex with GTP and *E. coli* elongation factor T (26).

7.3.2 Modified Nucleosides Located in the First Position of the Anticodon

A number of hypermodified nucleosides are present in the first position of the anticodon, and they are, therefore, presumably involved in codon-anticodon interaction. Inosine in this position is an example of a wobbling base, pairing with U, C, or A in the third position of the codon sequence (27–29). This is exemplified by the case of eucaryotic tRNA^{Ile}, since in yeast or mammalian cells only species of tRNA^{Ile} that contain inosine have so far been isolated (14, 15). Uridin-5-oxyacetic acid, which is present in *E. coli,* recognizes A, G, or U in the third position of the codon when assayed by binding of tRNA to ribosomes (30). These modified nucleosides presumably recognize more than two codon sequences. Uridin-5-oxyacetic acid appears to be unique to *E. coli* tRNA. Recently 5-methoxyuridine was found instead of uridin-5-oxyacetic acid in *B. subtilis* tRNA (9, 31). Organisms may be classified into three groups depending on whether they contain uridin-5-oxyacetic acid, 5-methoxyuridine, or inosine as a wobbling base in the antcodons of tRNA for valine, serine, and alanine.

In contrast to the wobbling base, the presence of a 2-thiouridine derivative functions to achieve exclusive base pairing with A, but not with G, in the third letter of the codon sequence, as clearly proved in the case of yeast tRNA$_3^{Glu}$ that contains 5-(methoxycarbonylmethyl)-2-thiouridine, in in vitro hemoglobin biosynthesis (32). Preferential recognition of A in the first position of the anticodon was also shown for other species of tRNA containing 2-thiouridine, such as *E. coli* tRNA^{Glu} (33) and tRNA^{Gln} (34), and rabbit (reticulocyte and liver) tRNA^{Lys}, tRNA^{Glu}, and tRNA^{Gln} (35, 36). Reeves and Roth isolated a recessive U-G-A *E. coli* suppressor strain that is defective in a tRNA-methylase (37). The enzyme might be related to the biosynthesis of the modified nucleoside in the first position, possibly 5-methylamino-methyl-2-thiouridine. *E. coli* cells contain two isoaccepting species of tRNA^{Ile}, tRNA$_{major}^{Ile}$ and tRNA$_{minor}^{Ile}$, in a ratio of about 19 to 1 (16). Transfer tRNA$_{major}^{Ile}$, which contains guanosine in the first position of anticodon, recognizes A-U-U and A-U-C, while tRNA$_{minor}^{Ile}$, which contains an unidentified modified nucleoside, recognizes only A-U-A (16). This minor tRNA^{Ile} presumably regulates the rate of protein synthesis, since only particular proteins can use the A-U-A codon as a code word (38).

At one time it appeared that the function of modified nucleosides in the first position of the anticodon was clearly understood. However, it has recently been shown that rabbit liver tRNA^{Val}, which contains inosine in the first position of the anticodon, recognizes all four codons for valine with preference for G-U-G (39). This unusual codon recognition by rabbit liver tRNA^{Val} cannot be explained by base pairing of inosine with a wobbling mechanism, as proposed by Crick (an I·G pair is not allowed). Recognition of the third letters of the codon by tRNA^{Val} may be influenced by the total three-dimensional structure of the anticodon region. Similar findings with an in vitro protein-synthesizing system have recently been reported by Mitra et al. (40). They showed that tRNA^{Val} from yeast and *E. coli* recognizes the four valine codons in MS2 coat protein messenger, whether tRNA^{Val} contains inosine, uridin-5-oxyacetic acid, or guanosine. It seems that the presence of inosine is not the only requirement for wobbling. *E. coli* tRNA^{Trp}, which contains C in the first position of the anticodon and recognizes only G in the third position of the codon, can be altered to recognize and suppress the U-G-A codon as well as U-G-G, by a single nucleotide change in the D-stem that is far from the anticodon region (41).

None of the tRNAs so far sequenced contains normal U or A in the first position of the anticodon; the U is always modified either to uridin-5-oxy-acetic acid or to 5-methoxyuridine or 2-thiouridine derivatives, and likewise A is replaced by I. It appears that in nature it is not possible to have normal U or A in that position, and if present these are presumably lethal to cells. [An exceptional case is *Staphylococci* tRNA$_1^{Gly}$ used for peptidoglycan synthesis, but not for protein synthesis (42). Other exceptions are $tRNA_1^{Leu}$ from baker's yeast (151) and $tRNA_2^{Gly}$ from *E. coli* (43). These tRNAs contain U in the first position of the anticodon. Nucleotide sequences of these two tRNAs were determined by postlabeling methods and Sanger's technique, respectively. *E. coli* suppressor $tRNA_1^{Tyr}$ su$_{oc}^+$ that was previously claimed to have unmodified U in the first position in fact contains a modified U (44).] Another possible function of 2-thiouridine derivatives is to prevent miscoding in protein synthesis. In the case of $tRNA^{Leu}$ (specific for codons for U-U-A and U-U-G), $tRNA^{Gln}$, $tRNA^{Lys}$, $tRNA^{Glu}$, and $tRNA^{Arg}$ (specific for codons for A-G-A and A-G-G), pairing of U in the first position of the anticodon with U or C may cause miscoding that would be lethal to cells. Modification of uridine to the 2-thiouridine derivatives may prevent this mispairing. 2-Thiouridine derivatives have been found only in some tRNAs, but they are present in diverse cells of species ranging from mammals to bacteria. In the case of *E. coli* no $tRNA^{Glu}$ other than that containing mnm^5s^2U in the anticodon is isolated. This suggests that mnm^5s^2U recognizes G as well as A in the wobbling position, although a $mnm^5s^2U \cdot G$ base pair is weaker than a $mnm^5s^2U \cdot A$ base pair.

A modified nucleoside, designated Q, having a 7-deazaguanosine nucleus and cyclopentene diol as a side chain is present in the first position of the anticodon of *E. coli* $tRNA^{Tyr}$, $tRNA^{His}$, $tRNA^{Asn}$, and $tRNA^{Asp}$, which all recognize $X \cdot A \cdot {}^U_C$ codons (45). Q and its derivative Q* have also been found in *Drosophila,* wheat germ, mammals, and other animal tissues in tRNA for the same amino acids but not in yeast (46, 47). Q has a wobbling property similar to G's, although it has a greater affinity for U than for C (45). Thus the function of Q in protein biosynthesis is uncertain at present.

To understand the function of modified nucleosides in the wobbling position more clearly it would be useful to isolate a mutant that lacks these modified nucleosides. Björk has recently isolated an *E. coli* mutant that lacks 5-methylaminomethyl-2-thiouridine (48).

7.3.3 Modified Nucleosides Present in Other Places

Modified nucleosides located in regions other than the anticodon area are not randomly distributed in tRNA but are generally located in specific positions in the cloverleaf structure. These modified nucleosides may be classified into three groups depending on their location and species specificity: (1) modified nucleosides present in most organisms and most aminoacylating species, such as ribothymidine, pseudouridine, and dihydrouridine; (2)

modified nucleosides present in a variety of organisms but specific for certain types of amino acids, such as 7-methylguanosine and 3-(3-amino-3-carboxypropyl)uridine; (3) modified nucleosides specific for tRNA from particular organisms, such as 4-thiouridine in *E. coli* and other bacterial cells; 1-methyladenosine in many eucaryotes and some microorganisms; and N^2-methylguanosine, N^2, N^2-dimethylguanosine, and 5-methylcytosine present in most eucaryotes.

There is also suggestive evidence for certain functions of these modified nucleosides: (1) involvement of ribothymidine and pseudouridine in the binding of tRNA to ribosomes (49); (2) stabilization of the conformation of tRNA (50–52); (3) enhancement of resistance of tRNA molecules to attack by RNase; and (4) enhancement of specificity for recognition of aminoacyl-tRNA synthetase (53). However, there is also much evidence to the contrary. First, tRNA completely lacking ribothymidine has been isolated from a mutant of *E. coli* that grows similarly to the wild-type strain of *E. coli* (54). Moreover tRNA isolated from this mutant functions normally both in aminoacylation and protein synthesis. A pure species of tRNA^{Ile}, isolated from *Mycoplasma* sp. (kid), also has activity both as an amino acid acceptor and in transfer, although it completely lacks ribothymidine (55). Purified methyl-deficient tRNA$_f^{Met}$ of *E. coli* functions normally in amino-acylation, formylation, codon recognition, and interaction with initiation factor (56). *E. coli* tRNA$_1^{Val}$ in which T and Ψ in the GTΨ-loop are replaced by 5-fluorouridine can function in aminoacylation (57) and in vitro protein synthesis (58). No ribothymidine has been found in mouse myeloma tRNA$_1^{Val}$ (58) or in wheat embryo tRNA^{Gly} or tRNA^{Thr} (60). Transfer RNA^{Trp} and tRNA^{Pro}, which are used as primers for the reverse transcriptase of Rous sarcoma virus and murine leukemia virus, respectively, contain pseudouridine instead of ribothymidine (61, 62). It is also known that mutants of *E. coli* completely lacking 7-methylguanosine (63) or 4-thiouridine (64) in tRNA can grow normally. Some species of micro-organisms, such as *Micrococcus luteus* (65) or *Mycobacterium smegmatis* (66), contain scarcely any ribothymidine in total tRNA.

A mutant of *E. coli* that lacks ribothymidine appears to be normal with respect to its growth rate and enzyme induction. However, when the mutant is grown with the original wild-type strain of *E. coli*, the mutant cells are overcome by those of the wild-type strain after several generations (67). This clearly shows that the presence of ribothymidine is necessary or at least advantageous for *E. coli* cells when they are grown in a natural environment. Modified nucleosides may slightly enhance tRNA function, and their effect may be too small to detect by a biochemical assay in vitro; nevertheless they may give a slight advantage that is obligatory for maintaining the organisms in a natural environment.

7.4 Possible Role of Modified Nucleosides in the Regulatory Function of tRNA

Ames and his co-workers have shown that $tRNA^{His}$, isolated from a mutant of *Salmonella* (his T) in which histidine biosynthesis is not repressed by the addition of histidine, contains the sequence U-U in the region of the anticodon stem and loop, while $tRNA^{His}$ from wild-type cells contains the sequence Ψ - Ψ in this region (68). Histidine tRNA from the his T mutant functions normally in aminoacylation and transfer (68). Histidyl-$tRNA^{His}$ of the mutant is presumably inactive in control of transcription of the histidine operon. It is possible that lack of the Ψ - Ψ sequence induces a small local conformational change in $tRNA^{His}$ that prevents hisitidyl-$tRNA^{His}$ from interacting with a regulatory component, although at present there is no experimental evidence for this (69). Leucine tRNA, isolated from the same mutant, also lacks two pseudouridine residues in the same region of the anticodon area. The mutant is also derepressed with respect to leucine starvation (70). These observations are widely recognized as proof of a positive role of modified nucleosides in the regulatory function of tRNA. However, it can be argued that the phenomena observed are artificial because they are observed only in a laboratory mutant. If insertion of Ψ - Ψ into tRNA occurs only at a certain stage of cell growth related to the regulation of cell function, it can be said that pseudouridine functions positively in a regulatory process. However, $tRNA^{His}$ with Ψ - Ψ may have been formed at an early stage of evolution of the organism, and a component that controls the function of the histidine operon may happen to recognize a conformation of tRNA that is altered by lack of the Ψ - Ψ sequence.

The modified nucleoside Q may be important for the regulatory function of eucaryotic tRNAs. The amount of Q relative to unmodified G in $tRNA^{Tyr}$, $tRNA^{His}$, $tRNA^{Asn}$, and $tRNA^{Asp}$ varies at different stages in the life cycle of *Drosophila*, suggesting that the presence of Q in tRNA is related to differentiation of the cells (46). Inhibition of the enzymatic activity of tryptophan pyrolase by *Drosophila* $tRNA^{Tyr}$ from a vermilion mutant may depend on whether the $tRNA^{Tyr}$ contains Q or guanosine in the first position of the anticodon (71), although some results contradict this possibility (72). A derivative of Q, called Q*, which was isolated from mammalian tRNAs, was recently characterized as Q containing mannose (manQ) or galactose (galQ) at the 4-position of cyclopentene diol (11). This most complex hypermodified nucleoside is specifically present in a particular tRNA, namely manQ in tRNA^{Asp} and galQ in tRNA^{Tyr} (73). Transfer RNA^{Asp} and tRNA^{Tyr} react specifically with concanavalin A and *Ricinus communis* lectin, respectively (72). The presence of such hypermodified nucleosides in an exposed region of the specific tRNA molecule probably makes the molecule a good receptor of components in reactions related to regulation of the cell function.

7.5 Biosynthesis of the Modified Nucleosides in tRNA

All the modified nucleosides in tRNA are thought to be synthesized by modification of a parental nucleotide residue after synthesis of the polynucleotide linkage of the tRNA molecule. The only exception to this general biosynthetic process may be the Q nucleoside. The earlier work on the isolation and specificity of tRNA modification enzymes has been described in a review by Söll (74). Methylation of tRNA and biosynthesis of modified nucleosides in precursor molecules have also recently been reviewed (75–78). Thus only recent developments and essential problems are discussed here.

When pulse-labeled tRNA precursors are analyzed, some modified nucleosides are found to have been formed already in the tRNA precursor (79–85). However, no modified nucleosides are found in the extra segment, although they are found in the tRNA moiety of the precursor. Modified nucleosides do not seem to have any role in the biosynthesis of the precursor molecule. It is still not clear whether formation of modified nucleosides proceeds in a sequential fashion or at random (86). In both mammalian and bacterial tRNA precursors, different modified nucleosides predominate at different stages (80–82), but this differential modification may be due to changes in the relative activities of the different enzymes causing tRNA modification.

The original nucleosides corresponding to the modified nucleosides have been determined by analyzing the sequences of unmodified tRNA or tRNA precursors (79, 87), by determining the sites of modification of heterologous tRNA modified in vitro (88–90), and by measuring the reduced level of modified nucleosides in tRNA containing 5-fluorouracil (57, 91). It was shown that Q and m$_2^2$G are derived from G, that m^{1}A is derived from A, and that T, D, Ψ, s^{4}U, uridin-5-oxyacetic acid, and mnm^{5}s^{2}U are derived from

U. Enzymic synthesis in vitro of modified nucleosides has also shown that the precursors of m^5C, acp^3U, m^2G, m^1G, m^7G, m^2G, i^6A, and t^6A are C, U, G, G, G, A, and A, respectively (74, 77, 92). It is not clear how enzymes convert only one particular nucleotide residue to the corresponding modified nucleoside, although many tRNA-modification enzymes have now been isolated and purified and their strict specificities have been demonstrated in vitro.

These tRNA-modification enzymes probably recognize the oligonucleotide sequence around the particular nucleotide residue that they modify. This conclusion is based on studies on heterologous tRNA methylation using rat liver m^1A-methylase and purified individual $E.$ $coli$ tRNAs as acceptor molecules (90). Since $E.$ $coli$ tRNAs contain normal A instead of m^1A in the GTΨC-loop, A in that position is available for methylation, to form m^1A, by m^1A-methylase from rat liver. The extent of methylation is greatly influenced by the nucleotide sequence in the GTΨC-loop. The most active tRNAs are always found to contain the G-T-Ψ-C-G-A-A-U-C sequence in the GTΨC-loop, but nucleotide sequences in other parts of these tRNAs vary considerably. It is also evident that tRNA-modification enzymes recognize the conformation or three-dimensional structure, as well as the nucleotide sequence of tRNA (93); a fragment obtained by a single cleavage of $E.$ $coli$ $tRNA_f^{Met}$ is only methylated by rat-liver tRNA methylase when it has been reconstituted by recombination with the other part of the tRNA fragment. It was also shown recently that base substitution in T4 phage–specific tRNA by mutation reduces the amounts of many modified nucleosides in this tRNA at the precursor stage (94).

Some modified nucleosides such as Ψ, D, m^5C, m^2G, m^1G, and m^1A are located in several places in the tRNA molecule in the cloverleaf structure. It is very likely that a different modification enzyme is required for each location. Hence $tRNA^{His}$ or $tRNA^{Leu}$ from His T mutant of $Salmonella$ may lack Ψ-Ψ in the anticodon stem and loop, but will contain Ψ in the G-T-Ψ-C region. (68, 70). m^1A-Methylase, which catalyzes methylation of the A-residue in the C-C-A stem only, has been isolated from the mitochondria and chloroplasts of $Phaseolus$ $vulgaris$ (95). Two distinct m^1G-methylases are found in $Saccharomyces$ $cerevisiae$ (96). They probably catalyze methylation of a guanylate residue at two different sites, respectively. However, it is still unknown whether a specific enzyme is required for each location of D, m^3C, m^5C, and m^2G.

A variety of tRNA-modification enzymes has been isolated and purified; the most highly purified so far are several tRNA methylases that use

S-adenosylmethionine as a methyl donor (77), but no enzymes have yet been purified to homogeneity. Isolating pure enzymes is difficult because of their low content in the cells and their instability. Pure species of tRNA-modification enzymes are needed for studies on the molecular mechanism of their interaction with tRNA. Isolation of m^1A-methylase form a heat-resistant bacterium *Thermus thermophilus* (97) using affinity column chromatography with S-adenosylmethionine as ligand (98) may be one way to obtain a pure enzyme.

The following are recent topics concerning tRNA-modification enzymes. The enzymes isolated include not only numerous tRNA methylases (77), but also the enzymes responsible for the syntheses of pseudouridine (103, 104, 105), N^6-isopentenyladenosine (99, 100), 4-thiouridine (101, 102), 5-meth-oxycarbonylmethyl)-uridine (106, 107), 3-(3-amino-3-carboxypropyl)uridine (acp^3U) (92), t^6A (108, 109), and manQ (110). The acceptor used in studies of the biosynthesis of t^6A was undermodified tRNA, isolated from cells of an *E. coli* mutant (rel⁻, thr⁻) grown with a suboptimal concentration of threonine. The carbonate ion was found to be the moiety of the keto group attached to the N^6 position of the purine (108, 109). In the synthesis of acp^3U, the 3-amino-3-carboxypropyl group is derived from S-adenosylmethionine (92). The involvement of the 3-amino-3-carboxypropyl group of S-adenosylmethionine was also shown for the synthesis of the Y-base (111). GDP-mannose was found to be the donor molecule for addition of the mannose moiety in the synthesis of manQ from the Q nucleoside (110). In the biosynthesis of T in *B. subtilis, Streptococcus faecalis* and other gram-positive bacteria, 5,10-methylenetetrahydrofolic acid was found to be the donor of the methyl group, whereas S-adenosylmethionine was the donor in most other organisms (65, 112–114). No T-methylases have yet been isolated from mammalian cells, perhaps because T-methylase is unstable in mammalian cells or because T-methylase recognizes only tRNA precursors as methyl acceptors.

Until recently all known tRNA modifications were one-way reactions, namely conversion of a normal nucleoside to a modified nucleoside; the normal nucleoside present in tRNA is converted to a modified nucleoside by addition of a side chain. However, a new type of tRNA modification was recently found to occur during formation of the nucleoside Q. Farkas first discovered the unique reaction called guanylation (115) in which an enzyme isolated from rabbit reticulocytes catalyzes the incorporation of guanine into specific tRNA (116). We found that when the rabbit reticulocyte enzyme is used, the modified base Q in *E. coli* tRNA can be specifically replaced by guanine without breaking the polynucleotide chain (117). This is a reverse reaction, converting a modified base in tRNA to a normal base. It is also a unique reaction in that it involves exchange of the bases by cleavage of the N-C glycoside bond without cleavage of the phosphodiester bond. The enzyme also catalyzes the exchange of guanine with guanine residues located in the position where the Q-base is located, namely guanine in the first position of the anticodon of $tRNA^{Tyr}$, $tRNA^{His}$, $tRNA^{Asn}$, and $tRNA^{Asp}$. The real function of the guanine-insertion reaction in a homologous system is still unknown. The enzyme isolated from rabbit reticulocytes seems to catalyze only an exchange of guanine with guanine in a homologous system (118). However, it has been shown that in morphogenesis of *Drosophila,* the guanine-insertion enzyme is only active at the stage where the conversion of Q to G proceeds in tRNA, suggesting that the enzyme is involved in conversion of Q-base to normal G in tRNA without new transcription of tRNA molecules (119). The guanine-insertion enzyme has also been found in other organisms such as *E. coli*, chick embryos (120), and Ehrlich ascites tumor cells (121). A preparation of the enzyme, extensively purified from *E. coli*, catalyzes exchange only of guanine with guanine, not of guanine with Q (120). The *E. coli* enzyme is probably involved in the biosynthesis of Q in tRNA. During conversion of guanine to Q, the carbon at position 8 of guanine is expelled along with the nitrogen at position 7 (122). Such a drastic modification of the heterocyclic nucleus may be difficult in a polynucleotide chain; thus the Q-base or its precursor may first be synthesized separately and then inserted into the tRNA molecule by the guanine insertion reaction.

Enzymes that liberate isopentenyl groups from tRNA have been isolated from bacteria and mammalian tissue (123). Cells may have the ability in vivo to demodify species of tRNA containing isopentenyladenosine residue.

7.6 Isoaccepting tRNA

There seem to be more than 40 tRNA species in a given organism, and there are generally multiple tRNA species for each amino acid, named isoaccepting species. The multiple peaks of amino acid acceptor activity in the chromatographic profile of tRNA are often regarded as an indication of the presence of isoaccepting tRNAs. Although identification of isoaccepting tRNAs in this way is convenient and simple, caution is necessary because artifacts may cause further peaks. For example, lack of terminal adenosine (124), conversion of the native form to a denatured form (125, 126), and limited digestion of tRNA (125, 127, 128) all change the elution profile of tRNA. Conversely in some cases isoaccepting tRNAs are not separated from each other by fractionation procedures.

Isoaccepting tRNA species can be classified into several categories. The first class includes tRNAs having different primary structures. These tRNAs are transcribed from different tRNA genes. This class of isoacceptors is further separated into two groups: (1a) tRNAs with different anticodon structures, which thus recognize different codon sequences, and (1b) tRNAs with the same anticodon structure but different nucleotide sequences in other parts. The second class consists of multiple tRNA species formed by different extents of posttranscriptional modification.

The presence of isoaccepting species belonging to class 1a may be important in regulating protein synthesis. Increase in the biosynthesis of specific proteins is correlated with increase in the amounts of particular tRNA species (129, 130). In vitro protein biosynthesis is stimulated by the addition of tRNA, which is used for decoding the corresponding gene (131, 132). A minor species of $tRNA^{Ile}$ of E. coli that recognizes only A-U-A is important for regulating protein synthesis (38). Isoaccepting species of this class generally have quite different nucleotide sequences in other regions as well as the anticodon itself, unlike suppressor tRNAs obtained by mutation in the laboratory (14, 15). It is likely that these isoaccepting tRNA species originated early in the evolutionary process. Methionyl-tRNAs, that is, initiator $tRNA^{Met}$ ($tRNA_f^{Met}$) and internal $tRNA^{Met}$ ($tRNA_m^{Met}$), also belong to this class, although they have almost the same anticodon structure.

Isoaccepting tRNAs belonging to class 1b have fewer base alterations than tRNAs in class 1a. E. coli cells contain two forms of $tRNA^{Tyr}$, $tRNA_1^{Tyr}$ and $tRNA_2^{Tyr}$, which differ in only two nucleotide residues in the extra loop, but which can be separated from each other by DEAE-Sephadex A-50 column chromatography (133). Isoaccepting tRNA species with such minor differences cannot usually be separated by column chromatography or other tRNA-fractionation procedures. The presence of these isoaccepting species is usually detected only during nucleotide sequence analysis of tRNA. For example, E. coli $tRNA_f^{Met}$ consists of two isoaccepting species, one having m^7G and the other having A in the same position in the extra loop. These two species of $tRNA_f^{Met}$ have not been easily separated from each other (134), but it is evident that they are transcribed from different genes: two $tRNA_f^{Met}$ genes are found in the E. coli gene map, and each gene codes for one species of $tRNA_f^{Met}$ (135). The significance of the existence of isoaccepting tRNAs of this category is unknown. It seems likely that the

tRNA gene was duplicated during evolution and that the resulting nucleotide alteration does not affect tRNA function and has been maintained.

Isoaccepting tRNAs of the second class that are formed by different
degrees of modification are often detected by a shift of chromatographic
profile, since the presence or absence of a particular modified nucleoside
greatly changes the properties of tRNA. For example, tRNA^{Tyr} detected
during sporulation of *B. subtilis* contains N^6-isopentenyladenosine (i^{6}A),
whereas vegetative cells possess tRNA^{Tyr} containing 2-methylthio-N^6-iso-
pentenyl-adenosine (ms^{2}i^{6}A) (136). Thus these isoaccepting species of
tRNA^{Tyr} can easily be separated by reversed-phase partition column
chromatography, since i^{6}A and ms^{2}i^{6}A have different lipophilic properties.
On the other hand isoaccepting tRNAs with a difference in minor
modification, such as U $\longrightarrow$ D, U $\longrightarrow$ T, G $\longrightarrow$ Gm, cannot be separated by
column chromatography; they can be detected only in sequence studies on
mixtures of two isoaccepting species. Undermodified tRNAs are formed in
methyl-deficient conditions (56, 137–139), on amino acid starvation (140,
141), on addition of chloramphenicol (142), or under other growth conditions (87, 143–145).

7.7 Isoaccepting tRNA Species in Eucaryotic Cells

New isoaccepting tRNA species are often observed in particular tissues, in
cells at different stages of differentiation, in tumor tissues, in transformed
cells, or in cells grown under different culture conditions. (for reviews see
refs. 125, 129, 146, 147). There are numerous reports on the appearance of
isoaccepting tRNA species, detected by change in the chromatographic
profile of amino acid acceptor activity. However, only a few have clearly
shown whether the isoaccepting species detected is due to a new transcript
of a gene or to a different degree of modification. The amount of
methylated nucleosides in tRNA and the activity of tRNA methylases is
generally high in tumor tissues (146, 148, 149), but there is no way to distinguish whether this change in the tRNA population is due to hypermethyl-
ation of preexisting tRNA or the appearance of new tRNA transcripts
containing higher amounts of methylated nucleosides. Moreover it is difficult to isolate a pure species of tRNA from tumor cells in large quantity for
primary sequence analysis. However, recent developments in nucleotide
sequence determination using the postlabeling technique (150, 151) have
made it possible to determine the sequence of tRNA with less than 5μg of
material.

Tumor tissues are known to contain a new isoaccepting species of tRNA^{Lys} (tRNA$_2^{Lys}$) that is scarcely detectable in normal cell lines. The amount of tRNA$_2^{Lys}$ in dividing cells is proportional to the rate of cell division (152, 153). Knowing the primary structure of tRNA, Gross showed that the primary sequence of tRNA$_4^{Lys}$ from SV40-transformed mouse fibroblasts is identical with that of tRNA$_2^{Lys}$, which is the major isoaccepting species of normal cells; the only difference between the two tRNAs is that tRNA$_4^{Lys}$ is not modified in several positions (154). This incomplete modification of tRNA$_4^{Lys}$ is not simply due to the fast growth of tumor cells, because 2'-O-methylribothymidine in tRNA$_2^{Lys}$ is partly replaced by Ψ in tRNA$_4^{Lys}$.

The new tRNA^{Phe} species that appears in tumor tissues is due to lack of modified base Y in the position next to the anticodon (155, 156). This was found indirectly by examining the change in the chromatographic profile after chemical modification of tRNA with acid. Gonano and his co-workers have reported that rat embryonic tissue and rat hepatoma cells contain a specific tRNA^{Phe} (157). They also stated that the nucleotide sequence of embryonic tRNA is different from that of adult tRNA^{Phe}, but they did not give any experimental values to support this statement.

Changes in the chromatographic profiles of tRNA^{Tyr}, tRNA^{His}, tRNA^{Asn}, and tRNA^{Asp} have been observed in various eucaryotic cells (46, 158–160). White et al. have shown that the appearance of two isoaccepting species for each of the four amino acids in *Drosophila* is due to the extent of modification of the first position of the anticodon of these tRNAs, that is, modification of normal guanosine to the modified nucleoside Q (46). Q and Q* (a derivative of Q having either mannose or galactose) are found in mammalian tRNAs (47). ManQ (Q with mannose) is present only in tRNA^{Asp}, and galQ (Q with galactose) in tRNA^{Tyr} (73). It is very likely that the presence or absence of modified nucleoside Q or Q* also affects the chromatographic profiles of the four tRNAs in mammalian tissues. Transfer RNA^{Tyr}, tRNA^{His}, tRNA^{Asn}, and tRNA^{Asp} containing normal guanine instead of Q or Q* can easily be detected using the guanine-insertion enzyme isolated from *E. coli* cells (161). It was shown that some tumor cells, such as rat ascites hepatoma cells, contain tRNA^{Asp} having normal guanine in the anticodon, although this tRNA is not detected in normal rat liver. Therefore it is possible that the isoaccepting species of tRNA^{Asp}, which appeared in SV40-transformed cells and other tumor tissues and was detected in earlier work from the chromatographic profile (159, 162, 163), is tRNA^{Asp} containing normal guanosine instead of manQ.

It is still uncertain whether a new tRNA transcript appears at a specific stage of cell differentiation or carcinogenesis. Tener and his co-workers showed that *Drosophila* tRNA genes are present in many chromosomes (164). It is possible that specific tRNA genes are transcribed at certain stages of cell growth, causing changes such as those observed in isozyme patterns. More sequence work on mammalian tRNAs is definitely needed to provide information on the real nature of isoaccepting species and their importance in cell function in eucaryotes.

7.8 Conclusion

Many of the modified nucleosides present in tRNA have now been characterized, and their locations in the tRNA molecule have been determined. Many enzymes involved in the biosynthesis of modified nucleosides have also been isolated. Various results suggest that modified nucleosides are involved in codon recognition or other functions of tRNA in protein biosynthesis, but more experiments are still required on the exact roles of modified nucleosides in tRNA function. A fascinating idea is that modified nucleosides have a role in the regulatory function of tRNA, although clear evidence to prove this is still lacking. One reason that studies on modified nucleosides are so attractive is that the function of tRNA in the regulatory process is still not understood, and this problem opens up a vast and interesting field of research on tRNA.

References

1. Ginsberg, T., Rogg, H., and Staehelin, M. (1971). Eur. J. Biochem. *21*, 249–257.

2. Nishimura, S. (1972). Prog. Acid Res. Mol. Biol. *12*, 49–85.

3. Dunn, D. B., and Hall, R. H. (1975). In Handbook of Biochemistry and Molecular Biology, 3rd ed., Nucleic Acids, vol. 1, G. Fasman, ed. (Cleveland, Ohio: CRC Press), pp. 65, 217.

4. Madison, J. T., and Holley, R. W. (1965). Biochem. Biophys. Res. Commun. *18*, 153–157.

5. Egami, F., Takahashi, K., and Uchida, T. (1964). Prog. Nucl. Acid Res. Mol. Biol. *3*, 59–101.

6. Kimura-Harada, F., von Minden, D. L., McCloskey, J. A., and Nishimura, S. (1972). Biochemistry *11*, 3910–3915.

7. Gray, M. W., (1974). Anal. Biochem. *62*, 91–101.

8. Fujimoto, M., Kuninaka, A., and Yoshino, H. (1974). Agr. Biol. Chem. *38*, 777–783.

9. Murao, K., Hasegawa, T., and Ishikura, H. (1976). Nucl. Acids Res. *3*, 2851–2860.

10. Kasai, H., Ohashi, Z., Harada, F., Nishimura, S., Oppenheimer, N. J., Crain, P. F., Liehr, J. G., von Minden, D. L., and McCloskey, J. A. (1975). Biochemistry *14*, 4198–4208.

11. Kasai, H., Nakanishi, K., Macfarlane, R. D., Torgerson, D. F., Ohashi, Z., McCloskey, J. A., Gross, H. J., and Nishimura, S. (1976). J. Am. Chem. Soc. *98*, 5044–5046.

12. McCloskey, J. A. (1974). In Basic Principles in Nucleic Acid Chemistry, vol. 1, P. O. P. Ts'o, ed. (New York: Academic Press), 3.

13. Schulten, H.-R., and Beckey, J. D. (1973). Org. Mass Spectrom *7*, 861–867.

14. Barrell, B. G., and Clark, B. F. C. (1974). Handbook of Nucleic Acid Sequences (Oxford: Joynson-Bruvvers).

15. Sodd, M. A. (1975). In Handbook of Biochemistry and Molecular Biology, 3rd ed., Nucleic Acids, vol. 2, G. Fasman, ed. (Cleveland, Ohio: CRC Press), p. 423.

16. Harada, F., and Nishimura, S. (1973). Biochemistry *13*, 300–307.

17. Files, J. G., Weber, K., and Miller, J. H. (1974). Proc. Nat. Acad. Sci. USA *71*, 667–670.

18. Brambilla, R., Rogg, H., and Staehelin, M. (1976). Nature *263*, 167.

19. Ishikura, H. Personal communication.

20. Gefter, M. L., and Russell, R. L. (1969). J. Mol. Biol. *39*, 145–157.

21. Miller, J. P., Hussain, Z., and Schweizer, M. P. (1976). Nucl. Acids Res. *3*, 1185–1201.

22. Pongs, O., and Reinwald, E. (1973). Biochem. Biophys. Res. Commun. *50*, 357–363.

23. Fittler, F., and Hall, R. H. (1966). Biochem. Biophys. Res. Commun. *25*, 441–446.

24. Thiebe, F., and Zachau, H. G. (1968). Eur. J. Biochem. *5*, 546–555.

25. Kimball, M. E., and Söll, D. (1974). Nucl. Acids Res. *1*, 1713–1720.

26. Ghosh, K., and Ghosh, H. P. (1972). J. Biol. Chem. *247*, 3369–3375.

27. Crick, F. H. C. (1966). J. Mol. Biol. *19*, 548–555.

28. Nirenberg, M., Caskey, T., Marshall, R., Brimacombe, R., Kellog, D., Doctor, B., Hatfield, D., Levin, J., Rottman, F., Pestka, S., Wilcox, M., and Anderson, F. (1966). Cold Spring Harbor Sym. Quant. Biol. *31*, 11–24.

29. Söll, D., Jones, D. S., Ohtsuka, E., Faulkner, D., Lohrmann, R., Hayatsu, H., and Khorana, H. G. (1966). J. Mol. Biol. *19*, 556–573.

30. Takemoto, T., Takeishi, K., Nishimura, S., and Ukita, T. (1973). Eur. J. Biochem. *38*, 489–496.

31. Albani, M., Schmidt, W., Kersten, H., Geibel, K., and Lüderwald, I. (1976). FEBS Letters *70*, 37–42.

32. Sekiya, T., Takeishi, K., and Ukita, T. (1969). Biochim. Biophys. Acta *182*, 411–426.

33. Ohashi, Z., Saneyoshi, M., Harada, F., Hara, H., Nishimura, S. (1970). Biochem. Biophys. Res. Commun. *40*, 866–872.

34. Yaniv, M., and Folk, W. R. (1975). J. Biol. Chem. *250*, 3243–3253.

35. Hilse, K., and Rudloff, E. (1975). FEBS Letters *60*, 380–383.

36. Rudloff, E., and Hilse, K. (1975). Hoppe-Seyler's Z. Physiol. Chem. *356*, 1359–1367.

37. Reeves, R. H., and Roth, J. R. (1975). J. Bacteriol. *124*, 332–340.

38. Jou, W. M., Montagu, M. V., and Fiers, W. (1976). Biochem. Biophys. Res. Commun. *73*, 1083–1093.

39. Jank, P., Shindo-Okada, N., and Nishimura, S., and Gross, H. J. Nucl. Acids Res. (submitted).

40. Mitra, S. K., Lustig, F., Åkesson, B., Largerkvist, U., and Strid, L. (1977). J. Biol. Chem. *252*, 471–478.

41. Hirsh, D. (1971). J. Mol. Biol. *58*, 439–458.

42. Roberts, R. J. (1974). J. Biol. Chem. *249*, 4787–4796.

43. Roberts, J. W., and Carbon, J. (1974). Nature *250*, 412–414.

44. Altman, S. (1976). Nucl. Acids Res. *3*, 441–448.

45. Harada, F., and Nishimura, S. (1972). Biochemistry *11*, 30–35.

46. White, B. N., Tener, G. M., Holden, J., and Suzuki, D. T. (1973). J. Mol. Biol. *74*, 635–651.

47. Kasai, H., Kuchino, Y., and Nishimura, S. (1975). Nucl. Acids Res. *2*, 1931–39.

48. Björk, G. R., and Kjellin-Stråby, K. (1976). In Abstracts of Tenth International Congress of Biochemistry, Hamburg, p. 33.

49. Sprinzl, M., Wagner, T., Lorenz, S., and Erdmann, V. A. (1976). Biochemistry *15*, 3031–3039.

50. Igo-Kelmers, T., and Zachau, H. G. (1971). Eur. J. Biochem. *18*, 292–298.

51. Wintermeyer, W., and Zachau, H. G. (1975). FEBS Letters *58*, 306–309.

52. Watanabe, K., Oshima, T., Nishimura, S. (1976). Nucl. Acids Res. *3*, 1703–1713.

53. Roe, B., Michael, M., and Dudock, B. (1973). Nature New Biol. *246*, 135–138.

54. Svensson, I., Isaksson, K., and Henningsson, A. (1971). Biochim. Biophys. Acta *238*, 331–337.

55. Johnson, L., Hayashi, K., and Söll, D. (1970). Biochemistry *9*, 2823–2831.

56. Marmor, J. B., Dickerman, H. W., and Peterkofsky, A. (1971). J. Biol. Chem. *246*, 3464–3473.

57. Horowitz, J., Ou, C.-N., and Ishaq, M. (1974). J. Mol. Biol. *88*, 313.

58. Ofengand, J., Bierbaum, J., Horowitz, J., Ou, C.-N., and Ishaq, M. (1974). J. Mol. Biol. *88*, 313–325.

59. Piper, P. W. (1975). Eur. J. Biochem. *51*, 295–304.

60. Marcu, K., Mignery, R., Reszelbach, R., Roe, B., Sirover, M., and Dudock, B. (1973). Biochem. Biophys. Res. Commun. *55*, 477–483.

61. Harada, F., Sawyer, R. C., and Dahlberg, J. E. (1975). J. Biol. Chem. *250*, 3487–3497.

62. Peters, G. G., Harada, F., and Dahlberg, J. E., Personal communication.

63. Marinus, M. G., Morris, N. R., Söll, D., and Kwong, T. C. (1975). J. Bacteriol. *122*, 257–265.

64. Ramabhadran, T. V., Fossum, T., and Jagger, J. (1976). J. Bacteriol. *128*, 671–672.

65. Delk, A. N., Romeo, J. M., Nagle, D. P., Jr., and Rabinowitz, J. C. (1976). J. Biol. Chem. *251*, 7649–7656.

66. Ramakrishnan, J., Taya, Y., and Nishimura, S. Unpublished results.

67. Björk, G. R., and Neidhardt, P. C. (1975). J. Bacteriol. *124*, 99–111.

68. Singer, C. E., Smith, G. R., Cortese, R., and Ames, B. N. (1972). Nature New Biol. *238*, 72–74.

69. Deeley, R. G., Goldberger, R. F., Kovach, J. S., Meyers, M. M., and Mullinix, K. P. (1975). Nucl. Acids Res. *2*, 545–554.

70. Allaudeen, H. S., Yang, S. K., and Söll, D. (1972). FEBS Letters *28*, 205–208.

71. Jacobson, K. B. (1971). Nature New Biol. *231*, 17–19.

72. Mischke, D., Kloetzel, P., and Schwochau, M. (1975). Nature *254*, 79–82.

73. Okada, N., Shindo-Okada, N., and Nishimura, S. (1977). Nucl. Acids Res. *4*, 415–423.

74. Soll, D. (1971). Science *173*, 293–299.

75. Schäfer, K. P., and Söll, D. (1974). Biochimie 56, 795–804.

76. Altman, S. (1975). Cell 4, 21–29.

77. Nau, F. (1976). Biochimie 58, 629–645.

78. Nishimura, S. (1974). In MTP International Review of Science, Biochemistry (series one), Biochemistry of Nucleic Acids, K. Burton, ed., (London: Butterworths; Baltimore: University Park Press), 6, 289.

79. Altman, S., and Smith, J. D. (1971). Nature New Biol. 233, 35–39.

80. Sakano, H., Shimura, Y., and Ozeki, H. (1974). FEBS Letters 48, 117–121.

81. Munns, T. W., and Sims, H. F. (1975). J. Biol. Chem. 250, 2143–2149.

82. Chen, G. S., and Siddiqui, M. A. Q. (1975). J. Mol. Biol. 96, 153–170.

82. Chang, S., and Carbon, J. (1975). J. Biol. Chem. 250, 5542–5555.

84. Barrell, B. G., Seidman, J. G., Guthrie, C., and McClain, W. H. (1974). Proc. Nat. Acad. Sci. USA 71, 413–416.

85. Guthrie, C. (1975). J. Mol. Biol. 95, 529–547.

86. Davis, A. R., and Nierlich, D. P. (1974). Biochim. Biophys. Acta 374, 23–37.

87. Goodman, H. M., Abelson, J. N., Landy, A., Zadrazil, S., and Smith, J. D. (1970). Eur. J. Biochem. 13, 461–483.

88. Baguley, B. C., Wehrli, W., and Staehelin, M. (1970). Biochemistry 9, 1645–1649.

89. Kuchino, Y., and Nishimura, S. (1970). Biochem. Biophys. Res. Commun. 40, 306–318.

90. Kuchino, Y., and Nishimura, S. (1974). Biochemistry 13, 3683–3688.

91. Kaiser, I. I. (1972). J. Mol. Biol. 71, 339–350.

92. Nishimura, S., Taya, Y., Kuchino, Y., and Ohashi, Z. (1974). Biochem. Biophys. Res. Commun. 57, 702–708.

93. Kuchino, Y., Seno, T., and Nishimura, S. (1971). Biochem. Biophys. Res. Commun. 43, 476–483.

94. McClain, W. H., and Seidman, J. C. (1975). Nature 257, 106–110.

95. Dudois, E., Dirheimer, G., and Weil, J. H. (1974). Biochim. Biophys. Acta. 374, 332–341.

96. Smolar, N., Hellman, V., and Svensson, I. (1975). Nucl. Acids Res. 2, 993–1004.

97. Taya, Y., and Nishimura, S. Unpublished results.

98. Nishimura, S. In The Biochemistry of Adenosylmethionine, F. Salvatore, ed. (New York: Columbia University Press), in press.

99. Kline, L. K., Fittler, F., and Hall, R. H. (1969). Biochemistry 8, 4361–4371.

100. Rosenbaum, N., and Gefter, M. L. (1972). J. Biol. Chem. 247, 5675–5680.

101. Lipsett, M. N. (1972). J. Biol. Chem. 247, 1458–1461.

102. Harris, C. L., and Titchener, E. B. (1971). Biochemistry 10, 4207–4212.

103. Schaefer, K. D., Altman, S., and Söll, D. (1974). Proc. Nat. Acad. Sci. USA 70, 3626, 3630.

104. Cortese, R., Kammen, H. O., and Spengler, S. J., Ames, B. N. (1974). J. Biol. Chem. 249, 1103–1108.

105. Mullenback, G. T., Kammen, H. O., and Penhoet, E. E. (1976). J. Biol. Chem. 251, 4570–4578.

106. Bronskill, P., Kennedy, T. D., and Lane, B. G. (1972). Biochim. Biophys. Acta 262, 275–282.

107. Kennedy, T. D., and Lane, B. G. (1975). Can. J. Biochem. 53, 1–10.

108. Körner, A., and Söll, D. (1974). FEBS Letters 39, 301–306.

109. Elkins, B. N., and Keller, E. B. (1974). Biochemistry 13, 4622–4627.

110. Okada, N., and Nishimura, S. (1977). Nucl. Acids Res. 4, 2931–2937.

111. Münch, H.-J., and Thiebe, R. (1975). FEBS Letters 51, 257–258.

112. Romeo, J. M., Delk, A. S., and Rabinowitz, J. C. (1974). Biochem. Biophys. Res. Commun. 61, 1256–1261.

113. Kersten, H., Sandig, L., and Arnold, H. H. (1975). FEBS Letters 55, 57–60.

114. Schmidt, W., Arnold, H.-H., and Kersten, H. (1977). J. Bacteriol. 129, 15–21.

115. Hankins, W. D., and Farkas, W. R. (1970). Biochim. Biophys. Acta 213, 77–89.

116. Farkas, W. R., and Singh, R. (1973). J. Biol. Chem. 248, 7780–7785.

117. Okada, N., Harada, F., and Nishimura, S. (1976). Nucl. Acids Res. 3, 2593–2603.

118. Farkas, W. R., and Chernoff, D. (1976). Nucl. Acids Res. 3, 2521–2547.

119. Farkas, W. R. Personal communication.

120. Okada, N., and Nishimura, S. Unpublished results.

121. Itoh, Y. H., Itoh, T., Haruna, I., and Watanabe, I. (1977) Nature 267, 467.

122. Kuchino, Y., Kasai, H., and Nishimura, S. (1976). Nucl. Acids Res. 3, 393–398.

123. McLennan, B. D. (1975). Biochem. Biophys. Res. Commun. 65, 345–351.

124. Lebowitz, P., Dohn, F. C., Jr., Richards, H. H., Neelon, F. A., and Cantoni, G. L. (1967). J. Biol. Chem. *242*, 4523–4527.

125. Sueoka, N., and Kano-Sueoka, T. (1970). Prog. Nucl. Acid Res. Mol. Biol. *10*, 23–55.

126. Kowalski, S., and Fresco, J. R. (1971). Science *172*, 384–385.

127. Nishimura, S., and Novelli, G. D. (1964). Biochim. Biophys. Acta *80*, 574–586.

128. Yudelvich, A. (1971). J. Mol. Biol. *60*, 21–29.

129. Littauer, U. Z., and Inouye, H. (1973). Ann. Rev. Biochem. *42*, 439–470.

130. Garel, J.-P. (1976). Nature *260*, 805–806.

131. Anderson, W. F. (1969). Proc. Nat. Acad. Sci. USA *62*, 566–573.

132. Anderson, W. F., and Gilbert, J. M. (1969). Biochem. Biophys. Res. Commun. *35*, 456–461.

133. Nishimura, S., Harada, F., Narushima, U., and Seno, T. (1967). Biochim. Biophys. Acta *142*, 133–148.

134. Dube, S. K., Marcker, K. A., Clark, B. F. C., and Cory, S. (1969). Eur. J. Biochem. *8*, 244–255.

135. Ikemura, T., and Ozeki, H. Personal communication.

136. Menichi, B., and Heyman, T. (1976). J. Bacteriol. *127*, 268–280.

137. Isham, K. R., and Stulberg, M. P. (1974). Biochim. Biophys. Acta *340*, 177–182.

138. Stern, R., Gonano, F., Fleissner, E., and Littauer, U. Z. (1970). Biochemistry *9*, 10–18.

139. Keisel, N., and Vold, B. (1976). J. Bacteriol. *126*, 294–299.

140. Fournier, M. J., and Peterkofskey, A. (1975). J. Bacteriol. *122*, 538–548.

141. Kitchingman, G. R., and Fournier, M. J. (1976). Biochem. Biophys. Res. Commun. *73*, 314–322.

142. Mann, M. B., and Huang, P. C. (1974). Biochemistry *13*, 4704–4710.

143. White, B. N. (1975). Biochim. Biophys. Acta *395*, 322–328.

144. Singhal, R. P., and Vold, B. (1976). Nucl. Acids Res. *3*, 1249–1261.

145. Harris, C. L., Marashi, F., and Titchenev, E. B. (1976). Nucl. Acids Res. *3*, 2129–2142.

146. Borek, E., and Kerr, S. J. (1972). Adv. Cancer Res. *15*, 163–190.

147. Symposium on tRNA and tRNA modification in differentiation and neoplasia (1970). Cancer Res. *31*, 591–721.

148. Craddock, V. M. (1970). Nature *228*, 1264–1268.

149. Randerath, E., Chia, L. S. Y., Morris, H. P., and Randerath, K. (1974). Cancer Res. *34*, 643–653.

150. Simsek, M., RajBhandary, U. L., Boisnard, M., and Petrissant, G. (1974). Nature *247*, 518–520.

151. Randerath, K., Chia, L. S. Y., Gupta, R. C., Randerath, E., Hawkins, E. R., Brum, C. K., and Chang, S. H. (1975). Biochem. Biophys. Res. Commun. *63*, 157–163.

152. Ortwerth, B. J., and Liu, L. P. (1973). Biochemistry *12*, 3978–3984.

153. Juarez, H., Juarez, D., Hedgcoth, C., and Ortwerth, B. J. (1975). Nature *254*, 359–360.

154. Gross, H. J. Personal communication.

155. Grunberger, D., Weinstein, I. B., and Mushinski, J. F. (1975). Nature *253*, 66–67.

156. Katze, J. R. (1975). Biochim. Biophys. Acta *407*, 392–398.

157. Gonano, F., Pirro, G., and Silvetti, S. (1973). Nature New Biol. *242*, 236–237.

158. Baliga, B. S., Borek, E., Weinstein, I. B., and Srinivasan, P. R. (1969). Proc. Nat. Acad. Sci. USA *62*, 899–905.

159. Sekiya, T., and Oka, T. (1972). Virology *47*, 168–180.

160. Katze, J. R. (1975). Biochim. Biophys. Acta *383*, 131–139.

161. Okada, N., Shindo-Okada, N., and Nishimura, S. Unpublished results.

162. Briscoe, W. T., Griffin, A. C., McBride, C., and Bowen, J. M. (1976). Cancer Res. *35*, 2586–2593.

163. Briscoe, W. T., Syrewicz, J. J., Marshall, M. V., and Griffin, A. C. (1975). Biochim. Biophys. Acta *383*, 441–445.

164. Tener, G. M. (1976). In Abstracts of Tenth International Congress of Biochemistry, Hamburg, p. 32.

165. Kasai, H., Murao, K., Nishimura, S., Liehr, J. G., Crain, P. F., and McCloskey, J. A. (1976). Eur. J. Biochem. *69*, 435–444.

166. Kasai, H., Ohashi, Z., Harada, F., Nishimura, S., Oppenheimer, N. J., Crain, P. F., Liehr, J. G., von Minden, D. L., and McCloskey, J. A. (1975). Biochemistry *14*, 4198–4208.

8 CONFORMATIONAL CHANGES OF tRNA

D. M. Crothers and P. E. Cole

8.1 Introduction

Transfer RNA plays a unique role in protein synthesis by providing the crucial link between the base sequence of a triplet codon and the insertion of a specific amino acid into a growing polypeptide chain. At least one tRNA is present on the ribosome during all the stages of polypeptide synthesis, which involves a complex series of events whose detailed molecular mechanisms generally remain unknown. Certainly it is unreasonable to suppose that tRNA is only a passive adapter between mRNA and the peptidyl transferase site, given the complexity of both tRNA and the task it must accomplish. It is probable that tRNA structure, now so elegantly defined by crystallographic work on yeast tRNA^{Phe} (1–5), may be substantially altered during residence on the ribosome. Indeed there is growing evidence for a series of cyclic conformational changes involving pairing between the several species of RNA present on active bacterial ribosomes. Besides the obvious interaction between the mRNA codons and the two tRNA anticodons, proposals have included binding between tRNA and 5S RNA (6), 5S RNA and 23S RNA (7), 23S RNA and 16S RNA (8), 16S RNA and tRNA (9), and 16S RNA and mRNA during initiation (10, 11). There may well be other undiscovered pairing interactions or other conformational changes in which tRNA participates.

Our knowledge of the molecular mechanisms of functional RNA conformational changes is in its infancy, usually limited to direct (11) or indirect (6–10) indications that a specific pairing occurs. In contrast, study of conformational changes of isolated tRNA molecules is a well-developed field because of the much greater simplicity of the problem. Even though there is now no way of knowing whether the conformational changes observed in solution are functionally relevant, the solution studies provide an essential basis for approaching the more difficult problem of functional processes. For example it can be shown that tRNA tertiary structure is capable of transient, localized opening (12, 16), which is opposed by the addition of Mg^{2+} ions (12). In addition the tertiary structure undergoes a concerted disruption (13–16), coupled to release of bound Mg^{2+} (17, 18), leaving only cloverleaf double helices as stable structural elements. The transient opening of structure could help initiate localized pairing inter-actions, for example between 5S RNA and the TΨC-loop of tRNA. Alter-natively the general opening of tertiary structure could provide tRNA with the flexibility that has been proposed as a means of facilitating transloca-tion (16, 19). At present we simply do not know which of the many interactions that stabilize tRNA structure in solution remain intact during its function.

It would be a mistake to infer that all the solution conformational changes of tRNA are now well understood. We do have a general knowledge of the molecular mechanism of thermal unfolding, including such details as the thermodynamic (20) and kinetic (15, 16) characteristics of individual steps. In specific cases we know the permissible limits of solution variables such as salt concentration (14) and pH (21) that leave the native structure intact. However, some vital areas still need intensive investigation. For example, there is evidence that aminoacylation of tRNA causes signifi-cant conformational changes far removed from the 3'-terminus (22), although other measurements reveal no important structural alteration (23). The nature of any change that may be induced remains a mystery, along with the mechanistic problem of how the allosteric effects of acylation could be transferred through the relatively flexible phosphodiester bonds at the single-stranded -C-C-A terminus to other regions of the molecule at which the influence of acylation apparently is felt. More than a decade of work has gone into our present understanding of conformational changes of tRNA in solution; probably at least another decade will be required before the job can be considered complete.

8.2 Conformational Changes of Isolated tRNA

8.2.1 A Phase Diagram Defines the Limits of Stability of Particular Conformational Forms

The native form of tRNA is generally defined as the stable structure that can be aminoacylated by an appropriate synthetase, in a buffer that typically contains 10 mM Mg^{2+}, roughly 10–100 mM Na^+, at neutral pH and 37°C. Years of experiments have demonstrated that change of the solution conditions can alter tRNA conformation, an effect that can conveniently be summarized in the form of a conformational phase diagram (14), which shows the stability limits of particular forms of the molecule. Any study of tRNA under nonphysiological conditions should be preceded by establishment of such a diagram for the specific tRNA and conditions used (pH, temperature, solvent), so that the relationship between the conformation actually studied and the native structure can be assessed.

A common phase diagram in the study of polynucleotides describes the variation of the transition midpoint temperature (T_m) with salt concentration for conversions such as those occurring between the single, double, and triple helical structures that result from mixing poly A and poly U (24). A similar approach is possible for tRNA conformational equilibria. Figure 8.1 shows typical melting curves for $tRNA_f^{Met}$ (E. coli), measured at the absorbance band of the single 4-thiouridine (s^4U) residue contained by that tRNA. The increased absorbance as temperature rises reflects the loss of stacking of s^4U that accompanies unfolding of the ordered structure. Notice that the character of the transition curves is different in high and low Na^+ concentration. At high concentrations a large and narrow lower-temperature transition is followed by a smaller absorbance change at 335 nm occurring at higher temperatures. When the salt concentration is lowered, the high-temperature transition is sharper and much larger than the broad, low-temperature melting process.

The phase diagram shown in figure 8.2 was established by a systematic search for variable tRNA conformations as a function of salt concentration and temperature. When interpreting such a diagram, one must have clearly in mind the definition of a *distinguishable conformation,* which in turn determines when a phase change has occurred. Whether a conformational change can be demonstrated depends on the experimental method used, so the definition of a separate region of the phase diagram is in a sense an operational one. However, one can choose experimental techniques that

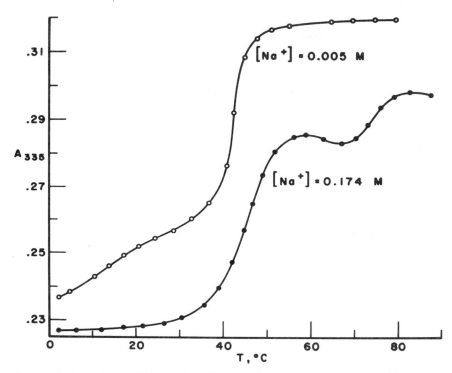

Figure 8.1 Comparison of high and low salt absorption melting curves of tRNA$_f^{Met}$ (*E. coli*) measured at 335 nm. Absorbance measurements refer to tRNA samples at identical concentrations; the absorbance difference between low and high salt was determined by a salt jump experiment. The buffers contained 1 mM phosphate, 0.1–10 mM sodium cacodylate, Na$_2$ EDTA (0.05–1 mM), and sufficient NaClO$_4$ to achieve the desired Na$^+$ concentration. From Cole et al. (14).

reflect the major tRNA conformational changes, giving the phase diagram a more general meaning than might otherwise be attributed to a single experimental approach.

The two variables in a nucleic acid phase diagram are generally temperature and counterion concentration. Experiments can be done holding one of the variables constant. Thermal melting curves at constant salt concentration yield well-defined melting points when measured at 335 nm, as seen in figure 8.1. These melting points provide the operational definition for the cooperative transitions between phases I and II, II and IV, and III and IV in figure 8.2. The nature of the structural difference between the forms is left for later consideration.

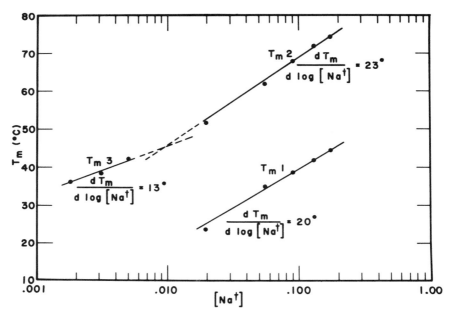

Figure 8.2 Variation of the three transition temperatures (measured at 335 nm) for tRNA $_f^{Met}$ (*E. coli*) with the logarithm of sodium ion concentration. The buffers are those described in figure 8.1. From Cole et al. (14).

The basis for distinguishing phase changes that occur when the salt concentration is changed at constant temperature also requires an operational definition. In this case the experiment is performed by adding salt to a tRNA solution held at constant temperature. By definition a phase boundary has been crossed whenever the tRNA structure responds to the change in salt concentration with a process slower than a few seconds and characterized by a large activation energy. As we will see, the effect of this definition is to draw a phase boundary between two conformations only when a change of salt concentration produces a conformation containing structural elements incompatible with the starting conformation. For example, in crossing the boundary from region III to II by adding salt, it is first necessary to dissociate base pairs or other structural interactions in form III before the bonds that stabilize conformation II can be formed.

Conversion between incompatible forms has a large activation energy because the ordered structure is stabilized by stacking interactions, whose dissociation requires heat (24). Studies on helix-coil transitions in oligonucleotides reveal that formation of helix from the single strands is a fast process with either zero or slightly negative activation energy. In contrast, disruption of structure has a large activation energy, which is approximately equal to the heat of dissociation of the structure (25, 26). Thus the heat required to dissociate the interactions that must be broken to convert from III to II appears in the activation energy for the process. Consequently, when salt is added to form III, a slow rate of conversion with a large activation energy is found for the reaction to conformation II. For example, the time required to convert $tRNA_f^{Met}$ from form III to I is about 1000 sec at 13°C, with an activation energy of 61 kcal/mol. In contrast, form II, which differs from form I only in the cooperative loss of (tertiary) structural interactions, converts to form I in a few milliseconds, with an activation energy very close to zero. Therefore conversion of II to I requires only *formation* of tertiary structure, not dissociation of incompatible structure. In summary, cooperative thermal melting transitions provide lines on the phase diagram, and we infer additional boundaries between incompatible structural forms from the interconversion kinetics.

These abstract considerations are more readily understood when the identity of the conformations is examined (figure 8.3).

Region I, found at high Na^+ and low Na^+, is called "native," even though Mg^{2+} is absent. A main basis for this identification is the failure to find any slow kinetic steps when Mg^{2+} is added to convert I to physiological buffer conditions under which its biological activity can be checked (14). (Exceptions in which Mg^{2+} is required for maintaining the native structure are considered in section 8.2.8.) In addition, NMR spectroscopic evidence indicates only very slight structural differences between tRNA in high Na^+ concentration with and without Mg^{2+} (27). Finally, as discussed in detail later (section 8.2.2), the observation of noncooperative bonding of Mg^{2+} to tRNA in form I indicates that binding is not coupled to a major conformational change (18, 28), in contrast to the cooperative binding observed for form III (29). Of course these considerations do not imply that Mg^{2+} binding to "native" tRNA is without structural effect. It may, for example, slightly alter the tertiary structure and enhance stacking interactions in regions such as the anticodon loop. This "fine tuning" of tRNA structure may be of considerable significance for biological function.

Form II differs from form I by the cooperative loss of tertiary structural interactions, the evidence for which we will take up later (section 8.2.4). Therefore this region of the phase diagram is labeled *cloverleaf or close variants*. As temperature is increased through region II, the cloverleaf

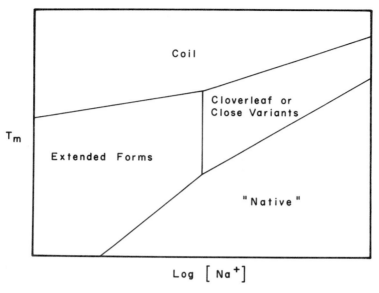

Figure 8.3 Simplified diagram identifying the conformational states corresponding to various regions of the phase diagram of figure 8.2. From Cole et al. (14).

double helices gradually melt, leaving only the single-stranded, nonhydrogen-bonded (coil) structure in region IV.

Form III is called *extended forms* because it results when the counterions that screen the polynucleotide phosphates from each other are reduced in concentration. As a result the electrostatic free energy of the compact tRNA structure rises, until an alternate, expanded or extended, structure becomes preferred because of its lower electrostatic free energy. One indication of this process is the smaller slope of the plot of T_m against log [Na⁺] for the extended forms, compared to cloverleaf. The slope of the lines is determined by the number of counterions released per calorie of heat in the transition (30, 31). Therefore a smaller slope for conversion of III to the coil implies a smaller reduction in the number of ions bound when III converts to coil than when II converts to coil (assuming the heat of melting is no larger at low than at high salt). This can result because III is more extended than II and thus binds fewer counterions.

Cole et al. (14) proposed low-salt structural models for a series of *E. coli* tRNAs, and Reeves et al. (37) suggested an extended conformation for yeast tRNA^Ala at low salt. However, these have not been investigated further and thus remain conjectural. Unfortunately a systematic phase diagram has not been published for that most-studied of tRNAs, the phenylalanine-specific tRNA from yeast, so the lower limit of Na^+ concentration at which the cloverleaf structure is stable without Mg^{2+}, relative to an extended form, is not clearly defined for that tRNA. NMR spectroscopy, now the method of choice for detailed conformational studies of tRNA, may be difficult to apply to the problem of low-salt structures because at the concentration of tRNA required for observation, the necessary counterions may already exceed the upper concentration limit for stability of the extended forms (32). This can be true even if the concentrated tRNA solution is dialyzed against a buffer in which tRNA in dilute solution is known to be in the extended form. In thermodynamic terms the influence is exerted by the effect of the (concentration-dependent) Donnan potential on the relative free energies of macro-ion conformations that differ in counterion binding and hence in charge.

8.2.2 Native tRNA Has a Special Affinity for Mg^{2+} and Other Divalent Ions
The binding of Mg^{2+} and other divalent ions by tRNA has been studied for many years (17, 18, 28, 29, 33–53). This subject has recently become much better understood in its outline and can be readily summarized with reference to the phase diagram in figure 8.3. There are two approaches to the study of Mg^{2+} binding: either the experiment begins with a tRNA solution, lacking Mg^{2+}, which corresponds to the native region of the phase diagram, or the starting sample does not possess the native tertiary structure as, for example, in the cloverleaf or extended-forms regions of the phase diagram. Binding isotherms for Mg^{2+} with native tRNA are not cooperative (17, 18, 45, 49); there are a few (1 to 5) sites with greater affinity than the remaining 20 to 25 sites. The actual number of "strong" sites is still disputed and may depend on conditions. The "weak" sites are roughly comparable in affinity to the Mg^{2+}-binding sites on standard double helical nucleic acids. Binding of Mg^{2+} to nonnative tRNA is cooperative (29, 36, 40–42, 48) because Mg^{2+} acts as an allosteric effector to convert tRNA from one conformation to another (49), and more than one Mg^{2+} ion usually is required for the conversion. However, as we will discuss below, caution should be exercised in interpreting the cooperative binding curves in terms of the number of special binding sites provided by the native structure (49). Possible structural origins of the Mg^{2+}-binding sites of potentially high affinity are revealed by analysis of ion-binding sites in crystals of yeast tRNA^{Phe} (51, 52).

Cooperative binding reactions can be distinguished by the shape of a Scatchard (54) plot of the binding isotherm. According to Scatchard's analysis, if each tRNA contains binding sites of association equilibrium constant k, the ratio of r (the number of ions bound per tRNA) to m (the free concentration of ions) is

$$r/m = k(n - r). \tag{8.1}$$

Therefore a plot of r/m versus r should be straight, with slope $-k$ and intercept $r = n$ when $r/m \rightarrow 0$. When the line is not straight, at least one of the assumptions is incorrect.

Deviations from linear isotherms are classified according to the direction of curvature. As shown in figure 8.4, cooperative binding produces curves that are concave downward (figure 8.4b), often with a local maximum, whereas anticooperative binding causes the opposite curvature (figure 8.4c). Curves of the latter shape can result either because the bound molecules or ions repel each other (a true anticooperative effect) or because the binding sites do not all have the same affinity. In general there is no way to distinguish these two possibilities for interpreting figure 8.4c simply on the basis of the shape of the isotherm. Specific examples illustrating these general principles follow.

Römer and Hach (18) and Stein and Crothers (17) independently demonstrated that the binding isotherm of Mg^{2+} with native tRNA are not cooperative. Their results, compared in figures 8.5a and 8.5b, respectively, show similarities but also some differences. The affinity of tRNA^{Phe} (yeast)

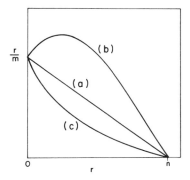

Figure 8.4 (a) Linear Scatchard plot, slope $= -k$, intercept $r = n$ at $r/m = 0$. (b) Cooperative binding. (c) Anticooperative binding. From Bloomfield et al. (24).

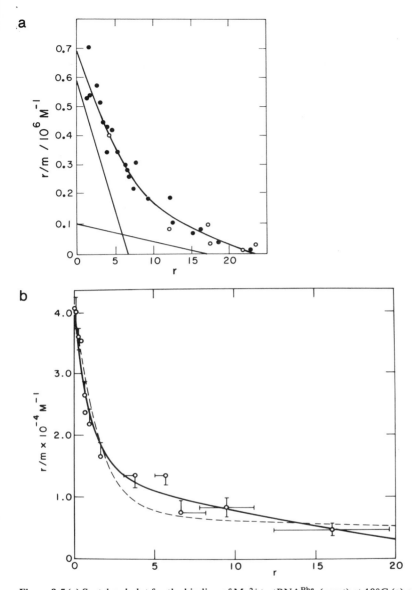

Figure 8.5 (a) Scatchard plot for the binding of Mg^{2+} to $tRNA^{Phe}$ (yeast) at 10°C (o) and 30°C (•) in 0.01 M sodium cacodylate, 0.022 M NaCl, pH 6.0. The points are experimental values determined by fluorescent titrations with 8-hydroxyquinoline 5-sulfonic acid. The curve represents a fit by two classes of noninteracting binding sites. The number n_1 of strong sites is 6.5 ± 3 with binding constant $k_1 = 9 \times 10^4 M^{-1}$. The number of weak sites is $n_2 = 17 ± 5$ with binding constant $k_2 = 6 \times 10^3\ M^{-1}$. From Römer and Hach (18). (b) Scatchard plot of Mg^{2+} binding to $tRNA_{f1}^{Met}$ in 0.17 M Na^+, phosphate-cacodylate buffer (pH 7.0, 4°C). The data were obtained by equilibrium dialysis and atomic absorption analysis. The solid curve is the best fit to equation 8.2 for $n_1 = 1$, $k_1 = 2.9 \times 10\ M^{-1}$, $n_2 = 26$, $k_2 = 4.2 \times 10^2 M^{-1}$. The dashed curve is the best fit to equation 8.2 with $n_1 = 2$. From Stein and Crothers (17).

for Mg^{2+}, as measured by r/m when $r \to 0$, is more than an order of magnitude larger than the affinity of $tRNA_f^{Met}$ (*E. coli*), probably because the Na^+ concentration is higher for $tRNA_f^{Met}$ (0.17 M) than for $tRNA^{Phe}$ (0.03M) and because Na^+ competes with Mg^{2+} in binding to nucleic acids. Both groups interpreted their results in terms of two classes of binding sites, of numbers n_1 and n_2 and affinities k_1 and k_2. Under these conditions equation (8.1) becomes

$$\frac{r}{m} = \frac{k_1 n_1}{1 + k_1 m} + \frac{k_2 n_2}{1 + k_2 m} \cdot \tag{8.2}$$

The main difference between the results observed for the two tRNAs, aside from the difference in affinities, is the number n_1 of strong sites: Stein and Crothers (17) found it to be 1–2; Römer and Hach (18), 6.5 ± 3. Both groups neglected a possible additional factor that could affect the shape of the binding isotherm, namely the electrostatic repulsion between bound Mg^{2+} ions (55), which would cause anticooperative curvature as shown in figures 8.4 and 8.5. This effect is more likely to be serious in the experiments on $tRNA^{Phe}$, because of the lower Na^+ concentration and higher number of ions bound in the experimental region of interest for determining n_1 ($r \leq 10$ for $tRNA^{Phe}$; $r \leq 2$ for $tRNA_f^{Met}$). Consequently n_1 is probably significantly less than 6–7, possibly as small as 1. More extensive experiments at varying Na^+ concentrations, along with theoretical calculations to take account of ion-ion repulsions (analogous to neighbor-exclusion effects; ref. 56) are clearly called for.

Cooperative binding of Mg^{2+} and other divalent ions to tRNA has more commonly been observed than noncooperative binding, and it has not gone without notice that the cooperativity arises from conversion of the low-salt (29) or partly melted form (18) to the native structure. However, the potential ambiguity in interpreting the cooperative binding isotherms has been underemphasized. For example, figure 8.6 shows similarly shaped cooperative binding isotherms for (a) Mn^{2+} binding to $tRNA^{Phe}$ (29) in 0.1 M triethanolamine, a buffer in which the native tertiary struture has melted (18), and (b) for Mg^{2+} binding to the denatured form of $tRNA^{Phe}$ (49), which is converted to the native form by Mg^{2+} addition. The interpretations given to the two experiments are significantly different, with appreciably differenct conclusions, in spite of the similar appearance of the binding isotherms. Schrier and Schimmel considered that there are n_1 cooperative binding sites, whose interaction exponent a indicates the number of Mn^{2+}

ions that must bind simultaneously. With n_2 independent sites the Scatchard plot is given by the expression (29)

$$\frac{r}{m} = \frac{n_1 k_1^a m^{a-1}}{1 + k_1^a m^a} + \frac{n_2 k_2}{1 + k_2 m}. \tag{8.3}$$

Analysis of the data in figure 8.6a leads to $n_1 = 5$, $k_1 = 2.7 \times 10^5 \, \mathrm{M^{-1}}$, $a = 2.3$ for the cooperative sites. The total number of noninteracting sites is approximately 20, and two affinity classes are required to fit the data.

In contrast, Bina-Stein and Stein (49) considered that Mg^{2+} acts to convert the denatured (D) form to the native (N) state because N binds more Mg^{2+} than D does:

$$D + xMg^{2+} \rightleftharpoons N-(Mg^{2+})_x. \tag{8.4}$$

The number x of Mg^{2+} ions required for the conversion depends on the equilibrium constant L between N and D in absence of Mg^{2+},

$$L = \frac{[D]}{[N]},$$

and on the difference in Mg^{2+} affinity between N and D. Bina-Stein and Stein (49) found that they could fit the data with values, measured independently for the N form, of $n_1 = 1$, $k_1 = 7.5 \times 10^4 \, \mathrm{M^{-1}}$, $n_2 = 36$, $k_2 = 8.3 \times 10^2 \, \mathrm{M^{-1}}$. The D-form lacks the single strong Mg^{2+} site and has an equal number, $n_3 = 36$, of weak sites as the N-form but these are of slightly weaker affinity, $k_3 = 4.3 \times 10^2 \, \mathrm{M^{-1}}$.

The significance of these results is that there is not necessarily a separate class of ~5 cooperative binding sites. The single strong site and all of the ~36 weak sites participate in forcing the equilibrium in equation (8.4) to the right. While the allosteric model is not necessarily correct in detail, it is clearly reasonable and requires reexamination of the earlier interpretations of Mg^{2+} binding that were based on a class of cooperative binding sites. In summary it can be said that the number of Mg^{2+} sites with high affinity remains in dispute, but it is probably less than 5 and may generally be as small as 1. There is no reason to suppose that native tRNA binds Mg^{2+} cooperatively; the observed cooperative curves for binding to nonnative forms can be explained in terms of one (or a few) strong sites on native tRNA, with a large class of weaker sites that are slightly stronger on native than on nonnative tRNA.

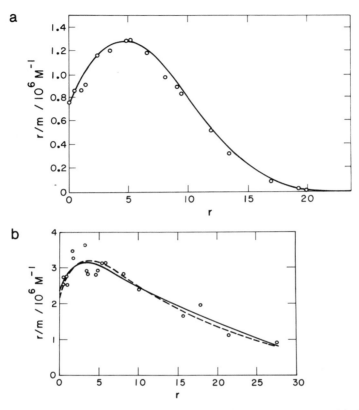

Figure 8.6 (a) Scatchard plot of manganese binding to tRNA[Phe] (yeast) in 0.1 M triethanolamine (pH 8.35) at 0°C. The data were obtained by equilibrium dialysis and atomic absorption measurements. Analysis of the least-squares fit solid line for the data yields, for the parameters in equation 8.3, $n_1 = 5$, $k_1 = 2.7 \times 10^5 \, M^{-1}$, $\alpha = 2.3$ for the cooperative sites. From Schreier and Schimmel (29). (b) Scatchard plot of Mg^{2+} binding to tRNA$_2^{Glu}$ (*E. coli*) in 0.1 M Na$^+$, pH 7.0, 34°C. The solid curve is the best fit to the allosteric model, using binding parameters $n_1 = 1$, $k_1 = 7.5 \times 10^4 \, M^{-1}$, $n_2 = 36$, $k_2 = 8.3 \times 10^2 \, M^{-1}$ for the native (N) form and $n_3 = 36$, $k_3 = 4.3 \times 10^2 \, M^{-1}$ for the denatured (D) form, and $L = 10$. The dashed line is the best fit to the cooperative binding model, with $k_1 = 6.7 \times 10^3 \, M^{-1}$, $n_1 = 4$, and $\alpha = 1.8$ for the cooperative sites. From Bina-Stein and Stein (49).

8.2.3 Mildly Acidic pH Can Convert tRNA to a Different Conformation

A theme currently characteristic of work on tRNA conformation is the search for alternative stable conformations of tRNA in order to explain the unusual effects observed, for example, on aminoacylation. Several lines of evidence indicate that alteration of solution pH from 7 to approximately 5–6 can alter tRNA conformation (21, 57–60). It is conceivable that the acidic-solution structure and native structures are close enough in free energy at neutral pH that other influences, such as aminoacylation, could shift the conformational balance toward the acid form.

Figure 8.7 shows a phase diagram for variation of the pH of solutions of $tRNA_i^{Tyr}$, at 0.17 M Na^+ without Mg^{2+} (21). Conformations I, II, and IV again represent the native, cloverleaf, and coil conformations, respectively, whereas III is a new conformation, recognized by its altered melting properties and slow rate of conversion to the native form. In the case of $tRNA_f^{Met}$ Bina-Stein and Crothers (57) were able to show that the acidic form is not chargeable and that the region of the molecule altered is the D-helix plus tertiary structure interactions. In acidic pH the altered tertiary structure of $tRNA_f^{Met}$ has a considerably higher T_m than does the native tertiary structure at neutral pH.

Steinmetz-Kayne et al. (59) have examined the NMR spectra of several tRNAs at acidic pH (5.5–7) in the presence of Mg^{2+}. The principal change is the appearance of two new tertiary proton resonances at 9.9 ppm, which the authors interpret as adenine amino protons in the A_9-A_{23}-U_{12} triple of yeast $tRNA^{Phe}$, stabilized by protonation of A_9 at N1. Notice that in this case pH change does not appear to cause a structural reorganization such as is observed in acidic pH without Mg^{2+}(21, 57); instead it affects the proton exchange rate of one of the tertiary interactions in the native structure.

Phase diagrams have not yet been measured for systematic variation of both pH and ionic strength of Mg^{2+} concentration. However, Lynch and Schimmel (58) showed that pH and Mg^{2+} binding are coupled. Using a fluorescent probe attached to the 3′-end of isoleucyl-tRNA^{Ile}, they found that the affinity for Mg^{2+} is reduced by a factor of 100 at pH 4.7 compared to pH 7. Since cooperative binding was studied under low-salt condition, they postulated special sites of protonation that stabilize the low-salt structures.

8.2.4 Pathways of Thermal Unfolding Are Known for Several tRNAs

Considerable effort has gone into study of the mechanism of thermal unfolding of tRNA, especially yeast $tRNA^{Phe}$ (15, 18, 20, 29, 61–73, 88, 89, 91), $tRNA^{Gly}$ (74), $tRNA^{Asp}$ (75, 76), $tRNA^{Ala}$ (77–79), and $tRNA^{Ser}$ (80),

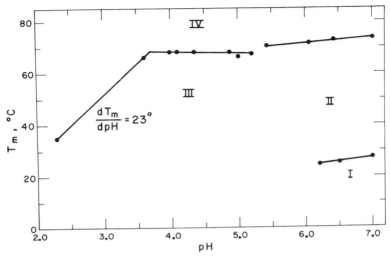

Figure 8.7 Variation of the transition temperatures T_m for $tRNA_I^{Tyr}$ with pH. The data were obtained at 150 mM Na$^+$, without Mg^{2+}, from thermal melting profiles at both 260 and 335 nm. From Bina-Stein and Crothers (21).

along with *E. coli* $tRNA_f^{Met}$ (14, 16, 17, 57, 81–84, 95), $tRNA^{Tyr}$ (14, 21, 32, 85–87), $tRNA^{Glu}$ (90, 92), $tRNA^{Val}$ (87, 94), and $tRNA^{Leu}$ (93). As a result more is known about the detailed mechanisms of folding these molecules than about any other class of macromolecules. Generally the conditions of the experiment are adjusted so that unfolding occurs over a wide temperature range and is therefore more sequential than concerted. In particular Mg^{2+} is deleted or added in very small amounts, thus destabilizing the tertiary structure. Under such conditions the first melting step includes loss of tertiary structure, an identification made originally by Fresco and his collaborators (13); sometimes unstable double helical arms melt together with the tertiary structure (15, 16). After the loss of tertiary structure the remaining double helices melt in the order of their stability, which is generally the order predicted (96–98) from their base composition and sequence.

The two techniques most successfully used to elucidate the pathway of thermal unfolding are T-jump relaxation kinetics and NMR spectroscopy of hydrogen-bonded protons. In the T-jump method the absorbance increase that accompanies melting is measured as a function of time. In this way a complex process can frequently be resolved into separate steps with different time dependences. The individual helical sections that make up a tRNA often melt separately, each giving rise to an exponential decay of the solution absorbance toward its new equilibrium value following the rapid temperature jump. In general the absorbance A approaches the equilibrium value A_{eq} by a sum of exponentials (for small perturbations):

$$A - A_{eq} = \Sigma A_i \, \exp(-t/\tau_i).$$

The constants τ_i are the relaxation times, and A_i are the absorbance amplitudes associated with each relaxation i (99, 100).

The amplitude of the relaxation corresponding to a particular structural section that melts as a unit reaches a maximum at the T_m of that transition, if we assume that the temperature jump size is held constant. Figure 8.8 shows an example from the extensive work of Maass, Riesner, Römer, and their collaborators on the unfolding of yeast tRNA^{Phe}; figure 8.8b shows the melting transitions for tRNA$_f^{Met}$ (E. coli). Included in the former diagram are the identifications assigned to each melting transition, on grounds that we will consider shortly.

When the NMR method is used to measure thermal unfolding, the primary quantity observed is increased exchange of the H-bonded ring $\geq$NH protons with solvent water as the structure melts. Of the $\geq$NH protons only those engaged in hydrogen bonds retain their chemical environment long enough to yield a high-resolution nuclear resonance. When the temperature is raised, the hydrogen bond becomes labile and the proton resonance is broadened or disappears (101). A simple model for the process is (101, 102)

$$\text{helix} \underset{k_{close}}{\overset{k_{open}}{\rightleftharpoons}} \text{coil} \xrightarrow{k_{ex}} \text{exchange}$$

in which only the nonbonded state (coil) can exchange with water. Usually the exchange rate (k_{ex}) is faster than the structure formation rate (k_{close}), so that the total exchange process is limited by k_{open} (16). When the helix lifetime (k_{open}^{-1}) becomes less than about 4 msec, the H-bonded proton resonances broaden and disappear, which can occur at a much lower temperature than the optically observed T_m (101, 16). Sometime k_{open}^{-1} is much greater than 5 msec at the optical T_m, in which case the resonance does not broaden but rather disappears in proportion to the fraction of the molecules that remain in the helix state (74, 101).

Figure 8.9 shows the unfolding mechanism deduced for E. coli tRNA$_f^{Met}$ (16), on which we can base our discussion of the principles governing this and analogous processes. The results refer to conditions (moderate salt con-

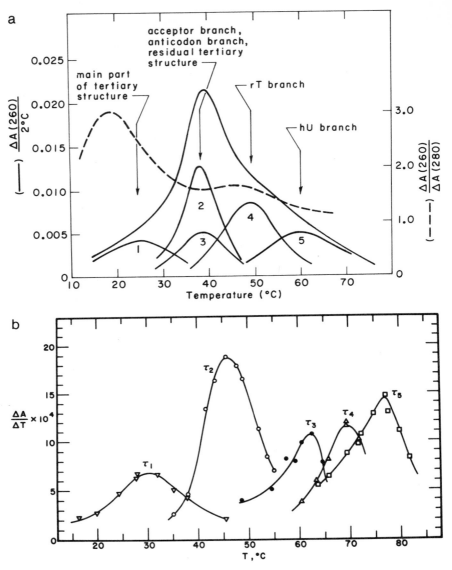

Figure 8.8 (a) Summary of the five differential thermal transition profiles of tRNA^{Phe} (yeast) in 0.01 M sodium cacodylate, 0.001 M Na EDTA, 0.02 M NaCl, pH 6.8. From Coutts et al. (69). (b) Differential thermal transition profiles for tRNA$_f^{Met}$ (*E. coli*) as determined directly from temperature-jump amplitude data. The amplitude for τ_3 was measured at 280 nm, while the other four effects were measured at 266 nm. The profile for τ_1 corresponds to tertiary structure D-helix melting in tRNA$_{f3}^{Met}$, while τ_2 is the tertiary structure D-helix melting for tRNA$_{f1}^{Met}$. The transitions τ_3, τ_4, τ_5 correspond to the melting of anticodon, TΨC, and stem helices, respectively, and are the same for tRNA$_{f1}^{Met}$ and tRNA$_{f3}^{Met}$. From Crothers et al. (16).

centrations, no Mg^{2+}) such that the tertiary structure interactions melt first, possibly accompanied by melting of the least stable of the cloverleaf stems. Several lines of evidence demonstrate that tertiary structure is being disrupted in the first melting step. Perhaps most definitive is the strong effect on the T_m of sequence changes in regions of the molecule not bonded in cloverleaf double helices. For example, replacement of m^7G by A in tRNA$_f^{Met}$ (E. coli) reduces the tertiary structure T_m by 15°C in 0.2 M Na$^+$ (see figure 8.8b; ref. 16), and substitution of G15 by A15 in E. coli tRNA$_{Su+}^{Tyr}$ dramatically affects the properties of the tertiary melting transition (86). Additional evidence for tertiary structure disruption is provided by the strong sensitivity of the earliest melting transition temperature to Mg^{2+} ions (14, 17, 18) and by the unusual dependence of T_m on the size of monovalent cations (68). Also exceptional is the destabilization of the first melting unit by ethidium binding (67), in contrast to the effect of ethidium on melting of known polynucleotide helices. In some cases tRNA molecules contain natural probes for tertiary interactions, such as s^4U in many E. coli tRNAs, which serve to monitor primarily tertiary rather than secondary (double helical) interactions (16, 103), because s^4U is located in a nonhelical region of the cloverleaf.

The problem of which double helices melt with the tertiary structure is usually answered by elimination: when no subsequent signal is found that corresponds to a particular helix, those base pairs must have melted in the first cooperative step. This is essentially the logic behind the assignment of loss of the D-helix to the first optical melting transition in figures 8.8b and 8.9. (Transient opening of hydrogen bonds, detected by the NMR method, precedes the first optical transition, which monitors equilibrium opening.)

Several lines of evidence have been used to support structural assignment of optical melting effects. For example the tRNA can be fragmented so that the T_m of a single hairpin helix can be evaluated (62, 63, 66). Also fluorescent chromophores, such as the Y-base in yeast tRNA^{Phe}, and a formycin-replaced A on the -C-C-A terminus have been used to monitor anticodon and acceptor stem regions (69). Alternatively the helix lifetime of a particular arm, deduced from the NMR method, can be correlated with the lifetime measured by relaxation kinetics, thus associating the optical signal with a helix whose identity is known from its NMR spectrum (16, 74, 91). Finally supporting arguments can be based on the apparent base composition as determined from the wavelength dependence of the absorbance change (15, 81), the melting enthalpy (16), and the observed time constant for helix formation to close a hairpin loop (16). (Small loops can

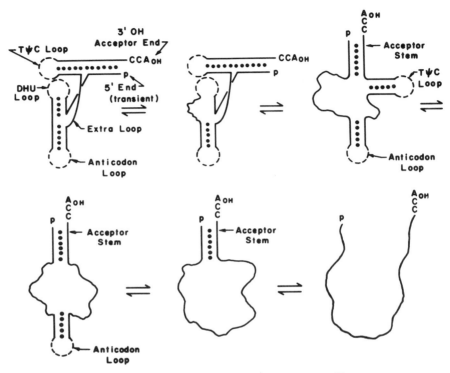

Figure 8.9 Schematic drawing of the molecular mechanism of tRNA$_f^{Met}$ thermal unfolding. The first optically detected melting step corresponds to the simultaneous disruption of "tertiary interactions" and the D- (denoted hUra) helix. Successive steps are then the anticodon helix, the TΨC-helix, and acceptor-stem melting.

NMR measurements indicate that some elements of the D- helix and/or tertiary structure can open transiently at temperatures below those of the first optical melting transition. From Crothers et al. (16).

be closed more rapidly than large ones because the bases that become bonded in the potential helix are closer together.)

In most cases the observed stability of tRNA hairpin helices is in good accord with predictions based on the properties of model oligonucleotides (96–98) when allowance is made for solution conditions (16). An exception is the remarkably high thermal stability deduced for the D-helix of yeast tRNA^{Phe}, which melts last (see figure 8.8a). Recently Hinz et al. (20) suggested that the assignments of melting the anticodon and D-helices should be interchanged based on calorimetric (20) and NMR (71, 90) evidence. Further, the chemical modification experiments of Rhodes (12) indicate that the D-helix melts relatively early. This reassignment would be more in line with predicted stabilities and with the observed order of melting of $tRNA_f^{Met}$, which has a D-helix identical in sequence but reversed in polarity compared with the D-helix in yeast $tRNA^{Phe}$. However, it would also require an explanation of the properties of the tRNA fragments on which the assignment in figure 8.8a is based (63, 66). If figure 8.8a is correct, some additional structural feature must be sought to account for the high stability of the D-helix in yeast $tRNA^{Phe}$.

8.2.5 Calorimetry and Unfolding Studies Allow Dissection of the Stabilization Energetics of tRNA

If cyclic switching of RNA-RNA interactions occurs on the ribosome, we must seek to understand the conformational energy balance that results. Assignment of an enthalpy change to melting of a given structural region is of particular interest in this context. Both T-jump melting curves and calorimetric measurements (20, 42, 104–108) permit enthalpies to be determined. The values found for melting of tertiary structure vary from about 25 to 60 kcal/mol. As an example, the heat for melting the $tRNA_f^{Met}$ tertiary structure plus D-helix in figures 8.8b and 8.9 is 52 kcal/mol, and the heats of melting the TψC, anticodon, and acceptor arms are 54, 58, and 70 kcal/mol, respectively (16). The total enthalpy of melting deduced from the sum of these individual optical transition heats, 234 kcal/mol, is significantly smaller than the recently measured total calorimetric heat of $tRNA_f^{Met}$ melting (108). Calorimetric measurements on other tRNAs also give high heats of melting compared to the optically determined heat for $tRNA_f^{Met}$: 314 kcal/mol for yeast $tRNA^{Ile}$ and 322 kcal/mol for yeast $tRNA^{Ser}$ (107).

The probable source of this discrepancy is neglect of the fast ($\tau < 5\,\mu sec$) unstacking transition in calculating the heats from the T-jump measurements. This effect, which arises from unstacking of bases in already melted regions, accounts for a substantial part of the optical signal and must be included to account for the total heat.

The outstanding improvement in calorimetric instrumentation achieved by Privalov has recently permitted heat uptake measurements of such precision that they can successfully be resolved into contributions from separate melting transitions (20). Figure 8.10 shows the heat capacity of yeast $tRNA^{Phe}$, along with its resolution into five components of enthalpies 45, 52, 76, 69, and 57 kcal/mol, respectively (total heat = 299 kcal/mol). The similarity of the individual components to those in figure 8.8a is unmistakable. However, no transition is included that corresponds to the fast un-

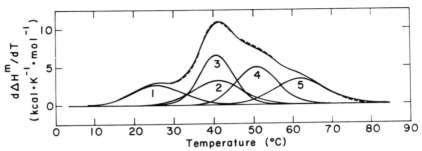

Figure 8.10 Decomposition of experimental molar excess heat capacity ΔH^m curve of tRNA ^{Phe} (yeast) into constituent functions. Equation of curve: $d\Delta H^m/dT = f(T)$. Dashed line segments indicate the experimental curve; the solid line marks the summation of constituent functions. Buffer is 10 mM sodium cacodylate, 1 mM EDTA, 20 mM NaCl, pH 6.8. From Hinz et al. (20).

stacking signal always observed in T-jump kinetics. Such a process must necessarily contribute to the heat capacity, so the resolution of the observed total heat uptake curve into components cannot be entirely correct because only the kinetically slower components are represented. It seems likely that a broad sixth transition curve corresponding to unstacking could be added to figure 8.10 and still allow excellent fit to the heat uptake data.

8.2.6 NMR Techniques Provide the Method of Choice for Future Studies of tRNA Conformational Changes

The increasing availability of high-resolution, high-sensitivity NMR spectrometers capable of measuring a variety of nuclei makes it likely that a number of tRNA conformational problems will be solved by this method. Four promising areas of analysis are identifiable at this time, depending on the nucleus. These include the nonexchangeable proton resonances of the modified nucleotides (such as the rT methyl resonance and the Y-base) and the exchangeable ring NH hydrogen-bonding protons, along with ^{31}P and ^{19}F NMR spectroscopy. Of these only the modified nucleotide nonexchangeable proton resonances are not currently plagued by the problem of resonance identification. In the other three areas unambiguous and agreed-on resonance assignment is the first priority before the phase of productive applications can begin.

Several revealing studies of proton NMR of nonexchangeable hydrogens in tRNA have appeared, for example, from the laboratories of Ts'o (70–72) and Schmidt (94). As an illustration Kan et al. (70–72) identified 12 methyl and 2 methylene resonances in yeast tRNA ^{Phe} and measured the temperature dependence of their chemical shifts, allowing identification of the thermal unfolding pattern of a number of separate regions of the molecule. Kastrup and Schmidt (94) studied the m⁶A, dihydrouridine, and rT resonances in *E. coli* tRNA^Val and were able to show that unfolding occurs in the order dhU, m⁶A, rT. The measured line width, along with direct measurement of the nuclear relaxation time T_1 can be expected to provide considerable information about the local freedom of motion in various portions of the molecule.

The main problem facing studies of the exchangeable proton resonances is definitive assignment. Figure 8.11 shows an example of the remarkable resolution now possible using correlation spectroscopy at 360 MHz with about 5 mg of tRNA and 15 min accumulation time (109, 110). Contrary to the early studies, it is now agreed that 3–8 tertiary H-bonded protons contribute to the resonances between −11 and −15 ppm, but the assignment of

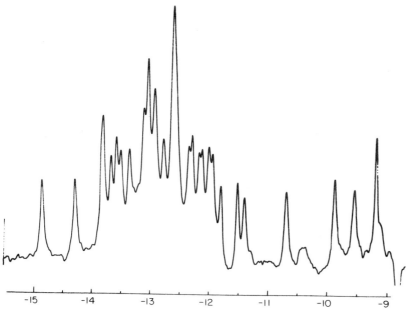

Figure 8.11 High-resolution proton NMR spectrum of tRNA^Val (*E. coli*) in the low field region, showing resonances from hydrogen-bonded protons; 7.0 mg of 99 percent pure tRNA dissolved in 190 μl of 10 mM cacodylate, 10 mM Na₄ EDTA, 9 mM MgCl₂, pH 7.0. 2500 Hz swept in 0.8 sec, with 0.2 sec delay; 1000 accumulated sweeps in 16 min. There are 21 resolved (or partially resolved) single proton peaks between −15 ppm and −9 ppm. The peaks at −13.8, −13.05, and −12.9 ppm contain 2 protons, and the peak at −12.6 ppm contains 4 protons, a total of 28 protons in this region. Spectrum courtesy of B. R. Reid.

these is still hotly debated. This subject has been extensively reviewed recently by various experts in the field (27, 111, 112).

Utilization of ^{31}P spectroscopy of tRNA (113–115) is even more hampered by the problem of resonance identification. As figure 8.12 shows, several ^{31}P resonances are shifted from the main envelope in tRNA (113). These are found to be sensitive to temperature and sometimes to Mg^{2+} addition, so they could be excellent probes for tRNA conformational changes. However, it is not at all obvious how definitive resonance assignments can be made.

An extremely promising NMR approach is ^{19}F spectroscopy (116). 5-Fluorouridine can be incorporated biosynthetically into tRNA, yielding functional molecules containing highly sensitive local probes. Figure 8.13 shows ^{19}F spectra of *E. coli* tRNA$_1^{Val}$, illustrating the extraordinary environmental sensitivity of the ^{19}F chemical shift. (All the ^{19}F nuclei are chemically identical.) Once assignments have been made for these resonances, very productive applications should be possible to conformational changes of tRNA both in free solution and bound to proteins.

8.2.7 There Remain Unexplained Observations Concerning the Conformation of tRNA

Even though the broad outlines of tRNA conformation in solution are understood, some puzzling observations about the conformation of tRNA still need explanation. For example, three separate laboratories have examined the NMR spectroscopy of the rT methyl resonance in Mg^{2+}-containing solutions and have concluded that there is more than one conformation of native tRNA. Both Reid (117) and Kastrup and Schmidt (94) studied tRNA$_1^{Val}$. Whereas the rT methyl resonance appeared as a single band at -1.7 to -1.8 at high temperature, it disappears and reappears at a higher field position as temperature is lowered. Reid assigned resonances at -1.0 and -1.2 in the low-temperature (native) spectrum to rT, whereas Kastrup and Schmidt (whose spectra were somewhat less resolved because of the lower frequency) assigned the -1.2 resonance (and a near neighbor at about -1.4) to rT and attributed the -1.0 resonance to contaminating tetraalkylammonium ion. Kan et al. (72), working with yeast tRNA^{Phe}, also ascribed two resonances (at -1.0 and -1.5) to rT. If Reid and Kan et al. are correct and the -1.0 resonance as well as the -1.2 or -1.5 resonances are both due to the rT methyl group in the native form, then there must be more than one native conformation of tRNA, with a conversion rate slower

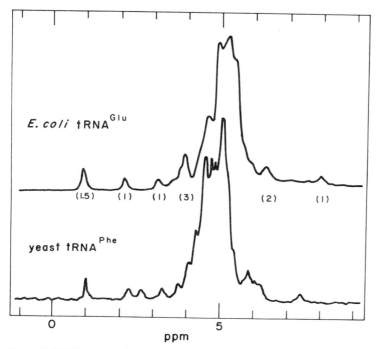

Figure 8.12 ^{31}P spectra of tRNA at 109 MHz in Fourier transform mode. The tRNA^{Glu} (*E. coli*) buffer contained 15 mM Mg^{2+}; the numbers in parentheses give the integrated intensity of the peaks, normalized to a total of 76 ^{31}P nuclei. The buffer for tRNA^{Phe} (yeast) contained an unknown quantity of Mg^{2+}, probably amounting to about 2.5 Mg^{2+}/tRNA. Accumulation time 10 hrs. From Gueron and Shulman (113).

than about 20 msec. This important observation clearly deserves to be confirmed or disproved by further work.

Another puzzling observation of considerable potential importance was reported by Olson et al. (118). Studying the diffusion coefficient of both bulk *E. coli* tRNA and purified yeast tRNA^{Phe} by the method of laser light scattering, they found an anomalous increase of 11 percent in the diffusion coefficient when the ionic strength was decreased from 0.2 to 0.1 M at 1 mM Mg^{2+}, pH 7.2, 20°C, extrapolated to zero tRNA concentration. No such anomalous behavior was observed at 10 mM Mg^{2+}, the other conditions remaining unchanged. Since the translational diffusion coefficient of a macromolecule is relatively insensitive to its shape, an 11 percent change would appear to require a substantial change in shape. Olson et al. suggest a hinge model, in which flexibility is gained at the lower salt conditions,

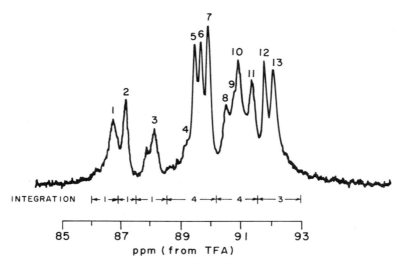

INTEGRATION

ppm (from TFA)

Figure 8.13 ^{19}F NMR spectrum of [5-FUra]tRNA$_1^{Val}$ (*E. coli*). The solutions contained approximately 2 mM [5-FUra]tRNA$_1^{Val}$ in 20 mM phosphate buffer, pH 6.4, 100 mM NaCl, and 10 mM MgCl$_2$ in D$_2$O; temperature 30°; frequency 93.6 MHz; accumulation time 9 hrs. The chemical shifts are given in parts per million from trifluoroacetate (TFA). From Horowitz et al. (116).

allowing the anticodon and acceptor stems to approach each other. Again this experiment suggests a second stable tRNA conformation in the presence of Mg^{2+}. No alternative explanation has yet been published, but it is clearly important that the matter be pursued and examined by other methods as well.

Another technique that has revealed unexplained conformational changes is measurement of the correlation time of spin labels attached to tRNA (119–121). A recent example is provided by the work of Caron and Dugas (119); they showed an apparent change in the slope of the temperature dependence of the correlation time of spin-labeled tRNA^{Glu}, tRNA$_f^{Met}$, and tRNA^{Phe} from *E. coli*, at about 32° in 10 mM Mg^{2+}, 20 mM Tris, pH 7.5. (Spin labels were attached at the anticodon loop, the X-base of the mini-loop, or s^{4}U8.) The molecular origin of the phenomenon is unknown.

Multiple tRNA conformational states are also indicated by the work of Rigler and Ehrenberg (122) on native yeast tRNA^{Phe}, which has ethidium covalently attached either at the dihydrouridine residues in the D-loop or at the position of the Y-base, with retention of biological activity. Even in the presence of up to 16 mM Mg^{2+} two components in the fluorescence lifetime decay curve were found at 10°C and three components at 35°C. This observation implies two different environments for the probe at 10°C and three at 35°C. The dynamics of interconversion between the states was studied by the fluorescence T-jump method, revealing, for example, two relaxation times of 10 and 100 msec at 23°C. The authors interpreted their results in terms of multiple conformational states, but they did not attempt to specify the molecular nature of these.

8.2.8 Denatured Forms Result from Hysteresis in Conformational Equilibria

Early work from the laboratories of Fresco, Sueoka, and Muench (13, 123–136) showed that tRNA can be converted to an inactive form, as judged by its failure to act as a substrate for aminoacylation. Usually denaturation is achieved by incubation in Mg^{2+}-free solution in low or moderate salt concentration buffer, but exposure to acidic pH can also produce nonchargeable species (21). The general physical basis for denaturation is understood: removal of Mg^{2+} shifts the relative conformational free energies so that an alternative structure is preferred over the native form. Even when conditions favorable to the native structure are restored, the molecule remains trapped in its metastable denatured state. In all cases there is a large activation energy for both denaturation and renaturation (~ 60 kcal/mol), indicating that both forms have bonds that must be broken to allow formation of the alternate conformer. Therefore elevated temperatures greatly accelerate equilibration of the two forms.

Recent physical studies of denatured conformers have concentrated on three species, $tRNA_3^{Leu}$ (137–142), $tRNA^{Trp}$ (143–145), and $tRNA_2^{Glu}$ (49, 92, 146). The widest variety of techniques has been employed to study $tRNA_3^{Leu}$, including chemical modification (138), tritium exchange (137), oligonucleotide binding (140), renaturation kinetics (142), and exchangeable proton NMR (93, 139, 141). All the results seem compatible with the model proposed by Kearns et al. (139). In this model the dihydrouridine and anticodon helices and one base pair from the TΨC-helix are disrupted in favor of adding five base pairs to the remaining TΨC-helix and forming a new helix having one G·C and two A·U pairs. However, the secondary structure proposed is clearly expected to be less stable than the cloverleaf, as judged by the standard rules (96–98), and there is no basis on which to assess the possible contributions of tertiary structure interactions to the stability of the denatured form.

In the model proposed by Jones et al. (144) for denatured $tRNA^{Trp}$ on the basis of NMR (144) and chemical modification (143) data, the TΨC-helix and acceptor-stem helix are intact, and the D-helix is disrupted while the anticodon helix is augmented. The availability of the anticodon loop in the model is clearly inconsistent with the observation of Buckingham (145) that denaturation interferes with complexation of tRNA^{Trp} to tRNA^{Pro}, whose anticodon is complementary to tRNA^{Trp}. Jones et al. (144) propose that their model explains the observation of Hirsh (147) that tRNA$_{Su^+}^{Trp}$ is not denaturable because the G24 $\longrightarrow$ A24 mutation produces an A·U pair in the suppressor tRNA, thus stabilizing the dihydrouridine helix against opening for denaturation. However, a G·U pair in the denatured conformer would also be converted to A·U if the model is correct, so the effects on native and denatured forms should cancel.

In contrast to tRNA^{Trp} and tRNA$_3^{Leu}$, tRNA$_2^{Glu}$ has a denatured conformer in which the anticodon loop is available for complementary oligomer or tRNA binding (146, 148). Bina-Stein et al. (92) showed that the denatured form contains three helices whose properties match those of the acceptor-stem, anticodon, and TΨC-helices and proposed that the structure contains those helices along with a disrupted dihydrouridine helix and an altered tertiary structure. Eisinger and Gross (146) reached a similar conclusion.

It is evident that the proposed structures for denatured tRNA are highly speculative. It is likely that all stable denatured conformers have an altered tertiary structure, which makes improbable all models based solely on double helices. However, our knowledge of tertiary interactions in RNA now derives from one molecular structure, so it is premature to guess the higher-order structures of denatured tRNAs.

8.2.9 Transfer RNA Molecules Can Interact to Form Dimers

Depending on their sequence, tRNA molecules can interact to form a_2 or $a\beta$ dimers, with a wide variation in stability. For example the association between two tRNAs with complementary anticodons (149) is stable only below about 30°C (148), whereas the metastable dimer of tRNA^{Tyr} (150) must be heated above 50°C in Mg^{2+}-containing buffer before it dissociates to the thermodynamically more stable monomer.

The $a\beta$ complex between two tRNAs having complementary anticodons, discovered originally by Eisinger (149), has been further characterized by Grosjean and his collaborators (148). The strength of the interaction between yeast tRNA^{Phe} (anticodon G$_m$-A-A) and E. coli tRNA^{Glu} (anticodon s^{2}U-U-C) is about six orders of magnitude greater than expected for the association between two trinucleotides, based on the properties of model oligonucleotides. Since the formation rate constant ($\sim 3 \times 10^6 M^{-1}$) is approximately the same as found for the combination of two oligonucleotides (25, 26), the anomalous complex stability must be due to a much slower dissociation rate than expected for complementary trinucleotides. Grosjean et al. (148) showed that no single factor is responsible for the stabilization; contributions have been detected from the modified purine adjacent to the 3′-side of the anticodon, the closure of the single-stranded anticodon region into a hairpin loop, and the "dangling end" effect, which refers to the stabilization of double helical oligomers by the presence at the ends of noncomplementary nucleotides (151).

The model proposed by Grosjean et al. (148) stacks both tRNA anticodon loops on the two purines at the 3′-side of the anticodon loop (figure 8.14). The trinucleotide anticodon-anticodon helix is thus stacked between two almost continuous helices; the increased stacking energy of such a structure was proposed as the source of the anomalously large enthalpy of stabilization of the complex. More recent work (152) has characterized the influence of wobble pairing and mismatching on the stability of the complex diagrammed in figure 8.14.

The a_2 dimers formed from yeast tRNA^{Ala} (153), E. coli tRNA^{Tyr} (150), and other tRNAs (154, 155) are much more stable than the $a\beta$ anticodon-anticodon complex but are also much slower to form. With the exception of the yeast tRNA^{Ala} dimer (153), most such aggregates are inactive. Usually they can be renatured by heat treatment (50°–65°C) in Mg^{2+}-containing buffers.

The a_2 dimer most studied by physical techniques is that from E. coli tRNA^{Tyr} (150, 32). Yang et al. (150) showed that the rate constant for dimer formation (from a monomer whose tertiary structure has melted) is three to four orders of magnitude slower than the helix formation rate constant observed for complementary single-stranded oligonucleotides. Since the rate constant for dimer formation is much slower when the tertiary structure is intact, it is evident that dimer formation requires disruption of the native tertiary structure. Yang et al. (150) found also that the dimer has 4 percent lower absorbance and an enthalpy lower by 135 kcal per mole of dimer than two monomers that have lost tertiary structure (and possibly the dihydrouridine helix). Hence the dimer has more stacking interactions than a monomer lacking tertiary structure.

Rordorf and Kearns (32) measured NMR spectra of tRNA^{Tyr} of monomer and dimer, comparing them under conditions in which the monomer tertiary structure should be intact. They concluded that the dimer contains 12

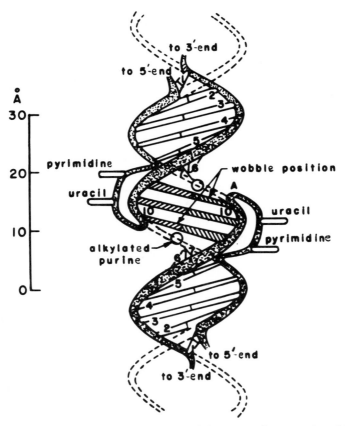

Figure 8.14 Schematic representation of the proposed structural model for the anticodon-anticodon interaction. The conformation of each anticodon loop is as proposed by Fuller and Hodgson (200) with the three bases of the anticodon stacking in a helical conformation with the two purines on their 5′-side. The two pyrimidines on the 5′-side of the loop are stacked together but separated from the other stacked bases by a short bend. The twofold rotational axis is perpendicular to the paper, through the bond joining the two middle bases of the anticodon. The three base pair double helix is sandwiched into an almost continuous double helix that includes the anticodon stems of both tRNAs. The stacking interactions generated in this structure can explain the stabilizing effect of the anticodon stems and of the modified purine residues. From Grosjean, Söll, and Crothers (148).

fewer base pairs than the sum of two native monomers. Given the demonstrated inaccuracies in determining the number of hydrogen-bonded base pairs from integration of poorly resolved NMR spectra (109), it is difficult to be confident that this conclusion is correct. In particular the method of Rordorf and Kearns may be subject to artifacts under conditions in which higher tRNA aggregates are formed, because nucleotides in the more mobile single-stranded regions of the higher aggregates could contribute to the nonexchangeable ring proton resonance used as an integration standard. However, the H-bonded proton resonances from the more immobilized double helices would be broadened and would not contribute to the low-field spectrum.

The models proposed by Yang et al. (150) and by Rordorf and Kearns (32) share the feature of dimerization through self-complementarity of the TΨC-loop in the sequence -T-Ψ-C-G-A-A-. Other tRNAs with this same sequence known to dimerize or aggregate are $tRNA_3^{Leu}$ (154), $tRNA_2^{Ser}$ (155), and $tRNA_2^{Glu}$ (92, 149). The last of these has been studied extensively by Eisinger and Gross (149), who showed that dimerization proceeds preferentially from the denatured form. Evidently the properties of both denatured and dimerized forms of tRNA depend on the sequence of each species; generalizations are thus difficult. It is unlikely that aggregated species have any functional significance. Whether the denatured conformation may exist at some stage of tRNA function is debatable.

8.3 The Functional Role of tRNA Conformational Changes

8.3.1 The Question of Whether Aminoacylation Induces a Conformational Change in tRNA Remains Unresolved

The existence of a conformational difference between aminoacyl- and nonaminoacyl-tRNA could be a basis for discrimination between the two by a protein, so a comparison of the two species free in solution is of considerable importance. The problem has been attacked by a battery of techniques—nuclease digestion (156), tritium exchange (157, 158), sedimentation (159), NMR (160, 176), ESR (161), Raman spectroscopy (162), fluorescence (163), circular dichroism (164–168), laser light scattering (22), low angle X-ray scattering (169), dye and steroid binding (170–172), Mn^{2+} binding (36), oligonucleotide binding (23, 171, 173, 174). Unfortunately the results are ambiguous and conflicting. Some approaches have indicated little or no conformational change (23, 156–158, 160, 162–167, 169, 174), while other data are consistent with a substantive structural alteration (22, 159, 161, 168, 170–173, 176), upon aminoacylation. For example it is generally agreed that oligonucleotides complementary to the TΨC- and D-loops do not binding to unacylated $tRNA^{Phe}$ (23, 172–175). Liesch and Uhlenbeck (23) have done an extensive analysis of equilibrium oligomer binding to the chemically stable phenoxyacetyl-$tRNA^{Phe}$ and found that the binding patterns are identical to those of unacylated and chemically deacyl-ated tRNA^{Phe}, except for a twofold enhancement of binding by oligomer complementary to the C-C-A terminus in the case of the peptidyl-like tRNA^{Phe}. Thus aminoacylation seems to have no effect on tRNA conforma-tion, as monitored by oligomer binding, except for a local C-C-A terminus perturbation by the N-blocked Phe moiety. Similarly Pongs et al. found no effect of aminoacylation on TΨC- and D-loop conformation using Phe tRNA^{Phe} with a CC-3'-NH$_2$A amide bond (stable against hydrolysis) to the amino acid (174). In conflict with these results Dvorak et al. (172) observed that oligonucleotides complementary to the TΨC-loop and D-loop do bind strongly to Phe-tRNA^{Phe}. The authors suggested that tertiary interactions involving these loops are destabilized or broken by aminoacylation. Dugas et al. (161) arrived at a conclusion similar to that of Dvorak et al. by study-ing the differences between tRNA^{Phe}, N-blocked Phe-tRNA^{Phe}, and chemi-cally deacylated tRNA^{Phe} using spin labels on the tRNAs as probes.

A possible source for the various disparities in these and other data is the preparation of the aminoacyl- or peptidyl-like tRNA. These species are typi-cally exposed to acid pH, ethanol precipitation, and low-salt dialysis during synthesis, conditions that can give rise to alternate metastable conforma-tional forms in the case of unacylated tRNA. An essential control seems to be missing in virtually all studies claiming to have observed a conforma-tional change upon aminoacylation: a comparative measurement must be made on a tRNA exposed to a mock aminoacylation treatment that lacks either aminoacyl ligase or amino acid. Evidence that chemical deacylation of tRNA reverts the aminoacyl or peptidyl conformation to that of unacyl-ated tRNA rules out nuclease degradation, but it does not eliminate con-formational changes for which the state of aminoacylation is irrelevant. For example, suppose that acid treatment or ethanol precipitation caused the aminoacylated tRNA sample to be trapped in a metastable conformation, not because of aminoacylation but because of changed solvent conditions. Incubation at 37°C for several hours to deacylate the tRNA could easily result in a renatured conformation. In other words it is essential to demonstrate that *only* acylation could be responsible for the altered conformational properties.

The laser light scattering experiments of Potts et al. (22) provide evidence in favor of a conformational change induced by aminoacylation. In their experiments aminoacylated tRNA^{Phe} samples initially at pH 4.5 and either 1 mM Mg^{2+} or 10 mM Mg^{2+} were allowed to deacylate at 20°C, pH 7. They found that the diffusion coefficient varied linearly with the extent of amino-acylation, suggesting a correlation between the two variables. Deacylation at 10 mM Mg^{2+} led to a 14 percent calculated increase in the translational diffusion coefficient, but there was no change in the same parameter for the sample in 1 mM Mg^{2+}. A substantial change in shape would be required to produce such a large change in diffusion coefficient. The authors suggest as one possible interpretation that aminoacylation disrupts tertiary base pairs between the D- and TΨC-loops and rearranges the L-structure to a less compact conformation. Since the authors did not report controls in which a tRNA sample was taken through a mock aminoacylation procedure, one cannot be absolutely certain that only aminoacylation could be responsible for the observations, even though the observed linear correlation between the diffusion coefficient and the degree of aminoacylation strongly suggests this interpretation.

Although early NMR studies on Phe-tRNA[Phe] (160) indicated that aminoacylation has no effect on the exchangeable proton spectrum, the recent improvements in resolution and sensitivity should make NMR of both proton and ^{19}F nuclei a powerful tool in resolving the question of a conformational difference between aminoacyl- and unacylated tRNA and in elucidating the structural details of that change if it exists. Recent prelim-inary NMR studies on Phe-tRNA[Phe] which has the chemically stable 3'-amide linkage to Phe replacing the 3'-ester linkage (176) suggest conforma-tional changes upon aminoacylation that involve yet-to-be-assigned hydro-gen bonds.

If future experiments indicate that the conformational changes are due to a real shift in the balance of tRNA conformational states mediated by the aminoacyl moiety rather than by an artifact of sample preparation, then the physical basis of the change needs to be explained. It is not obvious how the addition of an aminoacyl or peptidyl group on the C-C-A terminus could effect or stabilize structural changes as distant as the TΨC- or D-loops. Some workers (161, 169) have suggested that aminoacyl ligase mediates a conformational change in the tRNA during reaction. However, tritium exchange-out from a H^3-C8 purine-labeled Ileu-tRNA[Ileu]·Ileu ligase com-plex (177) and low-field NMR studies of tRNA[Glu]·Glu ligase complex (178) have shown no gross structural changes in tRNA conformation when it is bound to its cognate ligase. Further, this model does not suggest a molecular basis for maintaining the aminoacyl conformation nor does it mechanistically explain why *chemical* deacylation should mediate a reversion to the unacylated conformer. The possibility that the synthetase can mediate a conformational change merits further investigation, and studies with ^{19}F-labeled tRNA should be useful.

An alternate model that may account for aminoacylation effects has been presented by Reid (117). From NMR data implying the existence of two tRNA conformational states in Mg^{2+} and a consideration of the ligase·tRNA complex geometry, Reid has proposed a preexisting equilibrium between two tRNA conformations in solution: the L-shaped conformer (predominant) that corresponds to the crystal structure and a U-shaped conformer where the anticodon loop interacts with the X-C-C-A terminus (similar to the second Mg^{2+} conformation of Ford and co-workers; see ref. 118). Reid suggests that synthetase recognizes only the U conformation. Once charged, that U conformer may be trapped by additional amino acid stabilization or it may interconvert to the U-L equilibrium. Further work is needed to establish the validity of the Reid hypothesis.

In other work related to aminoacyl-tRNA conformation, a comparison of the conformation of tRNA^{Phe} and the elongation factor complex of T_u·Phe-tRNA^{Phe}·GTP showed no gross structural alteration of the tRNA as assayed by NMR (179), tritium exchange (180), and oligonucleotide binding (23). The conclusion from nuclease digestion of the Phe-tRNA^{Phe}·T_u·GTP complex that the TΨC- and D-loops are not in contact with T_u (181), in combination with the preceding results, implies that the interaction between the TΨC-loop and the D-loop is not disrupted in the T_u complex.

8.3.2 There is Evidence for a Conformational Change of Aminoacyl-tRNA and Pairing with 5S RNA at the A-Site of the Bacterial Ribosome

Crystallographic analysis (1, 3), chemical modification (182), and oligonucleotide binding (175, 183) have demonstrated that the TΨC-loop of isolated tRNA is masked and unreactive. Thus the observation that the oligonucleotide T-Ψ-C-G inhibits nonenzymatic binding (184, 185) and T_u-dependent binding to the *E. coli* ribosomal A-site (186, 170) implies indirectly that the tRNA must partially unfold to allow the TΨC-loop to interact with the ribosome.

The recent studies of Sprinzl, Erdmann, and co-workers (187) with yeast tRNA^{Phe} fragments indicate that enzymatic A-site binding also is inhibited by D-loop fragments, whereas tRNA^{Phe} fragments do not influence the enzymatic binding of initiator (or peptidyl) tRNA to the P-site of the *E.coli* ribosome. In addition a 3'-fragment with the TΨC-stem but lacking the terminal A_{OH} is a severalfold less efficient inhibitor than that same fragment containing the A_{OH}. (The existence of a C-C-A ribosomal binding site is confirmed by the observation that model aminoacyl oligonucleotides such as CpA-Phe inhibit both enzymic and nonenzymic A-site binding; ref. 189.) The interpretation that the terminal A enhancement of inhibition is an allosteric effect of the C-C-A binding region on the TΨC binding region remains a suggestion, however, until detailed binding studies are done. Most significantly the authors propose that at least the tertiary interactions G18-Ψ55, G19-C56, and T54-m′A58 are disrupted to free the D- and TΨC-loops for ribosomal interactions in the A-site, while peptidyl-tRNAs are probably refolded at the P-site (187, 190).

Whether interactions in addition to those involving the T-Ψ-C-G sequence are broken upon A-site binding remains to be determined. The functionality of tRNA photocross-linked between sU8 and C13 in protein synthesis (191) does imply, however, that the U8·A14 tertiary and G22·C13 secondary bondings remain unchanged during function. Sigler has proposed that G15-C48, G22-m^{7}G46, A9-A23, and the rest of the secondary structure may also remain intact on the ribosome (2). Transient opening of portions of the tertiary structure can be induced in isolated tRNA (12), and such a phenomenon may occur on the ribosome.

Even though tRNA fragments have not been found to inhibit P-site binding, it remains possible that some regions of the tRNA are unfolded but not bound to the ribosome. In such a case the corresponding oligonucleotides would not be inhibitory. Unfortunately a recent study of P-site binding with fluorescent techniques does not resolve this question. Robertson et al. have looked at the polarization of P-site-bound species of tRNA^{Phe} fluorescently labeled with ethidium bromide at either the anticodon or the D-loop (192). Their results indicate that in the absence of the appropriate mRNA (poly U), the D-loop and anticodon are considerably more flexible than when messenger is present. In the presence of poly U the emission polarization of both probes increases. The favored interpretation is that the anticodon is highly immobilized by base pairing with the poly U; the D-loop ethidium probe is stacked better, or the whole loop is less flexible due to direct interaction of the ribosome. Alternately a rigid tRNA flexibly attached in the absence of poly U could become immobilized to the P-site by poly U binding without a D-loop change. Either explanation is feasible. In addition it is not clear from the data that the measurements in the absence of poly U were done with enough excess ribosomes, a condition necessary to avoid an apparent depolarization effect that is simply due to an average of free and bound tRNA states.

At which stage the T-Ψ-C conformational change occurs during the sequence of molecular events termed A-site binding remains unclear. 5S ribosomal RNA has, however, been proposed as a pairing acceptor for the exposed TΨC-loop on the basis of sequence complementarity (193). It is further implicated by the work of Erdmann and his colleagues, who found that T-Ψ-C-G binds not only to the 50S subunit (186) but also to a complex of 5S RNA with 50S ribosomal proteins (6, 194). Further, chemical modification of the sequence C-G-A-A-C in 5S RNA impaired T-Ψ-C-G binding to the 5S RNA·protein complex, and reconstitution of the modified 5S RNA into 50S subunits resulted in particles with low activity (194). Thus the switching from a TΨC-loop/D-loop pairing within the tRNA to a TΨC-loop/5S RNA pairing in the A-site is a feasible portion of cyclical RNA conformational changes in ribosome function. Of course the present data do not prove such a switching, and the alternative interaction of the TΨC-loop with 23S RNA, for example, remains an untested possibility.

It has also been suggested from kethoxal modification of 30S subunits in the presence and absence of tRNA (9) that tRNA may interact with 16S rRNA. Studies on tRNA binding to 70S ribosomes reconstituted with modified 16S RNA from kethoxal-treated 30S subunits would clarify the ambiguities in this interpretation.

Extending the concept of a tRNA/5S RNA interaction, Gassen and co-workers have hypothesized that the interaction of a tRNA anticodon with its appropriate codon induces an allosteric change in the structure of aminoacyl-tRNA, thus exposing the T-Ψ-C-G sequence for interaction with the 5S RNA of the 50S subunit (195). Using equilibrium oligonucleotide binding, they showed that C-G-A-A will bind to a complex formed by poly U–programmed 30S subunits + Phe-tRNA^{Phe}·T$_u$·GTP. In addition they showed by the filter-binding method that C-G-A-A binds to Pro-tRNA^{Pro} that is enzymically bound to poly C–programmed ribosomes; disturbingly, Lys-tRNA^{Lys} bound to poly A + 30S subunits did not bind C-G-A-A. Their interpretation was that C-G-A-A binds to the TΨC-loop that has been exposed by the codon-anticodon interaction and that, in the absence of 50S subunits, is still available for oligomer binding (195, 196). In later experiments Gassen et al. find that C-G-A-A will bind to a complex of the "codon" U$_8$ + tRNA^{Phe}, to Phe-tRNA^{Phe}·T$_u$·GTP + U$_8$ as well as to Phe-tRNA^{Phe} enzymically and nonenzymatically bound to U$_8$-programmed 30S subunits (196, 197). The Mg^{2+} dependence of C-G-A-A binding to the U$_8$·tRNA^{Phe} complex in the presence and the absence of 30S subunits is apparently sigmoidal with a midpoint at 6 mM Mg^{2+}(196, 197). The authors conclude that the anticodon-codon interaction and Mg^{2+} have sole control in causing a conformational change in the TΨC-loop 60–70 Å away.

This case is weakened by a lack of proof that C-G-A-A binds to the T-Ψ-C-G sequence in tRNA. At least one alternate explanation is not eliminated by the data: C-G-A-A could complex with the remainder of the U_8 dangling from the anticodon, thus forming a helical stack of at least six base pairs. The possibility that C-G-A-A binding is monitoring a Mg^{2+}-induced change in the anticodon loop is interesting in light of (a) a putative Mg^{2+} site in the loop from crystal analysis (52) and (b) the observations of Knapp and Uhlenbeck (198) that the mere binding of U-C-A to the anticodon of $tRNA^{Phe}$ is enhanced twofold in the range of 5–8 mM Mg^{2+}. In apparent contradiction to the proposal (196, 197) that codon-anticodon interaction can mediate an allosteric conformational change in the TΨC-loop, Liesch and Uhlenbeck (23) have observed that oligomer binding to the anticodon of $tRNA^{Phe}$ does not change the binding pattern of oligonucleotides complementary to other regions of the tRNA (23).

8.3.3 Changes in the Conformation of the Anticodon Loop on the Ribosome Have Been the Subject of Much Discussion and a Recent Fluorescence Study

The central feature in Woese's mechanical model for translocation is the switching between two anticodon conformations (199). One conformation of the seven-base loop (the FH conformation, after its original proposal by Fuller and Hodgson, ref. 200) is a 3'-stacked conformation with the anticodon and two purines to its 3'-side stacked in a pseudohelical array so as to extend the stem helix on the 3'-side. In the other conformation, the hf or 5'-stacked conformation, the five bases on the 5'-side (anticodon and two pyrimidines) form a stack. The complex of mRNA with two anticodons, one in the FH and the other in the hf conformation, is illustrated in figure 8.15. Woese proposes that switching between the A- and P-sites is accomplished by flipping the two anticodons through a cycle of FH→hf and hf → FH changes (presumably mediated by peptidyl chain transfer). Such a mechanism leaves the tRNAs stationary on the ribosome (rather than physically translocated from the A-site to the P-site), and the mRNA is pulled through the ribosome by the double reciprocal rachetlike flipping of anticodon conformations. Kim (4) has proposed that tRNAs with different anticodon conformations (hf or FH) in A- and P-sites would minimize the extent of tRNA movement needed for translocation. From electron-microscopic localization of ribosomal proteins implicated in tRNA binding, Lake

(201) has hypothesized that aminoacyl-tRNA initially binds, with its anti-codon in a 5′-stack, to a recognition site (R-site) distal from A- and P-sites. With no motion of the mRNA, a substantial movement of the aminoacyl-tRNA to the A-site is supposedly caused by a conformational change of the anticodon to a 3′-stack, and a U_{33}-U_{33} pairing then occurs between the A-site anticodon and 3′-stacked P-site anticodon. As a final example, the speculation of Crick et al. (202) on the evolutionary origin of protein synthesis assumes that the anticodon loop of a primitive tRNA can take up two conformations, the hf and FH. Pairing to mRNA is via the five-base stack with the aminoacyl-tRNA in an hf state; peptide attachment causes a flip to the FH conformation.

The possible conformational states of anticodon loops in isolated tRNA and in codon-anticodon complexes have been widely studied (1, 5, 61, 70, 72, 146, 148, 152, 175, 183, 203–209). The 3′-stacked loop is observed in both crystal forms (1, 5). Fluorescence studies on the tRNA[Phe] Y-base in crystals and in solution indicate the same environment in both cases (205); NMR studies on tRNA[Phe] are also consistent with a 3′-stacked conformation (70, 72). In addition the sensitivity of the Y-base to environmental changes (203) shows an anticodon loop flexibility in solution. The suggestion of a 5′-stacked solution conformation comes from complementary oligonucleo-tide binding studies. Residues on the 3′-side of the loop appear inaccessible to complementary oligomer binding even when modified bases are not present (175, 183, 206), whereas tetramers and pentamers complementary to the anticodon and its adjacent 5′-pyrimidine bases bind more strongly than the trimeric anticodon complement (175, 183, 204, 206). This binding asymmetry has been attributed to an energetically favored 5′-stack in solution (204). An alternate explanation is that the addition of purines to the 3′-end of the trimeric anticodon complement enhances oligomer binding because of "dangling end" (151) stacking stabilization rather than additional base pair formation. This view is supported by data on dangling end stabilization (151) and by recent experiments of Pongs and Bald (209) showing that U-U-C-purine binds more strongly to tRNA[Phe] than U-U-C does, even though the purine does not hydrogen bond. Further work is needed to confirm that a 5′-stack is generated by oligomer binding in solution. In the case of two tRNAs with self-complementary anticodons, the large interaction enthalpy resultant from complex formation and the stabilizing effect of modified 3′-purines are consistent with a 3′-stacked anticodon (148).

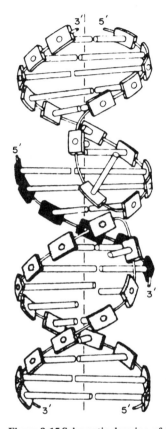

Figure 8.15 Schematic drawing of the conformation of two anticodon hairpins bound to adjacent mRNA triplets. The A-site anticodon (upper) is shown with a 5′-stack (hf) whereas the P-site anticodon loop is in the 3′-stack (FH) conformation (lower). The result is an interrupted helix containing six nucleotides from the mRNA. From Lake (201).

Recent fluorescence experiments of Fairclough and Cantor in collaboration with Zachau and Wintermeyer (210) yield further insight into the anticodon-codon conformation on the ribosome. At high Mg^{2+} and low temperature two poly U–programmed ribosomal binding sites for unacylated tRNA were studied using the normal Y-base containing tRNA^{Phe} and tRNA$^{Phe}_{pf}$, which has the highly fluorescent proflavin moiety substituted for Y37. The work of deGroot et al. indicates that under these conditions the strongest binding site for unacylated tRNA corresponds to the P-site (211); the second site studied may be the functional A-site, but confirming evidence is still needed. Singlet-singlet energy transfer measurements on poly U–programmed ribosomes with two tRNAs per ribosome gave a distance of 18 ± 4 Å between the dyes located at identical positions (3′-adjacent to the anticodon) in each of the two bound tRNAs. This distance, along with an estimate of the angle between the two proflavin residues at the P-site and the second site, constrains the two bound anticodons to a geometry close to that of an RNA-11 double helix. It is thus plausible that the anticodons of the two ribosome-bound tRNAs form a helical complex with *adjacent* U-U-U triplets of the poly U messenger. Further, the fluorescent spectral properties of the P-site bound anticodon are completely different from those of the second site anticodon, indicating that the Y-base position of the two anticodons has a very different environment in the two sites. The data are most easily rationalized by a 3′-anticodon stack (FH conformation) in the P-site and a 5′-stacked conformation (hf) in the second site, as illustrated in figure 8.15.

8.3.4 Although the Various tRNA Conformational States on the Ribosome Are Being Deciphered, the Detailed Mechanisms of tRNA Conformational Conversions in Protein Synthesis Are Still the Subject of Models and Future Studies

The first stage in elucidating the dynamics of tRNA conformational changes in protein synthesis is a complete characterization of the stable tRNA conformations that exist on the ribosome. The minimum and vastly oversimplified sequence of events in an elongation cycle of translation is diagrammed in figure 8.16. As recently discussed by Johnson et al. (212, 213) and Lake (201), there may be at least three tRNA binding conformations: R, A, and P (see figure 8.16). The question of whether A-site and R-site tRNA conformations are similar or whether they represent a change in position and anticodon conformation of the entire tRNA as suggested by Lake (201) remains to be answered. (According to Ofengand's competition experiments (189), the T Ψ C-loop apparently interacts with the ribosome in both R-sites and A-sites.)

The completion of a comprehensive description of tRNA structural states on the ribosome clearly requires substantial additional information, and a wide array of experimental approaches will be needed to achieve this goal. Transfer RNA mutants, either natural or created by well-defined chemical or enzymatic manipulations, will be helpful in establishing the various binding site requirements. The availability of functional, homogeneous tRNA species containing two fluorescent probes will allow conformational

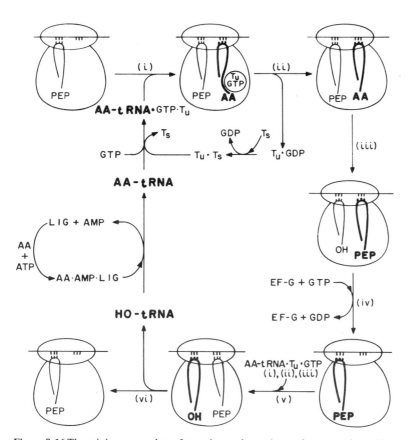

Figure 8.16 The minimum number of steps in an *elongation* cycle as experienced by an individual tRNA molecule (designated by the heavy black line) (212). Abbreviations: LIG aminoacyl-tRNA ligase; T_u, G, and T_s are elongation factors; AA-tRNA aminoacyl-tRNA; PEP-tRNA peptidyl-tRNA; HO-tRNA unacylated tRNA. The ribosome must in step i with Pep-tRNA in the P-site recognize and bind the correct AA-tRNA to its codon; the resultant localization of the AA-tRNA on the ribosome is called the recognition or R-site. Step ii positions the tRNA to accept the growing peptide chain; the resultant binding site of AA-tRNA is functionally defined as the A-site; the GTP hydrolysis and T_u release probably allow transacylation from the 2′ to 3′ OH of AA-tRNA to occur. In step iii transpeptidation occurs, allowing the formation of a peptide bond between the acceptor amino acid and the nascent protein chain; structurally all that is known is that the 3′-ends of the A-site and P-site tRNAs are localized at the peptidyl transferase center. Step iv is translocation of the newly formed peptidyl-tRNA to the P-site so that it will be able to donate the polypeptide chain to the next AA-tRNA. In step v the formation of another peptide bond is catalyzed, and step vi is the release of the deacylated tRNA. From Johnson et al. (212).

differences among the sites to be detected by singlet-singlet energy transfer. Tritium exchange-out from the C8 position of purine nucleotides in tRNA purines may be helpful in detecting microenvironmental changes, and RNA-RNA cross-linking agents that are nondestructive to the ribosome should be extremely powerful in localizing tRNA-rRNA and rRNA-rRNA interactions that may be involved in cyclic conformation changes.

Of course the ultimate aim is to establish the energetics and kinetics of conversion between various tRNA conformational states on the ribosome and the relation of those conformational alterations to the temporal sequence of functional steps in protein synthesis. This final stage is characterized by an impressive dearth of data, but two observations are extremely interesting. First, the finding that the tRNA^{Gly} (Su$^+$) frameshift suppressor has an extra nucleotide in its anticodon loop suggests that translocation of this tRNA involves movement of four rather than the usual three mRNA bases. This result implies that mRNA movement is determined by tRNA movement and not exclusively by the ribosome (214, 215). The second observation emphasizes our limited understanding of tRNA-ribosome interactions. A G24 → A24 mutation which produces an A·U pair in the D-stem of tRNA^{Trp} (*E. coli*) allows tRNA^{Trp}, which normally decodes U-G-G, to suppress the nonsense codon U-G-A (147). Intriguingly both wild-type and mutant tRNA have the same affinity for U-G-A (145). Why a D-stem alteration should affect anticodon-codon recognition is not clear.

Rich (215) has suggested that tRNAs do not undergo substantial conformational changes on the ribosome; he allows for small changes such as altered folding of the T Ψ C-loop during translocation. Recently Kurland et al. (216) have proposed that the low error frequency in protein synthesis can be achieved by having ribosomes preferentially bind tRNA molecules in conformations that depend on correct codon-anticodon interaction. The authors suggest that a correct codon-anticodon interaction alters the conformation of the anticodon loop; conformational changes in other regions of the tRNA are thereby initiated, and further tRNA-ribosome interactions are allowed. The result is amplification of the free energy difference between cognate and noncognate tRNA binding. Thus a correct codon is an allosteric effector of tRNA conformational states stabilized by the ribosome. (These hypothetical codon-induced tRNA conformations can exist spontaneously, free in solution, but with a much lower probability than on the ribosome.) Kurland et al. suggest that their model can account for the tRNA^{Trp} U-G-A suppressor observations in the following way. If the mutation destabilizes the tRNA^{Trp} tertiary structure, the mutant tRNA may undergo codon-dependent conformational rearrangement on the ribosome in response to proper matching of only two rather than three anticodon loop bases.

In summary the available evidence suggests considerable formation and breakage of RNA-RNA interactions during protein synthesis; tRNA participates in these, at least to the extent of tRNA-mRNA and tRNA–5S RNA bonding. A definite mechanistic advantage could be gained if tRNA became flexible during its function on the ribosome, as illustrated in figure 8.17. The tRNA occupying the A-site is shown in a conformation that is different from the native structure observed in crystals and allows interaction between the T-Ψ-C region and 5S RNA. Further, this altered conformation provides the molecule enough flexibility to place the aminoacyl linkage close to the peptidyl residue on the P-site tRNA, allowing the peptide chain to be transferred directly from the P-site tRNA to the A-site tRNA,

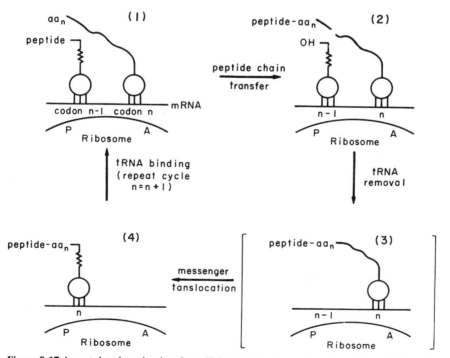

Figure 8.17 A postulated mechanism for utilizing the opening and closing of the tRNA molecule in translation.

converting state 1 to 2 (figure 8.17). As the spent tRNA is removed from the P-site, the peptidyl region of the A-site tRNA transfers to the portion of the P-site that is specific for binding the peptides and the tRNA acceptor stem. The step produces the hypothetical transient intermediate 3 which straddles the A- and P-sites. Conversion of this intermediate to state 4 is coupled to movement of the messenger RNA by three nucleotides and to restoration of the native tRNA tertiary structure. Binding of another tRNA leads to a repetition of the cycle.

The chief mechanistic advantage gained from the model illustrated in figure 8.17 is independence of movement of the two main functional parts of tRNA, the anticodon and the acceptor stem. In this way the tRNA remains bound to the ribosome through at least one of its parts. Further, the re-formation of the native tertiary structure provides a simple contractile element that serves to drive translocation.

A favorable free energy change for each step is a crucial feature for rapid operation of a cyclic set of conformational changes such as that shown in figure 8.17. Initial binding of tRNA to the A-site requires loss of some tRNA tertiary structure, whose free energy must be compensated by terms arising from interactions between tRNA and the components of the ribosome. Next, spontaneous peptide bond formation provides the free energy for converting state 1 to state 2. Removal of the spent tRNA is coupled to the free energy of GTP hydrolysis through EF-G and may also be assisted by occupancy of the P-site peptide-binding region by the hypothetical straddling intermediate 3. Finally spontaneous reformation of the tRNA tertiary structure drives the translocation step, producing state 4. This last step may also be coupled to or assisted by a flip in the anticodon conformation.

These considerations make it likely that the balance of RNA conformational energy is an important factor in protein synthesis. For example a tRNA whose tertiary structure is too stable will not undergo the tRNA binding step, which requires loss of structure. On the other hand if the tertiary structure is energetically too weak, an important factor for driving the translocation step may be lost. Leon et al. (86) showed that the A15 mutation in *E. coli* tRNA$_{Su+}^{Tyr}$ makes the tertiary structure free energy less favorable by 1–2 kcal. It is not yet known whether changes of this magnitude in the conformational energy balance can have an appreciable effect on protein synthesis. Many decades of work remain before the complex series of conformational changes occurring during ribosomal protein synthesis can be elucidated at the molecular level.

Acknowledgments

This work was supported by award of an NIH Career Development Award (GM00209) to P.E.C. and NIH research grant GM21966 to D.M.C. and GM21352 to P.E.C.

References

1. Jack, A., Ladner, J. E., and Klug A. (1976). J. Mol. Biol. *108*, 619–649.

2. Sigler, P. (1975). Ann. Rev. Biophys. Bioeng. *4*, 477.

3. Rich, A., and RajBhandary, U. L. (1976). Ann. Rev. Biochem. *45*, 805.

4. Kim, S.-H. (1978). This volume, ch. 9.

5. Quigley, G. J., Wang, A. H.-J., Seeman, N. C., Suddath, F. L., Rich, A., Sussman, J. L., and Kim, S.-H. (1975). Proc. Nat. Acad. Sci. USA *72*, 4866–4870.

6. Erdman, V. A. (1976). Prog. Nucl. Acid Res. Mol. Biol. *18*, 45–90.

7. Herr, W., and Noller, H. (1975). FEBS Letters *53*, 248.

8. van Duin, J., Kurland, C. G., Dondon, J., Grunberg-Manago, M., Branlant, C., and Ebel, J. P. (1976). FEBS Letters *62*, 111–114.

9. Woller, H. F., and Chaires, J. B. (1972). Proc. Nat. Acad. Sci. USA *69*, 3115–3121.

10. Shine, J., and Delgarno, L. (1974). Proc. Nat. Acad. Sci. USA *71*, 1342–1346.

11. Steitz, J., and Jakes, K. (1975). Proc. Nat. Acad. Sci. USA *72*, 4734–4738.

12. Rhodes, D. (1977). Eur. J. Biochem. *81*, 91–102.

13. Fresco, J. R., Adams, A., Accione, R., Henley, D., and Lindahl, T. (1966). In Cold Spring Harbor Symp. Quant. Biol. *31*, 527–537.

14. Cole, P. E., Yang, S. K., and Crothers, D. M. (1972). Biochemistry *11*, 4358.

15. Riesner, D., Maass, G., Thiebe, R., Philippsen, P., and Zachau, H. G. (1973). Eur. J. Biochem. *36*, 76–88.

16. Crothers, D. M., Cole, P. E., Hilbers, C. W., and Shulman, R. G. (1974). J. Mol. Biol. *87*, 63–88.

17. Stein, A., and Crothers, D. M. (1976). Biochemistry *15*, 160–168.

18. Römer, R., and Hach, R. (1975). Eur. J. Biochem. *55*, 271–284.

19. Crothers, D. M. (1975). In Functional Linkage in Biomolecular Systems, F. O. Schmitt, D. M. Schneider, and D. M. Crothers, eds. (New York: Raven Press), pp. 24–32.

20. Hinz, H.-J., Filimonov, V. V., and Privalov, P. (1976). Eur. J. Biochem. *72*, 79–86.

21. Bina-Stein, M., and Crothers, D. M. (1974). Biochemistry *13*, 2771–2775.

22. Potts, R., Fournier, M. J., and Ford, N. C., Jr. (1977). Nature *268*, 563–564.

23. Liesch, J. (1977). Ph.D. dissertation, U. of Illinois.

24. Bloomfield, V., Crothers, D. M., and Tinoco, I., Jr. (1974). Physical Chemistry of Nucleic Acids (New York: Harper and Row), chs. 6 and 7.

25. Pörschke, D., and Eigen, M. (1971). J. Mol. Biol. *62*, 361–381.

26. Craig, M. E., Crothers, D. M., and Doty, P. (1971). J. Mol. Biol. *62*, 383–401.

27. Reid, B. R., and Hurd, R. E. (1977). Accts. Chem. Res. (in press).

28. Stein, A., and Crothers, D. M. (1976). Biochemistry *15*, 157–160.

29. Schreier, A. A., and Schimmel, P. R. (1974). J. Mol. Biol. *86*, 601–620.

30. Record, M. T., Jr. (1967). Biopolymers *5*, 975.

31. Crothers, D. M. (1971). Biopolymers *10*, 2147.

32. Rordorf, B. F., and Kearns, D. R. (1976). Biochemistry *15*, 3320–3330.

33. Steiner, R. F., and Beers, R. F. (1961). Polynucleotides (New York: Elsevier).

34. Millar, D. B., and Steiner, R. F. (1966). Biochemistry *5*, 2289–2301.

35. Vournakis, J. N., and Scheraga, H. A. (1966). Biochemistry *5*, 2997–3005.

36. Cohn, M., Danchin, A., and Grunberg-Manago, M. (1969). J. Mol. Biol. *39*, 199–217.

37. Reeves, R. H., Cantor, C. R., and Chambers, R. W. (1970). Biochemistry *9*, 3993–4001.

38. Willick, G. E., and Kay, C. M. (1970). Biochemistry *10*, 2216–2222.

39. Danchin, A., and Guéron, M. (1970). Eur. J. Biochem. *16*, 532–536.

40. Sander, C., and Ts'o, P. O. P. (1971). J. Mol. Biol. *55*, 1–21.

41. Danchin, A. (1972). Biopolymers *11*, 1317–1333.

42. Rialdi, G., Levy, J., and Biltonen, R. (1972). Biochemistry *11*, 2472–2479.

43. Willick, G., Oikawa, K., and Kay, C. M. (1973). Biochemistry 12, 899–904.

44. Krakauer, H. (1974). Biochemistry *13*, 2579–2589.

45. Wolfson, J. M., and Kearns, D. R. (1974). J. Am. Chem. Soc. *96*, 3653–3654.

46. Jones, C. R., and Kearns, D. R. (1974). Proc. Nat. Acad. Sci. USA *71*, 4237–4240.

47. Kayne, M. S., and Cohn, M. (1974). Biochemistry *13*, 4159–4165.

48. Lynch, D. C., and Schimmel, P. R. (1974). Biochemistry *13*, 1841–1852.

49. Bina-Stein, M., and Stein, A. (1976). Biochemistry *15*, 3912–3917.

50. Rordorf, B. F., and Kearns, D. R. (1976). Biopolymers *15*, 1491–1504.

51. Jack, A., Ladner, J. E., Rhodes, D., Brown, R. S., and Klug, A. (1977). J. Mol. Biol. *111*, 315–328.

52. Holbrook, S. R., Sussman, J. L., Warrant, R. W., Church, G. M., and Kim, S. H. (1977). Nucl. Acid Res. (in press).

53. Leroy, J.-L., Guéron, M., Thomas, G., and Favre, A. (1977). Eur. J. Biochem. *74*, 567–574.

54. Scatchard, G. (1949). Ann. N.Y. Acad. Sci. *51*, 660.

55. Leroy, J.-L., and Guéron, M. (1977). Biopolymers *16*, 2429–2446.

56. Crothers, D. M. (1968). Biopolymers *6*, 575–584.

57. Bina-Stein, M., and Crothers, D. M. (1975). Biochemistry *14*, 4185–4191.

58. Lynch, D. C., and Schimmel, P. R. (1974). Biochemistry *13*, 1852–1861.

59. Steinmetz-Kayne, M., Benigno, R., and Kallenbach, N. R. (1977). Biochemistry *16*, 2064–2073.

60. Ladner, J. E., and Schweizer, M. P. (1974). Nucl. Acid Res. *1*, 183–192.

61. Robinson, B., and Zimmerman, T. P. (1971). J. Biol. Chem. *246*, 110–117.

62. Wintermeyer, W., Thiebe, R., Zachau, H. G., Riesner, D., Römer, R., and Maass, G. (1969). FEBS Letters *5*, 23–27.

63. Römer, R., Riesner, D., Maass, G., Wintermeyer, W., Thiebe, R., and Zachau, H. G. (1969). FEBS Letters *5*, 15–19.

64. Thiebe, R., and Zachau, H. G. (1970). Biochim. Biophys. Acta *217*, 294–304.

65. Römer, R., Riesner, D., and Maass, G. (1970). FEBS Letters *10*, 352–357.

66. Riesner, D., and Römer, R. (1973). In Physico-chemical Properties of Nucleic Acids, vol. 2, J. Duchesne, ed. (New York: Academic Press), p. 237–320.

67. Urbanke, C., Römer, R., and Maass, G. (1973). *33*, 511–516.

68. Urbanke, C., Römer, R., and Maass, G. (1975). Eur. J. Biochem. *55*, 439–444.

69. Coutts, S. M., Riesner, D., Römer, R., Rabl, C. R., and Maass, G. (1975). Biophys. Chem. *3*, 275–289.

70. Kan, L. S., Ts'o, P. O. P., von der Haar, F., Sprinzl, M., and Cramer, F. (1975). Biochemistry *14*, 3278–3291.

71. Kan, L. S., Ts'o, P. O. P., von der Haar, F., Sprinzl, M., and Cramer, F. (1974). Biochem. Biophys. Res. Commun. *59*, 22–29.

72. Kan, L. S., Ts'o, P. O. P., Sprinzl, M., von der Haar, F., and Cramer, F. (1977). Biochemistry *16*, 3143–3154.

73. Beltchev, B., Yaneva, M., and Staynov, D. (1976). Eur. J. Biochem. *64*, 507–510.

74. Hilbers, C. W., Robillard, G. T., Shulman, R. G., Blake, R. D., Webb, P. K., Fresco, R., and Riesner, D. (1976). Biochemistry *15*, 1874–1882.

75. Coutts, S. M., Gangloff, J., and Dirheimer, G. (1974). Biochemistry *13*, 3938–3948.

76. Robillard, G. T., Hilbers, C. W., Reid, B. R., Gangloff, J., Dirheimer, G., and Shulman, R. G. (1976). Biochemistry *15*, 1883-1888.

77. Riesner, D., Römer, R., and Maass, G. (1970). Eur. J. Biochem. *15*, 85–91.

78. Riesner, D., Römer, R., and Maass, G. (1969). Biophys. Biochem. Res. Commun. *35*, 369–376.

79. Römer, R., Riesner, D., Coutts, S. M., and Maass, G. (1970). Eur. J. Biochem. *15*, 77–84.

80. Pilz, I., Malnig, F., Kratky, O., and von der Haar, F. (1977). Eur. J. Biochem. *75*, 35–41.

81. Cole, P. E., and Crothers, D. M. (1972). Biochemistry *11*, 4368–4374.

82. Goldstein, R. N., Stefanovic, S., and Kallenbach, N. R. (1972). J. Mol. Biol. *69*, 217–236.

83. Kyogoku, Y., Inubushi, T., Morishima, I., Watanabe, K., Oshima, T., and Nishimura, S. (1977). Nuc. Acid Res. *4*, 585–593.

84. Wong, K. L., Wong, Y. P., and Kearns, D. R. (1975). Biopolymers *14*, 749–762.

85. Yang, S. K., and Crothers, D. M. (1972). Biochemistry *11*, 4375–4381.

86. Leon, V., Altman, S., and Crothers, D. M. (1977). J. Mol. Biol. *113*, 253–265.

87. Dourlent, M., Yaniv, M., and Helene, C. (1971). Eur. J. Biochem. *19*, 108–114.

88. Kearns, D. R., Wong, K. L., and Wong, Y. P. (1973). Proc. Nat. Acad. Sci. USA *70*, 3843.

89. Hilbers, C. W., Shulman, R. G., and Kim, S. H. (1973). Biochem. Biophys. Res. Commun. *55*, 953–960.

90. Hilbers, C. W., and Shulman, R. G. (1974). Proc. Nat. Acad. Sci. USA *71*, 3239–3242.

91. Römer, R., and Varadi, V. (1977). Proc. Nat. Acad. Sci. USA *74*, 1561–1564.

92. Bina-Stein, M., Crothers, D. M., Hilbers, C. W., and Shulman, R. G. (1976). Proc. Nat. Acad. Sci. USA *73*, 2216–2220.

93. Kearns, D. R., Wong, Y. P., Chang, S. H., and Hawkins, E. (1974). Biochemistry *13*, 4736–4746.

94. Kastrup, R. V., and Schmidt, P. G. (1975). Biochemistry *14*, 3612–3618.

95. Wilderauer, D., Gross, H. J., and Riesner, D. (1974). Nucl. Acid Res. *1*, 1165–1182.

96. Gralla, J., and Crothers, D. M. (1973). J. Mol. Biol. *73*, 497–511.

97. Borer, P. N., Dengler, B., Tinoco, I., Jr., and Uhlenbeck, O. C. (1974). J. Mol. Biol. *86*, 843–853.

98. Tinoco, I., Jr., Borer, P. N., Dengler, B., Levine, M. D., Uhlenbeck, O. C., Crothers, D. M., and Gralla, J. (1973). Nature New Biol. *246*, 40–41.

99. Crothers, D. M. (1971). In Procedures in Nucleic Acid Research, vol. 2, G. L. Cantoni and D. R. Davies, eds. (New York: Harper and Row), pp. 369–388.

100. Eigen, M., and De Maeyer, L. (1974). In Techniques of Chemistry, 3rd ed., vol. 6, part 2, G. G. Hammes, ed., (New York: John Wiley), pp. 63–145.

101. Crothers, D. M., Hilbers, C. W., and Shulman, R. G. (1973). Proc. Nat. Acad. Sci. USA *70*, 2899–2901.

102. Teitelbaum, H., and Englander, W. (1975). J. Mol. Biol. *92*, 55–78.

103. Shalitin, N., and Feitelson, J. (1976). Biochemistry *15*, 2092–2097.

104. Levy, J., Rialdi, G., and Biltonen, R. (1972). Biochemistry *11*, 4138–4144.

105. Bode, D., Schernau, U., and Ackermann, Th. (1974). Biophys. Chem. *1*, 214–221.

106. Brandts, J. F., Jackson, W. M., and Ting, T. Yao-Chung (1974). Biochemistry *13*, 3595–3600.

107. Filimonov, V. V., Privalov, P. L., Hinz, H. J., von der Haar, F., and Cramer, F. (1976). Eur. J. Biochem. *70*, 25–31.

108. Privalov, P. L. To be published.

109. Reid, B. R., Ribeiro, N. S., McCollum, L., Abbate, J., and Hurd, R. E. (1977). Biochemistry *16*, 2086–2094.

110. Hurd, R. E., Robillard, G. T., and Reid, B. R. (1977). Biochemistry *16*, 2095–2100.

111. Kearns, D. (1976). Prog. Nucl. Acid Res. Mol. Biol. *18*, 91–149.

112. Kearns, D. R., and Shulman, R. G. (1974). Accts. Chem. Res. *7*, 33–39.

113. Gueron, M., and Shulman, R. G. (1975). Proc. Nat. Acad. Sci. USA *72*, 3482–3485.

114. Weiner, L. M., Backer, J. M., and Rezvukhin, A. I. (1974). FEBS Letters *41*, 40–42.

115. Hayashi, F., Akasaka, K., and Hatano, H. (1977). Biopolymers *16*, 655–667.

116. Horowitz, J., Ofengand, J., Daniel, W. E., and Cohn, M. (1977). J. Biol. Chem. *252*, 4418–4420.

117. Reid, B. R. (1977). In Nucleic Acid-Protein Recognition, H. J. Vogel, ed. (New York: Academic Press), pp. 375–390.

118. Olson, T., Fournier, M. J., Langley, K. H., and Ford, N. C., Jr. (1976). J. Mol. Biol. *102*, 193–203.

119. Caron, M., and Dugas, H. (1976). Nucl. Acids Res. *3*, 35–47.

120. Schofield, P., Hoffman, B. H., and Rich, A. (1970). Biochemistry *9*, 2525–2533.

121. Sprinzl, M., Krämer, E., and Stehlik, D. (1974). Eur. J. Biochem. *49*, 595–605.

122. Rigler, R., and Ehrenberg, M. (1976). Quart. Rev. Biophys. *9*, 1–19.

123. Sueoka, N., Kano-Sueoka, T., and Gartland, W. T. (1966). Cold Spring Harbor Symp. Quant. Biol. *31*, 571.

124. Lindahl, T., Adams, A., and Fresco, J. R. (1966). Proc. Nat. Acad. Sci. USA *55*, 941–948.

125. Gartland, W., and Sueoka, N. (1966). Proc. Nat. Acad. Sci. USA *55*, 948–956.

126. Muench, K. H. (1966). Cold Spring Harbor Symp. Quant. Biol. *31*, 539–542.

127. Adams, A., Lindahl, T., and Fresco, J. (1967). Proc. Nat. Acad. Sci. USA *57*, 1684–1691.

128. Lindahl, T., Adams, A., and Fresco, J. R. (1967). J. Biol. Chem. *242*, 3129–3134.

129. Lindahl, T., et al. (1967). Proc. Nat. Acad. Sci. USA *57*, 178–186.

130. Adams, A., Lindahl, T., and Fresco, J. R. (1967). Proc. Nat. Acad. Sci. USA *57*, 1684–1691.

131. Ishida, T., and Sueoka, N. (1967). Proc. Nat. Acad. Sci. USA *58*, 1080–1087.

132. Ishida, T., and Sueoka, N. (1968). J. Biol. Chem. *243*, 5329–5336.

133. Ishida, T., and Sueoka, N. (1968). J. Mol. Biol. *37*, 313–316.

134. Gartland, W. J., et al. (1969). J. Mol. Biol. *44*, 403–413.

135. Ishida, T., and Sueoka, N. (1970). Biochim. Biophys. Acta *217*, 209–211.

136. Ishida, T., Snyder, D., and Sueoka, N. (1971). J. Biol. Chem. *246*, 5965–5969.

137. Webb, P., and Fresco, J. R. (1973). J. Mol. Biol. *74*, 387–402.

138. Hawkins, E. R., and Chang, S. H. (1974). Nucl. Acids Res. *1*, 1531–1538.

139. Kearns, D. R., Wong, Y. P., Hawkins, E., and Chang, S. H. (1974). Nature *247*, 541–543.

140. Uhlenbeck, O. C., Chirikjian, J. G., and Fresco, J. R. (1974). J. Mol. Biol. *89*, 495–504.

141. Rordorf, B. F., Kearns, D. R., Hawkins, E., and Chang, S. H. (1976). Biopolymers *15*, 325–366.

142. Hawkins, E. R., Chang, S. H., and Mattice, W. L. (1977). Biopolymers *16*, 1557–1566.

143. Greenspan, C. M., and Litt, M. (1974). FEBS Letters *41*, 297–301.

144. Jones, C. R., Kearns, D. R., and Muench, K. H. (1976). J. Mol. Biol. *103*, 747–764.

145. Buckingham, R. H. (1976). Nucl. Acids Res. *3*, 965–975.

146. Eisinger, J., and Gross, N. (1973). Biochemistry *14*, 4031–4041.

147. Hirsh, D. (1971). J. Mol. Biol. *58*, 439–458.

148. Grosjean, H., Söll, D. G., and Crothers, D. M. (1976). J. Mol. Biol. *103*, 499–519.

149. Eisinger, J. (1971). Biochem. Biophys. Res. Commun. *43*, 854–861.

150. Yang, S. K., Söll, D. G., and Crothers, D. M. (1972). Biochemistry *11*, 2311–2320.

151. Martin, F. H., Uhlenbeck, O. C. and Doty, P. (1971). J. Mol. Biol. *57*, 201–215.

152. Grosjean, H., de Henau, S., and Crothers, D. M. (1978). Proc. Nat. Acad. Sci. USA (in press).

153. Loehr, J. S., and Keller, E. B. (1968). Proc. Nat. Acad. Sci. USA *61*, 1115–1122.

154. Kowalski, S., and Fresco, J. R. (1971). Science *172*, 384–385.

155. Adams, A., and Zachau, H. G. (1968). Eur. J. Biochem. *5*, 556–558.

156. Hänggi, U. J., and Zachau, H. G. (1971). Eur. J. Biochem. *18*, 496–502.

157. Gantt, R. R., Englander, S. W., and Simpson, M. V. (1969). Biochemistry *8*, 475–482.

158. Englander, J. J., Kallenbach, N. R., and Englander, S. W. (1972). J. Mol. Biol. *63*, 153–169.

159. Chatterjee, S. K., and Kaji, H. (1970). Biochim. Biophys. Acta *224*, 88–98.

160. Wong, Y. P., Reid, B. R., and Kearns, D. R. (1973). Proc. Nat. Acad. Sci. USA *70*, 2193–2195.

161. Caron, M., Brisson, N., and Dugas, H. (1976). J. Biol. Chem. *251*, 1529–1530.

162. Thomas, G. J., Jr., Chen, M. C., Lord, R. C., Kotsiopoulos, P. S., Tritton, T. R., and Mohr, S. C. (1973). Biochem. Biophys. Res. Commun. *54*, 570–577.

163. Beres, L., and Lucas-Leonard, J. (1973). Biochemistry *12*, 3998–4002.

164. Hashizume, H., and Imahori, K. (1967). J. Biochem. *61*, 738–749.

165. Bernardi, A., and Cantoni, G. L. (1969). J. Biol. Chem. *244*, 1468–1476.

166. Adler, A. I., and Fasman, G. D. (1970). Biochim. Biophys. Acta *204*, 183–190.

167. Wickstrom, E. (1971). Biochem. Biophys. Res. Commun. *43*, 976–983.

168. Watanabe, K., and Imahori, K. (1971). Biochem. Biophys. Res. Commun. *45*, 488–494.

169. Ninio, J., Luzzati, V., and Yaniv, M. (1972). J. Mol. Biol. *71*, 217–229.

170. Tritton, T. R., and Mohr, S. C. (1973). Biochemistry *12*, 905–914.

171. Chin, R. C., and Kidson, C. (1971). Proc. Nat. Acad. Sci. USA *68*, 2448–2452.

172. Dvorak, D. J., Kidson, C., and Chin, R. C. (1976). J. Biol. Chem. *251*, 6730–6734.

173. Danchin, A., and Grunberg-Manago, M. (1970). FEBS Letters *9*, 327–330.

174. Pongs, O., Wrede, P., Erdmann, V. A., and Sprinzl, M. (1976). Biochem. Biophys. Res. Commun. *71*, 1025–1033.

175. Pongs, O., Bald, R., and Reinwald, E. (1973). Eur. J. Biochem. *32*, 117–125.

176. Kan, L. S., Ts'o, P. O. P., Sprinzl, M., van der Haar, F., and Cramer, F. (1976). Biophys. J. *16*, 11a.

177. Schoemaker, H. P., and Schimmel, P. R. (1976). J. Biol. Chem. *251*, 6823–6830.

178. Shulman, R. G., Hilbers, C. W., Söll, D., and Yang, S. K. (1974). J. Mol. Biol. *90*, 609–611.

179. Shulman, R. G., Hilbers, C. W., and Miller, D. L. (1974). J. Mol. Biol. *90*, 601–608.

180. Jekowsky, E. J. (1976). Ph.D. dissertation, Massachusetts Institute of Technology.

181. Jekowsky, E., Miller, D. L., and Schimmel, P. R. J. Mol. Biol. (in press).

182. Rhodes, D. (1975). J. Mol. Biol. *94*, 449–460.

183. Uhlenbeck, (1972). J. Mol. Biol. *65*, 25–41.

184. Ofengand, J., and Henes, C. J. (1969). J. Biol. Chem. *244*, 6241–6253.

185. Shimizu, N., Hayashi, H., and Miura, K. (1970). J. Biochem. *67*, 373.

186. Richter, D., Erdmann, V. A., and Sprinzl, M. (1973). Nature New Biol. *246*, 132–135.

187. Spinzl, M., Wagner, T., Lorenz, S., and Erdmann, V. A. (1976). Biochemistry *15*, 3031–3039.

188. Grummt, F., Grummt, I., Gross, H. J., Sprinzl, M., Richter, D., and Erdmann, V. A. (1974). FEBS Letters *42*, 15–17.

189. Ringer, D., Chladek, S., and Ofengard, J. (1976). Biochemistry *15*, 2759–2765.

190. Erdmann, V. A., Lorenz, S., Sprinzl, M., and Wagner, R. T. (1976). In Alfred Benzon Symposium *9*, Control of Ribosome Synthesis, N. C. Kjeldgaard and O. Maaloe, eds., (New York: Academic Press), pp. 427–436.

191. Yaniv, M., Chestier, A., Gros, F., and Favre, A. (1971). J. Mol. Biol. *58*, 381–388.

192. Robertson, J. M., Kahan, M., Wintermeyer, W., and Zachau, H. G. (1977).

Eur. J. Biochem. *72*, 117–125.

193. Forget, B. G., and Weissman, S. M. (1967). Science *158*, 1695–1699.

194. Erdmann, V. A., Sprinzl, M., and Pongs, O. (1973). Biochem. Biophys. Res. Commun. *54*, 942–948.

195. Schwarz, V., Lührmann, R., and Gassen, H. G. (1974). Biochem. Biophys. Res. Commun. *56*, 807–814.

196. Schwarz, V., Menzel, H. M., and Gassen, H. G. (1976). Biochemistry *15*, 2484–2490.

197. Schwarz, V., and Gassen, H. G. (1977). FEBS Letters *78*, 267–270.

198. Knapp, G. (1977). Ph.D. dissertation, U. of Illinois.

199. Woese, C. (1970). Nature *226*, 817–820.

200. Fuller, W., and Hodgson, A. (1967). Nature *215*, 817–821.

201. Lake, J. A. (1977). Proc. Nat. Acad. Sci. USA *74*, 1903–1907.

202. Crick, F. H. C., Brenner, S., Klug, A., and Pieczinik, G. (1976). Origins of Life *7*, 389–397.

203. Beardsley, K., Tao, T., and Cantor, C. R. (1970). Biochemistry *9*, 3524–3532.

204. Eisinger, J., and Spahr, P. (1973). J. Mol. Biol. *73*, 131–137.

205. Langlois, R., Kim, S.-H., and Cantor, C. R. (1975). Biochemistry *14*, 2554–2558.

206. Freier, S. M., and Tinoco, I., Jr. (1975). Biochemistry *14*, 3310–3314.

207. Yoon, K., Turner, D. H., and Tinoco, I., Jr. (1975). J. Mol. Biol. *99*, 507–518.

208. Yoon, K., Turner, D. H., Tinoco, I., Jr., von der Haar, F., and Cramer, F. (1976). Nucl. Acid Res. *3*, 2233–2241.

209. Bald, R., and Pongs, O. Unpublished data.

210. Fairclough, R. H. (1977). Ph. D. dissertation, Columbia University, New York.

211. de Groot, N., Panet, A., and Lapidot, Y. (1971). Eur. J. Biochem. *23*, 523.

212. Johnson, A. E., Fairclough, R. H., and Cantor, C. R. (1977). Nucleic Acid-Protein Recognition, H. J. Vogel, ed. (New York: Academic Press), pp. 469–490.

213. Johnson, A. E., and Cantor, C. R. To be published.

214. Riddle, D. L., and Carbon, J. (1973). Nature New Biol. *242*, 230–234.

215. Rich, A. (1974). In Ribosomes, M. Nomura, A. Tissières, and P. Lengyel, eds. (Cold Spring Harbor, New York: Cold Spring Harbor Laboratory), pp. 871–884.

216. Kurland, C. G., Rigler, R., Ehrenberg, M., and Blomberg, C. (1975). Proc. Nat. Acad. Sci. USA *72*, 4248–4251.

9 CRYSTAL STRUCTURE OF YEAST tRNA^{Phe}: ITS CORRELATION TO THE SOLUTION STRUCTURE AND FUNCTIONAL IMPLICATIONS

S.-H. Kim

9.1 Introduction

Among the various aspects of tRNA research in the last four years, one of the most satisfying advances has been made in the structural studies of tRNA by the X-ray crystallographic method. Since the revelation of the three-dimensional folding of the polynucleotide backbone of yeast tRNA^{Phe}, many more details are now known for the same tRNA in two different crystal forms, and they have been reviewed at various stages of progress (1–3). The crystallographic refinements of the structure of this tRNA are practically completed, and almost all the structural features of biological interest are already known.

This chapter attempts first to summarize the essential features of the crystal structure (the three-dimensional structure of a molecule in the crystalline state) of yeast tRNA^{Phe} and to point out the features that can be considered reliably determined from the crystallographic point of view; second, to correlate the crystal structure with the solution structure (the three-dimensional structure of a molecule in a solution state) of the tRNA implied by the vast amount of physical, chemical, and enzymatic data and to assess, although indirectly, how representative the crystal structure is; third, to extract the functional implications of the three-dimensional structural features of the tRNA.

To avoid lengthy and often confusing descriptions of structural features, many figures have been used throughout this chapter.

9.2 Crystal Structure of Yeast tRNA^{Phe}

The crystallographic studies of tRNA started in 1968, when many laboratories succeeded in obtaining large single crystals (4–7) as well as microcrystals (8) from various purified tRNA species and mixtures. Since then three important technical developments have led to the determination of the backbone structure of yeast tRNA^{Phe}: (a) the application of the vapor-diffusion method to quickly screen a large number of tRNA crystallization conditions with a minimum amount of material (6, 9); (b) the discovery of the crystallization conditions under which yeast tRNA^{Phe} forms crystals that give high resolution X-ray diffraction data (10–12); and (c) the discovery of the first heavy-atom compound, osmium trioxide·(pyridine)$_2$, that gave a reasonably good isomorphous replacement derivative (13). These developments were instrumental in obtaining the first glimpse of the three-dimensional structure of yeast tRNA^{Phe} in 1973, the folding of the polynucleotide backbone at 4 Å resolution (14).

There is a general tendency to freely accept X-ray crystallographic results of macromolecular structures despite the cautious remarks investigators often make. Such acceptance is usually safe for gross structural features such as backbone folding, and secondary structures, but not for features such as isolated hydrogen bonds, small differences in conformation, and light metal ions and water positions. These shortcomings can be partially overcome if there exist other compelling experimental results supporting such detailed features or if one can compare several structures of the same molecule determined and refined independently by different groups. With yeast tRNA^{Phe} both conditions apply.

9.2.1 Overall Structural Features

The backbone structure of yeast tRNA^{Phe} as determined by X-ray crystallographic studies at 4 Å resolution (14) revealed all the overall structural features of the molecule, which have subsequently been confirmed by higher resolution studies of the tRNA in the orthorhombic (15) as well as the monoclinic crystal forms (16, 17). They can be summarized as follows (see figures 9.1 and 9.2):

1. The molecule has an overall shape of the letter L.
2. The base-paired double-helical stems implied by the cloverleaf secondary structure are maintained in the three-dimensional structure, and they are all right-handed, antiparallel double helices.

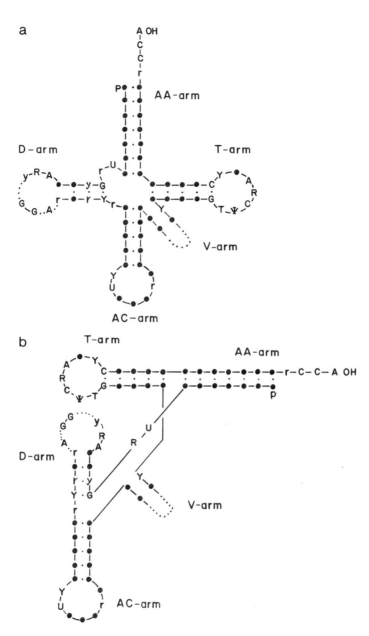

Figure 9.1 Generalized representation of tRNA in the cloverleaf (a) and the L arrangements (b). The bases common in all tRNAs participating in peptide elongation in ribosomes are indicated. Besides standard symbols, R and Y represent a purine and a pyrimidine, respectively, in all tRNAs, and r and y represent the same in most tRNAs. The dotted lines are three regions (α, β, and γ) in the polynucleotide chain where the number of nucleotides varies from one tRNA to another.

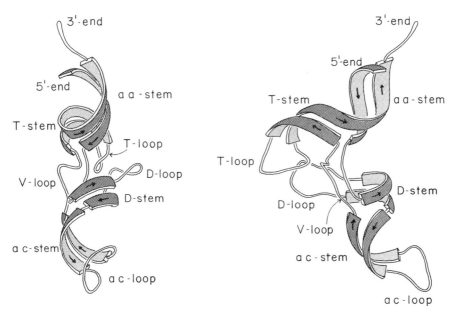

Figure 9.2 Sketch showing the flow of the polynucleotide backbone of yeast tRNA[Phe]. The tubes represent single-stranded regions and the "ribbons" represent the double helical stems with the arrows indicating the polarity of the backbone in the 5'-end → 3'-end direction.

3. The amino acid acceptor (aa) stem and T-stem form one continuous double helix with a gap, and the D and the anticodon (ac) stems form the other approximately continuous double helix with two gaps. These two continuous double helices form an L.

4. The 3'-end of the molecule is at one end and the anticodon at the other end of the L; the D- and T-loops form the corner of the L.

5. The molecule is flat with a thickness of about 20 Å (which corresponds to the width of a double helical RNA A-form), and the two axes of the L are about 70 Å long each.

A pair of generalized representations of tRNA are given in figure 9.1, and an artist's drawing of the backbone structure is shown in figure 9.2. Although the structural details are described in later sections, a set of photographs of a space-filling model is shown in figure 9.3 to convey the surface and volume features of the crystal structure of yeast tRNA[Phe].

9.2.2 Secondary Hydrogen Bonds

The term *secondary* hydrogen bonds (H-bonds) is used for the hydrogen

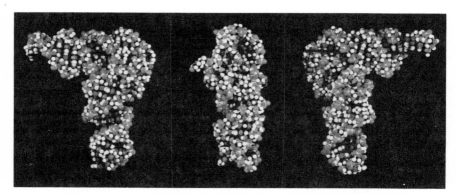

Figure 9.3 Three views of a space-filling model of yeast tRNA^{Phe} structure as determined by the X-ray crystallographic method. The model is successively rotated with 90° increment.

bonds in the complementary base pairs as predicted by the cloverleaf representation (18), the secondary structure.

All the secondary H-bonds implicit in the cloverleaf representation of this tRNA (see figure 9.4) have been observed in the three-dimensional structures from both the orthorhombic (15) and monoclinic crystalline forms (16, 17): seven base pairs in the aa-stem, four base pairs in the D-stem, five base pairs each in the ac-stem and the T-stem. All the secondary base pairs are of the Watson-Crick type as shown in figure 9.5a, b except a G·U pair in the aa-stem, which is of a "wobble" type as shown in figure 9.5c.

Due to the stereochemical constraints required for forming two or three H-bonds simultaneously in an approximately coplanar surface, the Watson-Crick base pairs in a double helical region are easy to interpret in the electron density map. Therefore the observations of this type are more reliable and less subjective from a technical point of view.

9.2.3 Tertiary Hydrogen Bonds between Bases

The term *tertiary* H-bonds is used to describe all H-bonds other than the secondary H-bonds. One of the most striking features of the secondary structure of tRNAs is the presence of *conserved* and *semiconserved* bases. Those in class 1 (D_4V_5) tRNAs, which have four base pairs in the D-stem and five nucleotides in the V-loop (for tRNA classification see refs. 20, 21) are indicated in figure 9.4. The X-ray crystallographic study of this tRNA revealed that most but not all of the conserved and semiconserved bases form H-bonds among themselves to establish a specific architectural framework of the tRNA structure (15, 16). These and a few additional tertiary

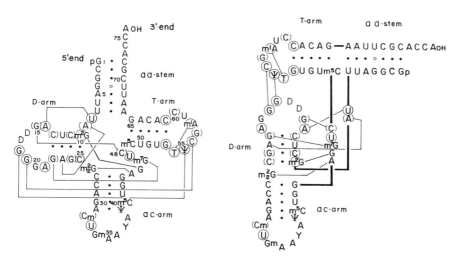

Figure 9.4 Nucleotide sequence of yeast tRNA^{Phe} (19) in the cloverleaf (left, 18) and L arrangements (right, 14). The bases that are conserved (circles) or semiconserved (parentheses) in D$_4$V$_5$ class (20), class 1 (21), tRNAs are indicated. The bases that form base-base tertiary H-bonds are connected by thin lines. Notice that most of the conserved and semiconserved bases are localized to the middle of the L in the right figure and are involved in forming the tertiary H-bonds.

H-bonds subsequently found (22, 23, 17, 24, 24a) are described in sequence starting from the 5′-end of the molecule.

1. U8 forms a base pair with A14 by a "reverse Hoogstein" scheme (figure 9.5d).

2. A9 forms a base pair with A23, which in turn is base-paired to U12 (figure 9.5e$_1$).

3. G10 forms a H-bond with G45 (figure 9.5k).

4. G15 and C48 make a base pair of a reverse Watson-Crick type (figure 9.5f$_2$).

5. N-2 (nitrogen at position 2) and/or N-1 of G18 makes one or two H-bonds to O-6 of Ψ55 (figure 9.5g).

6. G19 forms a bent Watson-Crick base pair with C56.

7. G22, which is base-paired to C13, makes two H-bonds with G46 (figure 9.5h$_2$).

8. G26 makes two H-bonds to A44 (figure 9.5j), where these two bases are not coplanar to each other, and the two methyl groups in G26 are not coplanar with the base of G26.

9. T54 makes a base pair with A58 (figure 9.5i).

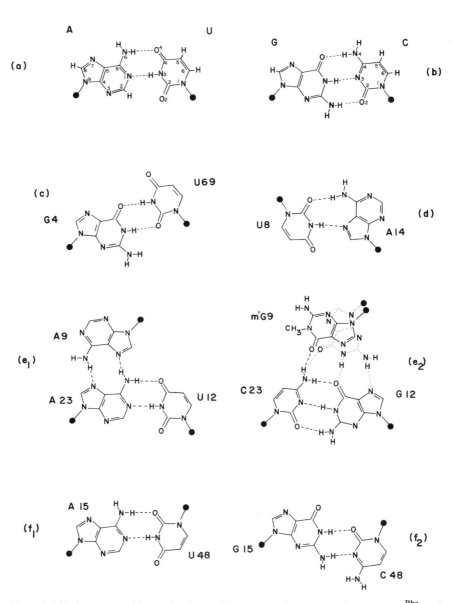

Figure 9.5 Various types of basepairs observed in the crystal structure of yeast tRNA^{Phe} and postulated for others in the D_4V_5 class. The filled black circles indicate the C-1' position of the riboses. In e_2 an alternative position for guanine is shown in dotted lines. The numbering of nucleotides is such that all the nucleotides in the stems have the same numbering as that of yeast tRNA^{Phe} regardless of the size of the D-loop.

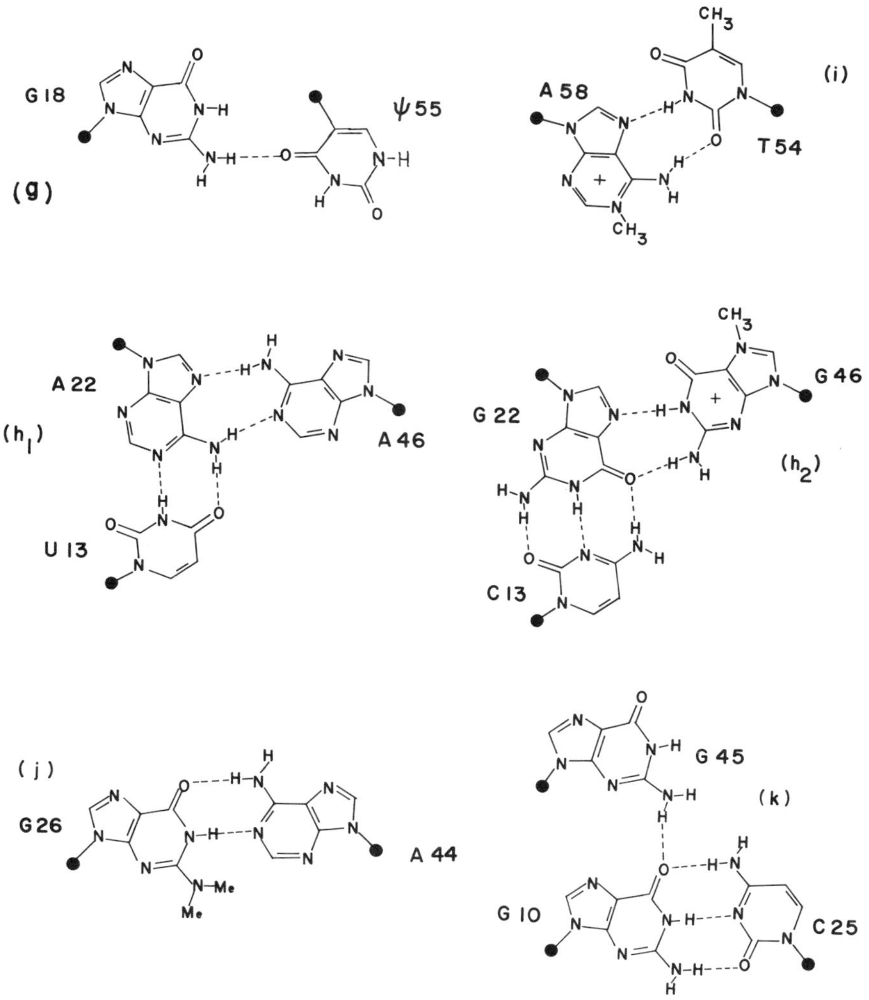

G 18

(g)

ψ 55

(i)

A 58

T 54

(h₁)

A 22

A 46

U 13

G 22

G 46

(h₂)

C 13

(j)

G 26

A 44

G 45

(k)

G 10

C 25

These tertiary H-bonds are all between the bases, and all except G19·C56 have H-bonding schemes different from the Watson-Crick type. They are schematically indicated in the cloverleaf and L diagram in figure 9.4, and their locations in the three-dimensional structure are shown in figures 9.6a, b. An additional possible H-bond between C32 and A38 (23, 17), and an unusual H-bond between O-2 (U33) and C-8 (A36) (24), have been assigned by some but not by others.

9.2.4 Tertiary Hydrogen Bonds Involving Riboses and Phosphates

There are many H-bonds stabilizing the tertiary structure of the molecule (22–24) in addition to those between the bases. Many of these utilize O-2′ of ribose and some bases as a hydrogen donor or acceptor and phosphate oxygen as a hydrogen acceptor. Since most of these H-bonds are much less stereospecific and isolated compared to the secondary or tertiary base pairs, they are technically more difficult to assign and subject to prejudice. In fact, although essentially all the secondary and tertiary base-pair assignments agree among four different laboratories presently engaged in the refinement of this structure (MIT and Duke groups for the orthorhombic crystal form; MRC and Wisconsin groups for the monoclinic form), the assignment of H-bonds involving the backbone (ribose and phosphates) varies considerably among them at the current stage of refinement. It is difficult to critically evaluate each assignment because arbitrary criteria are used in defining the H-bond, and there may exist some differences in H-bonds of *this* category in two different crystal forms. Thus it is difficult to compare them. The poor reliability in the assignment of this class of H-bonds is reflected in the poor agreement between the assignments by two groups refining on the same tRNA using the same data: of 33 H-bonds assigned for this class of H-bonds, less than half are agreed to by both groups (23, 24).

Among the H-bonds assigned by both groups, the following ones are especially interesting: (a) N-1 of a conserved base A21 forms an H-bond to O-2′ of U8; (b) N-3 of a conserved base Ψ55 makes an H-bond to the phosphate oxygen of A58, and similarly, N-3 of another conserved base U33 makes an H-bond to O-5′ or the phosphate oxygen of A36, stabilizing the sharp bends of the T-loop and the anticodon loop, respectively. The similarity of these two sharp bends suggests that such bends may be a general pattern in loop formation (25, 23); (c) the base G57 makes several H-bonds to riboses of residues 18, 19, and 55 to further stabilize the T-loop/D-loop interaction.

The preponderance of H-bonds involving O-2' suggests the advantage of RNA over DNA in forming structural nucleic acid molecules.

9.2.5 Two Base-Stacked Helical Columns

One of the most spectacular features of the molecule is that almost all the bases are stacked along two major helical axes (20). This is shown schematically in figure 9.6. One base-stacked column runs horizontally and the other vertically. Thus at the end of each stem the helix is augmented along at least one strand by the stacked bases.

This extensive stacking interaction is probably the primary stabilizing, although not specific, force in the crystal structure of tRNA and is expected to be a general feature for all tRNAs.

All four stems are right-handed, antiparallel, and double helical, and their backbone conformation is similar to that of the RNA-A form (26); the riboses are in the C3'-endo conformation, the bases are tilted with respect to the helical axes, the helical repeat angle is similar to the helices with 10 to 12 base pairs per turn, and the diameter of the helix is about 20 Å. Although all four stems are similar to the RNA-A type helix, there are considerable irregularities along the backbone of the helices. It is not clear whether the irregularities can be correlated with base-pair sequence.

9.2.6 Nucleotide Conformation

Four major nucleotide conformational elements that affect the polynucleotide conformation are the conformation of (a) the glycosyl bond χ, (b) the ribose, (c) the C-4'-C-5', δ, and (d) phosphodiester bonds, a and β (see figure 9.7).

Although all glycosyl bonds appear to have anticonformation of varying degrees, the glycosyl conformations for D16, D17, and U47, three pyrimidines that do not form any H-bonds involving bases and that all have C2'-endo riboses, and A76 are difficult to assign. A few show the high anticonformation. In all cases they are associated with C2'-endo riboses.

The ribose conformation of the tRNA turned out to be much more variable and complex than the simple C3'-endo or C2'-endo conformation previously assumed (27–29, 17). When we allow the ribose conformation to be refined against X-ray diffraction data using a least-squares method (30), we find that the riboses in yeast tRNA show considerable conformational variability. However, if one crudely classifies those riboses with C3'-endo and/or C2'-exo conformation as A-type riboses, and those with 2'-endo and/or 3'-exo conformation as B-type riboses (figure 9.7b), they can all be

a

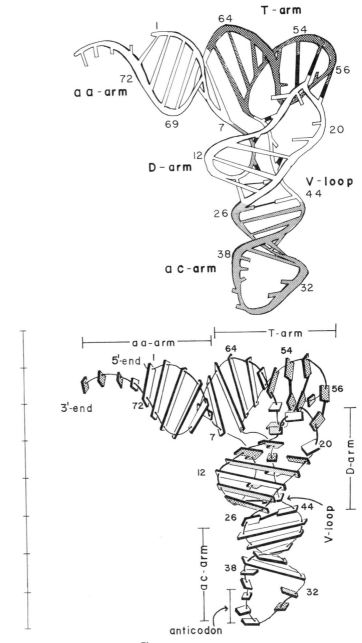

b

Figure 9.6 (a) Schematic drawing of yeast tRNA^{Phe} showing the ribose-phosphate backbone as a continuous tube, base pairs as long bars, and single bases as short bars. Tertiary H-bonds between the bases are indicated by solid dark rods. Adjacent arms are shaded differently to distinguish them easily. (b) Artist's drawing of yeast tRNA^{Phe} to show extensive base stacking in the structure. The secondary and the tertiary base pairs are shown as bent or fused slabs or by connecting two slabs with dark rods. Conserved and semiconserved bases are shaded dark and gray, respectively.

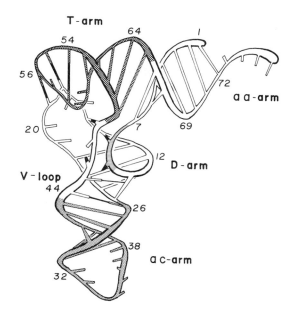

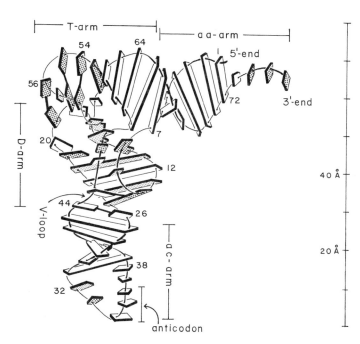

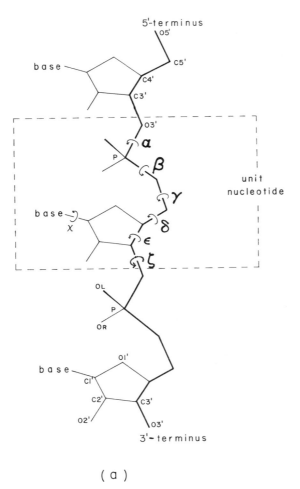

(a)

Figure 9.7 The nomenclature used in this article for the (a) various backbone conformation angles; (b) ribose conformations, where the five-membered ribose ring is viewed along a direction parallel to the plane defined by C-1', O-1', and C-4'; (c) glycosyl conformations; (d) δ conformation.

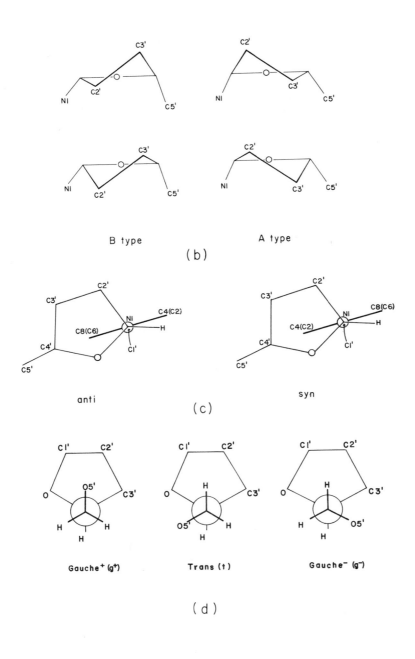

B type   A type

(b)

anti   syn

(c)

Gauche+ (g+)   Trans (t)   Gauche- (g-)

(d)

classified as belonging to the A-type except riboses from residues U7, A9, D17, G18, G19, A21, G46, C48, A58, and C60 by all four groups. One group recently added two more, D16 and U47, and deleted A21 from the list (24).

Since most of the nucleotides are in the two extended helical columns, most torsion angles are in the ranges expected for double helical RNA. Some nucleotides in the single-stranded regions, however, have different torsion angles. Details of these conformations are beyond the scope of this article.

9.2.7 Metal Binding Sites

It has been known for a long time that tRNA molecules assume a functional conformation in the presence of certain metal ions, most commonly magnesium ions (32). The effort to characterize and determine the nature, the site(s), and the number of sites of the metal binding has produced rather confusing and conflicting results (33 and references therein). Although the X-ray crystallographic studies on yeast tRNA^{Phe} appear unable to resolve the conflicting results of solution studies unambiguously, they shed some light on the subject. The difference $F_o - F_c$ and the sum $F_o + (F_o - F_c)$, electron density maps computed using a highly refined yeast tRNA^{Phe} structure (24) revealed several very well isolated electron density peaks, which could be interpreted as bound magnesium or hydrated magnesium ions (34). Among these the three most interesting sites are (a) the sharp turn formed between the aa-stem and the D-stem, where the magnesium hydrate ion is in a pocket formed by four phosphates from residues 8, 9, 11, and 12; (b) the corner of the D-loop, where the phosphate of residues 19 and the bases G20, U59, and C60 surround another magnesium hydrate ion; (c) the pocket in the ac-loop formed by the bases from C32, Y37, A38, and Ψ39 and the phosphate of Y37. The first site is also approximately the same site where samarium ions have been found in the tRNA crystals soaked with samarium acetates (35). In all four cases hydrated metal ions appear to be stabilizing the sharp corners and turns of the polynucleotide backbone.

9.2.8 Conformational Variability

The crystal structures determined by X-ray crystallographic methods are static structures that represent one of the most energetically favorable time-averaged conformations for a free, unbound tRNA under the particular crystallization conditions. It is generally assumed that when a tRNA binds interactively with other macromolecules, varying degrees of conformational changes in tRNA structure may be induced. Therefore one has to consider tRNA crystal structure as a flexible starting point for studies on tRNA function. Even without macromolecules, there is always a question whether the structure in crystalline form is the same as that in solution as a free molecule. In the case of tRNA, if one assumes the flexibility of the ac-loop and 3'-end, there is overwhelming evidence that the crystal structure of yeast tRNA^Phe is consistent with vast amounts of solution data.

Despite the overwhelming evidence on the similarity between tRNA structure in crystal and in solution, no crystal structure provides any direct information on the dynamic aspect of the molecule. However, by examining the crystal structure carefully, one can obtain some indications about possible conformational changes.

Amino acid arm The four nucleosides at the 3'-terminus are in a single strand and at least the last three are expected to be quite flexible in solution. The distance between C72 and A76 can be as long as about 30 Å when the backbone is extended. Even in the crystalline state residues 74, 75, and 76, as well as a few terminal base pairs, show very high thermal motion, as can be seen in figure 9.8.

Although the aa-stem maintains the same orientation with respect to the rest of the molecule in two crystal forms, this stem may have some limited small-bending flexibility.

Anticodon arm The anticodon loop is well isolated from the rest of the molecule, and it is probably safe to assume that its conformation may be variable to some extent. This can also be deduced from the higher thermal motion of the loop compared to the rest of the molecule (figure 9.8). The conformations of this loop in two different crystal forms are roughly the same in that five bases on the 3'-side of the loop are stacked (see figure 9.6b), but the extent of the base stacking appears to be slightly different, again an indication of the conformational variability of this part of the molecule. The sharp corner between residues 33 and 34 is accomplished by a rotation around the P-O5' bond of residue 34 and stabilized by a hydrogen bond between N3 of U33 and O5' or the phosphate oxygen of A36. However, the structure itself as well as model building suggest that it should be possible to alter the anticodon loop so that more bases are stacked on the 5'-side rather than on the 3'-side.

Although in two crystal forms this stem maintains the same orientation with respect to the rest of the molecule, the ac-stem may have limited small-

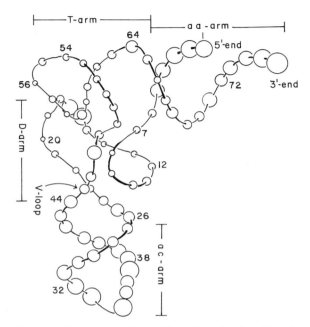

Figure 9.8 An average "thermal" motion of each residue of yeast tRNA^{Phe} in an orthorhombic crystal form is indicated (24), in a relative scale, by a circle proportional to the isotropic crystallographic temperature factor. The larger the circle, the higher the thermal vibration of that residue in the crystal. The anisotropy of the thermal motion is not shown. The term *thermal* is used to cover a wide variety of small motions. For example, in addition to the true thermal vibration, the high thermal motion at the 5'-end may be due to a partial opening and closing of the double helix, and that in the anticodon arm may be due to the flexing motion of the whole arm. The overall thermal behavior was found to be not correlated to the lattice contacts.

bending flexibility. This possibility is again reflected in the considerably high thermal motion of the ac-stem compared to that of the D-stem (figure 9.8).

D-arm/T-arm The D-loop and the T-loop are intricately interdigitated and hydrogen bonded to each other, suggesting that this part of the molecule, the corner of the L (see figure 9.6), is probably not flexible when a tRNA exists as a free molecule without interacting with other molecules. Such stability is also apparent from the small thermal motion of that region relative to other parts of the molecule except residues D16 and D17, which bulge out into solution and have high thermal motion (see figure 9.8). The relative orientation of the T-stem and D-stem is likely to remain fixed as long as D-loop/T-loop interaction is not broken by other environmental factors. Thus the D-arm/T-arm domain forms the structural core of the molecule. This apparent rigidity does not exclude the possibility of conformational changes on interaction with other macromolecules or organelles such as ribosomes. In fact careful refinement of crystallographic structure shows that the G19·C56 tertiary base pair is bent by about 30° from planarity (24) and appears to be strained as if it is ready to be opened easily.

9.3 Generalized Three-Dimensional Structure of tRNA

Since all tRNA sequences can be arranged in the secondary structure of the cloverleaf, and since most of the tertiary H-bonding between the bases found in the crystal structure of yeast tRNA^{Phe} involves the conserved or semiconserved bases, one can safely assume that those H-bondings, secondary as well as tertiary, will be preserved in all tRNAs (20, 36). Tertiary base pairs (two or more H-bonds between the two bases) that are expected to be common in class D_4V_5 (class I) tRNAs are the following:

1. 8·14 pair. In all tRNAs, this is U·A and expected to have the base-pairing scheme shown in figure 9.5d.
2. 9·23·12 triple. Either A·A·U or G·C·G, as shown in figure 9.5e$_1$, e$_2$. In both cases the relative positions of the riboses are similar; thus replacing one triple with the other will not change the backbone structure significantly at the region in the molecule. These three positions are a conserved coordinated triple.
3. 15·48 pair. Either A·U or G·C as shown in figure 9.5f$_1$, f$_2$. The difference will be a slight change in the backbone conformation of that region, a conserved coordinated pair.
4. 19·56 pair. In all tRNAs this pair is G·C, as shown in figure 9.5b.
5. 13·22·46 triple. Either C·G·G or U·A·A as shown in figure 9.5h$_1$, h$_2$. Here again the relative positions of the riboses in both triples are similar; therefore these form another conserved coordinated triple.
6. 54·58 pair. This is T·A in all tRNAs except the initiator tRNA from mammalian cells. The base-pairing scheme is shown in figure 9.5i.

These six types of base pairs or triples provide a very specific architectural network between helical stems that can be further stabilized by the extensive nonspecific stacking of the bases and by other H-bonds involving riboses and phosphates found in yeast tRNA^{Phe} structure. Thus one can consider that figures 9.2, 9.3, 9.6, and 9.8 represent all tRNAs in class $D_4 V_5$ (class I). Similar generalizations can be made for the other classes of tRNAs to a limited extent.

9.4 Correlation between Crystal Structure and Solution Structure of tRNA

The crystal structure of yeast tRNA[Phe] is generally in good agreement with the vast amount of experimental data obtained from solution studies of free tRNAs. Although most of these have been reviewed (2), they are summarized here with more recent results.

9.4.1 Small-Angle X-Ray Scattering

For particles in a nearly random arrangement, as in a gas or dilute solution, the angular dependence of the scattered X-ray intensity closely approximates that from a single particle averaged over all orientations. Usually the intensity decreases rapidly within a few degrees of scattering angles (thus small-angle). When scattered X-ray intensities are plotted against scattering angles on a log-log scale, the shape of the curve provides information about overall orientation-averaged size and often shape. The most reliable quantity obtainable from this method is the radius of gyration of a molecule.

The radii of gyration for the various tRNAs in aqueous solution have been measured to be around 23.5 Å–25 Å (37–41), which is in good agreement with the value of 24 Å calculated for yeast tRNA[Phe] using only the phosphate groups (S.-H. Kim, unpublished results).

9.4.2 Fluorescence Studies

The method of fluorescence energy transfer to estimate the distance is based on the observation that energy absorbed by a chromophore can be transmitted to another chromophore some distance away. In singlet-singlet transfer the return of the energy donor from the lowest excited state to the ground state is coupled to the excitation of the energy acceptor from the ground state to a higher singlet level. This type of energy transfer occurs over the appreciable distance of tens of angstroms and is suitable for the distance estimation in biological macromolecules (42) provided that there is a single donor and a single acceptor at specific sites on the molecule, that information on the relative orientation of the donor-acceptor pair is available, and that the distance at which singlet-singlet transfer is about 50 percent efficient is comparable to the magnitude of the distance measured.

Singlet-singlet energy transfer between the hypermodified base H37 and an energy acceptor such as acriflavin, proflavinyl acetic acid, hydrazide, or 9-hydrazinacridine at the 3′-end of yeast tRNA^{Phe} provided an estimate of the distance between these two points to be greater than 40 Å (43).

Recently similar experiments have been carried out with fluorescent groups attached to the 5′-end; at position 8 (between the aa-stem and T-stem), on dihydrouracil in the D-loop, on Ψ in the T-loop, and at the 3′-terminus of E. coli tRNA$_f^{Met}$, and on s^{2}U in the anticodon and at the 3′-end of E. coli tRNA^{Glu} (44). The distance have been estimated to be 24 Å between the 5′- and 3′-ends, 38 Å between position 8 and the 3′-end, 55 Å between Ψ and the 3′-end, 36 Å between Ψ and D, and greater than 65 Å between the anticodon and the 3′-end. The corresponding distances in yeast tRNA^{Phe} in the orthorhombic crystal are 16 Å, 42 Å, 52 Å, 25 Å, or 20 Å, and 67 Å. The distances involving the 3′-terminus require special consideration because the single-stranded stretch at the 3′-end of a free tRNA is expected to be flexible in solution. In addition the distance between the base at the 3′-end and the fifth base on the same strand is likely to range between 15 Å, when the bases are stacked, and 22 Å, when the backbone is in extended conformation. With this consideration and the recognition of the uncertainty about the exact position and orientation of the fluorescent group, one can say that the experimental results of this type do not contradict the crystal structure. However, the agreement is poor, either due to the assumptions one has to make in the distance estimate by this method or due to a genuine difference in the solution structure. The former is more likely in the light of other evidence. It is interesting, though, that there is good agreement for distances greater than about 40 Å.

Additional evidence for the similarity of the anticodon conformations in solution and in the crystal comes from the direct comparison of the fluorescence lifetimes, in crystal and in solution, of the hypermodified base that lies at the 3′-side of the anticodon of yeast tRNA^{Phe}. The fluorescence properties of this base are very sensitive to the environment, that is, to the conformation of the anticodon loop (43, 44), presumably because of the influence of two adjacent stacked bases (see figure 9.9a). The fluorescence lifetime of this base for tRNA in the orthorhombic crystal is identical with that in the solution from which these crystals were grown and very similar to that in the dilute solution of tRNA (46). Thus within the limits of the sensitivity of the method, the conformation of the anticodon loop around this hypermodified base is the same in the crystal and in solution.

9.4.3 Complementary Oligonucleotide Binding

In an aqueous solution of moderate ionic strength a short oligonucleotide of a specific sequence, as small as a trinucleotide, can bind specifically and strongly to an exposed, single-stranded region of RNA that contains a sequence complementary to the oligonucleotide. Short oligonucleotides have been employed as probes to search for the exposed single-stranded regions of tRNA (47, 48). Such experiments have been done on *all* regions for *E. coli* tRNA^{Tyr}, tRNA$_f^{Met}$; yeast tRNA^{Phe}, tRNA^{Leu}; and initiator tRNA^{Met} (49–52). The results agree with the crystal structure as to the availability of the ac-loop and the tetranucleotide at the 3′-end of tRNA if one assumes some flexibility of those regions. Four out of five tRNAs so examined indicate that the T-loops are not exposed (the exception is given in ref. 50), and this agrees with the crystal structure of yeast tRNA^{Phe}. The availability of parts of the D-loop and the V-loop as studied by this method varies among different tRNAs, and this may imply either some structural differences or competition for these regions of the molecule between complementary oligonucleotides and other parts of the molecule (51).

9.4.4 UV-Induced Cross Linking

Irradiation with near ultraviolet (UV) light (300–380 nm) induces a specific and quantitative photochemical cross link between 4-thiouracil at position 8 and the cytosine at position 13 in the nucleotide sequences of many procaryotic tRNAs (53, 54). Such photo cross linking suggests that those two bases are spatially close to each other in the tertiary structure of tRNAs. This was proven to be the case in the crystal structure of yeast tRNA^{Phe}, where C13 is stacked on U8, which in turn is base paired to A14. The relative orientation of the U8-C13 stack in yeast tRNA^{Phe} in the orthorhombic crystal form is shown in figure 9.9b, which is similar to that suggested earlier based on the model compound studies (55).

Further, the photo cross-linked *E. coli* tRNAs can be aminoacylated by the cognate synthetase and remain functional in protein synthesis though with a reduced rate (56). They also act as substrates for purified nucleotidyl transferase (56) and bind to the Tu factor (57). Such observations support the validity of this crystal structure as a biologically meaningful functional form.

9.4.5 Tritium Labeling

When nucleic acids or nucleotides are incubated in tritiated water to allow tritium-proton exchange to occur, and when free and loosely bound tritium

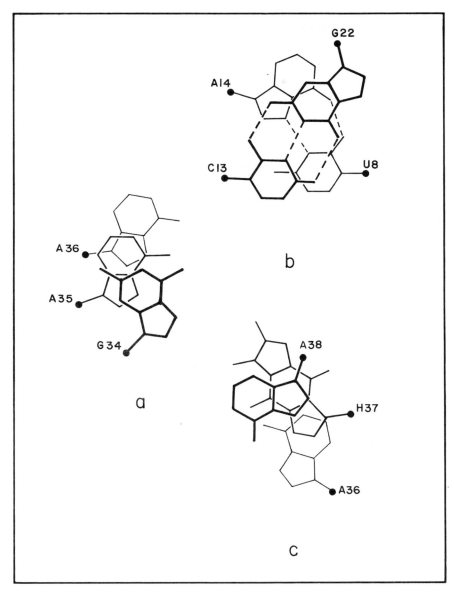

Figure 9.9 Base stacking patterns in the crystal structure of yeast tRNA[Phe] (24): (a) the anticodon triplet viewed perpendicular to the base A35; (b) the relative orientation of U8 and C13
viewed perpendicular to the base C13; (c) the hypermodified base H37 and its neighbors viewed
perpendicular to the H37 base plane.

is subsequently removed from them, the tritiums were found primarily bound to the C-8 position of purines (because the protons at the C-8 position are slightly acidic). Under the conditions where exchange-out does not occur, the nucleic acids can be digested with various nucleases, fragments, and nucleotides separated, and individual purines can be identified to measure specific activities. When the tritium labeling was done for a limited period, the exchange rate constants at 37°C for purines in the poly(A)·poly(U) duplex, duplex viral RNA and tRNA showed a very wide range of reduction compared to those for free mononucleotides or poly(A), presumably depending on the microenvironment around the C-8 of purines (58).

The tritium labeling on yeast tRNA^{Phe} (59) showed that among the 17 identified purines all except A76, G18, and G34 and/or A35 have reduced exchange rates compared to poly(A). This observation agrees in general with the crystal structure where all the identified purines except A76 are involved in secondary or tertiary base pairs or in stacking. The G34 and A35 are the two bases at the very end of the anticodon loop. Although they are stacked, they may be more easily available for tritium exchange. The G18 in the crystal structure is not only involved in tertiary H-bonding with ψ 55, but also is well stacked by other bases on both sides. However, the crystal structure shows that C-8 of G18 is well exposed to the solvent, not only because it is at the corner of the D-loop, but also because the ribose of G18 is in the 2'-endo conformation, which provides more space between C-8 and O-5'.

The tritium-labeling experiments on *E. coli* tRNA^{Ile} and tRNA$_2^{Tyr}$ show similar sensitivity (60).

Although one can correlate in an approximate way the tritium-labeling behavior with the crystal structure, the dependency of the exchange rate to the microenvironment around the C-8 position seems too sensitive to make a rigorous correlation with the crystal structure in terms of the secondary or tertiary H-bonding. However, the technique can be potentially useful in determining conformational differences too subtle to distinguish by other methods.

9.4.6 Base-Specific Chemical Modification

Before the crystal structure of tRNA was known, this method provided very high resolution information about the degree of exposure of certain bases (for a review see ref. 61). Several chemical reagents react specifically with certain bases under mild conditions if those bases are available. The reagents and specific bases attacked are listed in table 9.1. The basic procedure is as follows. Allow tRNA to react with one of the reagents in table 9.1; enzymatically digest the reaction product to obtain oligonucleotides; identify those oligonucleotides that have altered mobilities compared to the unreacted oligonucleotide pattern by column or gel chromatography; and finally identify which base is altered by further digestion of the altered oligonucleotides. Such experiments have been carried out with many pure tRNAs: *E. coli* tRNA$_2^{Glu}$, tRNA$_{Su3}^{Tyr}$, tRNA$_f^{Met}$, tRNA^{Leu}; yeast tRNA^{Val}, tRNA^{Ala}, tRNA^{Phe}, tRNA^{Tyr}, tRNA^{Ser}; mammalian initiator tRNA^{Met} (62–70).

Table 9.1 Base-specific chemical reagents

Reagents	pH Range for Reaction	Base Attacked	Reaction Product [a]
Hydroxymine, methoxyamine	7	Cytosine	
Carbodiimides	7	Uracil, guanine, hypoxanthine	
Monoperphthalic acid	7	Adenine (cytosine)	
Kethoxal	7	Guanine	
Borohydride	8–10	Dihydrouracil	$NH_2-C-N-CH_2-CH_2-CH_2-OH$
Bisulphite	5.8	Cytosine	

[a] As nucleoside in RNA (R = ribosyl).

A summary of these results for D₄V₅ class (class I) tRNAs is shown in figure 9.10. As can be seen from the figure, most of the bases (besides the bases in the stems) at the positions that are protected from chemical modification are involved in tertiary H-bonding in yeast tRNA^{Phe}. Among the remaining protected bases, the bases G4 and U69 constitute a wobble base pair; the base A21 forms an H-bond to the ribose of U8; the base G57 forms H-bonds to the riboses of G18, G19, and Ψ55; and U59 and C60 are buried inside the molecule.

In conclusion the results from the base-specific chemical modification are in extremely good agreement with the crystal structure at the individual base level (15, 16, 68, 71).

9.4.7 NMR Studies

There have been very extensive NMR studies on tRNA in the aqueous medium as well as in D_2O. The vast amount of NMR data so far obtained is by and large consistent with the crystal structure of yeast tRNA^{Phe} or the three-dimensional structures of other tRNAs predicted based on the crystal structure of yeast tRNA^{Phe}.

Low-field proton NMR studies in aqueous media Early observations showed that the ring NH protons from the individual base pairs in nonaqueous solvent give NMR peaks in the extreme low field of the proton spectrum (72). Subsequently the pioneering studies on tRNA showed that the exchange of the ring NH protons involved in base pairing is slow enough even in aqueous solvent to generate low-field NMR in the region between –11 and –15 ppm (73). Since each base pair contains one such proton (at N-1 of G in the G·C pair and at N-3 of U in the A·U pair), the low-field NMR studies on tRNA can provide relatively detailed structural information in solution especially because of the spreading of the resonance peaks due to the ring-current effects from adjacent bases.

In general three types of information are available. The atomic environment around the ring NH protons, the number of H-bonds involving the ring NH protons, and the lifetime of those H-bonds can be derived from the chemical shift positions, the integrated intensities, and the line widths of the proton NMR peaks, respectively. The low-field NMR studies on tRNA have recently been reviewed (74).

An approximate spectral assignment in this region was based on two lines of observations: (a) NMR spectra from tRNA fragments containing double-stranded stems show resonance peaks in this region and correlate with the

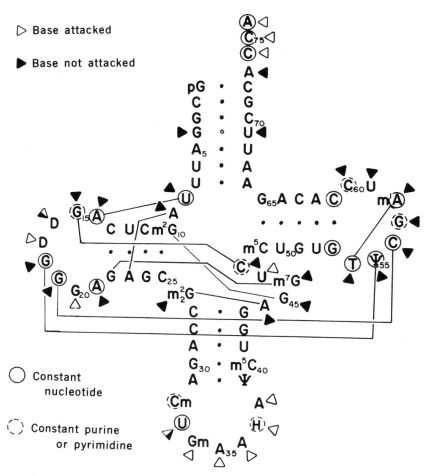

Figure 9.10 Summary of the results of the base-specific chemical modification experiments on many D_4V_5 class tRNAs is shown in relation to the nucleotide sequence of yeast tRNA^{Phe}. H = hypermodified guanosine. The circled bases are conserved, and the dashed circles indicate the semiconserved bases in all D_4V_5 class tRNAs participating in peptide elongation. The accessibility of various bases to chemical modification is indicated by open or filled triangles. The half-filled triangles indicate the bases with possible partial exposure. All the bases in the stems are not available for chemical modification. The tertiary H-bonds between the bases are indicated by connecting lines.

number of base pairs in the stems (75); (b) a semiempirical rule can be derived (76) that can predict approximately the NMR spectral positions from the theoretical calculation (77) of ring-current shifts, assuming that the bases are stacked as in the 12-fold RNA-A' double helix (a form very similar to RNA-A).

There still remain two areas of discrepancy among the workers in this field. One is the precise intrinsic positions of the ring NH proton resonances for A·U and G·C pairs in the absence of any neighboring ring current. The other is the method of integration of the resonance peaks to estimate the total number of secondary and tertiary base pairs. However, the following summary can be made. (a) All the Watson-Crick base pairs predicted from the cloverleaf secondary structures are present. (b) Initially it was concluded that these were the only base pairs in yeast $tRNA^{Phe}$ in solution (78). Subsequent interpretation of the NMR spectra suggests that there are about three more ring NH resonances attributable to the tertiary base pairs by one group (79) and about six by another group (80) for yeast $tRNA^{Phe}$. (c) The four helical stems are of the RNA-A type. (d) The amino acid stem is stacked on the T-stem. (e) The ends of the helical stems are usually stacked by the adjacent bases along at least one strand. (f) Similar results have been observed for many other tRNAs. These are all in good agreement with the crystal structure of yeast $tRNA^{Phe}$. A recent attempt to compare the calculated spectrum with the observed one showed the overall validity of this technique as well as the danger of overinterpretation (81, 82).

High-field proton NMR studies in deutrated water Yeast $tRNA^{Phe}$ contains 14 methyl groups and 4 methylene groups belonging to 12 modified nucleosides, and they are distributed throughout the molecule in all except the aa-arm. These methyl and methylene resonances were assigned by careful comparison of the resonances of the intact tRNA with that of the mononucleoside, mononucleotide, and appropriate oligonucleotide at various temperatures (83). By measuring the chemical shifts and line widths of these resonances as a function of temperature between 10°C and 80°C, the following interpretation has been made. (a) Methyl groups in the anticodon loop of yeast $tRNA^{Phe}$ (from Cm32, Gm34, and H37) are in a similar magnetic environment as those in mononucleotides and dodecamers containing nucleotides 31–42, the implication being that the whole anticodon loop does not associate with other parts of the molecule. (b) Methyl resonances m^5C40, m^5G10, and m^5A58, even though these bases are in the double-helical stems, do not show significant chemical shift changes in the absence of Mg^{++} ions as the temperature is raised, again implying that these methyl groups are not near diamagnetic regions in the molecule. However, methyl resonances from m^{5}C40 and m^{5}C49 experience thermal transition in the presence of Mg^{++} ions. (c) Methyl resonances from m$_2^2$G26 and T54 go through dramatic chemical shifts as well as line-width changes on heating, implying that they are within strong diamagnetic regions of the molecule and not free.

All these observations except the Mg^{++} effect on m^{5}C40 and m^{5}C49 can be understood in terms of the crystal structure of yeast tRNA^{Phe}. Thus the overall agreement between the crystal structure and the solution structure as observed by NMR is very good.

9.4.8 Conformational Changes
The conformation of tRNA is expected to change, during interaction with other proteins and ribosomes, to an unknown extent. Noticeable changes have been observed even for free tRNA under various solution conditions as detected by circular dichroism measurements (84) as well as laser light scattering (85). The latter measures the change in diffusion constants. One of the questions frequently asked was: Is there a significant conformational difference between tRNA and aminoacyl-tRNA? The experimental data or the interpretations of them from NMR (86), other spectroscopic methods (87, 88), and the oligonucleotide-binding technique (89) are conflicting. The crystal structure does not suggest any significant conformational changes in tRNA on aminoacylation other than in the immediate neighborhood of the 3'-end and possibly the 5'-end. There is, however, a structural feature suggestive of opening up the corner of the L formed by the T-loop and the D-loop in ribosome.

9.5 Functional Implications from the Crystal Structure

Throughout its life cycle a tRNA molecule goes through numerous interactions with proteins and other nucleic acids. In fact each tRNA can be considered a common substrate to more than a dozen proteins.

In an enzyme-substrate interaction knowledge of the three-dimensional structure of a free substrate of small molecular size often does not provide any significant information in understanding the mechanism of recognition or action. This is because of the large and many conformational changes a small molecule can go through at the expense of a small amount of energy with minimal, if any, conformational change in the enzyme structure. The situation is quite different when the substrate is as large as enzymes, as in the case of tRNA. It is therefore likely that knowledge of tRNA's three-dimensional structure can provide as much as half the information about enzyme-tRNA interaction. With this assumption as a starting point, this section will examine various functions of tRNA from the point of view of its three-dimensional structure. The following discussions are based mostly on the implications derived from crystal structure and are described here to sharpen our focus for more detailed future studies of tRNA functions. Most of the following topics have been discussed recently (90).

9.5.1 In Aminoacylation of tRNA

The fidelity in the transfer of genetic information at the translational level depends largely on the accuracy with which each tRNA is aminoacylated by its cognate aminoacyl-tRNA synthetase. The search for the common sites on tRNA, which the enzyme recognizes to distinguish cognate from non-cognate tRNA, has resulted so far in a confusing picture. Practically all parts of the tRNA molecule have been implicated as the possible recognition site(s) by one synthetase or another (reviewed in refs. 91, 92).

The search has been further complicated by the observation that misacyl-ation by wrong amino acids can occur not only in a homologous system but also in a heterologous system in vitro; for example, most of the *E. coli* tRNAs can be misacylated by one yeast aminoacyl-tRNA synthetase (93, 94). Such an observation suggests that there are two types of recognition involved in aminoacylation of tRNA: a general recognition, in which the enzymes recognize the common structural features of all tRNA, and a speci-fic recognition, in which the cognate tRNA is selected. It appears that the specific enzyme recognition sites on tRNA may vary from one tRNA to another.

There is a set of structural features of tRNA and the enzyme that suggests a model for the general recognition. Many aminoacyl-tRNA synthetases are composed of one or more large subunits, each of which contains two sub-stantially homologous amino acid sequences (95–98). For the rest, it has been shown that the functional enzymes are composed of an even number of small subunits. If one assumes that this sequence homology implies a two-domain structure for the large subunits, one can generalize that the structures of all aminoacyl-tRNA synthetases are likely to be composed of an even number of domains or subunits.

With these observations and assumptions one can ask whether there are any three-dimensional structural features of tRNA that suggest a possible mode of recognition between the enzyme and tRNA. One such mechanism has been proposed (99) for the general recognition based on the symmetry properties of tRNA and the synthetase; it assumes that two domains or sub-units of an enzyme are related by a pseudo twofold axis. A closer examina-tion of the crystal structure of yeast tRNA^{Phe} reveals the existence of a pseudo twofold axis as shown in figure 9.11. The symmetry-recognition hypothesis proposes that the pseudo twofold symmetry of an aminoacyl-tRNA synthetase is recognized by the pseudo twofold symmetry of tRNA as schematically shown in figure 9.12. Such a general recognition involves a large contact area between the tRNA and the synthetase, primarily along the inner side of the L, as also suggested by Rich and Schimmel (100), and provides the stability of the complex. Within the contact area the specific recognition can take place to distinguish the cognate from the noncognate tRNAs. A recent study by the tritium-labeling method (101) focused atten-tion to six bases on *E. coli* tRNA^{Ile}, a D$_4$V$_5$ (class 1) tRNA, as markers for the synthetase contact sites. The tritium exchange rates at the C-8 position of the following purines in the tRNA were found to be significantly retarded when the cognate synthetase was bound to it: A22, G23, G35, G50, A74 and 77. These six bases are located in approximately symmetric regions (see fig-ure 9.12e), which could be expected from figure 9.12c.

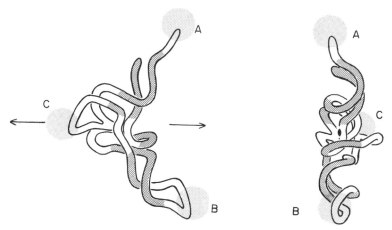

Figure 9.11 Two views of the tRNA backbone structure, 90° apart. The molecular pseudo two-fold axis is shown as arrows when in the plane of the page, or as ◗ when perpendicular to the plane of the page. The regions related by the symmetry are heavily stippled. The peptide elon-gation site, the codon recognition site, and the presumed 5S rRNA recognition site are lightly shaded and indicated by A, B, and C respectively.

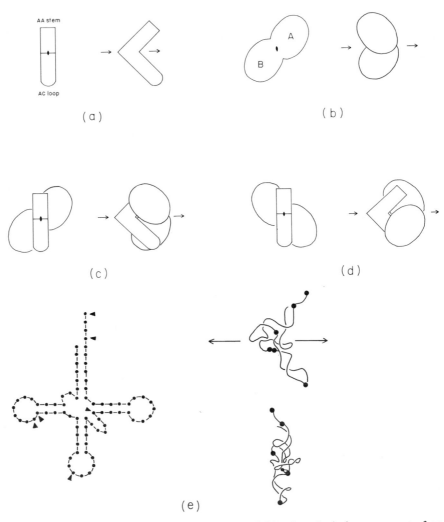

Figure 9.12 Two orthogonal views of (a) tRNA and (b) a hypothetical arrangement of an aminoacyl-tRNA synthetase, showing the location of the pseudo twofold symmetry axes. Parts (c) and (d) are two alternative proposed models for the tRNA·enzyme complex formed by coinciding the symmetry axes of both molecules. (e) The six locations on tRNA that have been implicated as the minimum synthetase contact points from the tritium-labeling experiment (101) are found to be arranged symmetrically around the molecular pseudo twofold axis.

9.5.2 In Protein Synthesis

The crystal structure of yeast tRNA^{Phe} reveals interesting functional designs of the molecule. Three important functions of tRNA in ribosomes are associated with three well-separated regions of the molecule (see figure 9.11); the 3'-end, where the peptide elongation is occurring (region A), is maximally separated from the anticodon (region B), where codon-anticodon interaction takes place with messenger RNA. Both regions, A and B, are again well separated from the presumed ribosomal 5S RNA interaction site C, thus preventing cross interference among three important function sites.

Five bases including the anticodon triplet and two bases on the 3'-side of it are stacked in a pseudohelical conformation (see figures 9.6 and 9.9c), and the sharp corner of the loop is stabilized by at least one hydrogen bond. If one assumes that part of the codon-anticodon interaction remains in the P-site and that the anticodon loop conformation as observed in the crystal structure remains approximately the same in either the A-site or the P-site of the ribosome, the structure has the following implications: (a) it is less probable that tRNAs in the A-site and the P-site will have the same anticodon loop conformation, because that requires a large rotational movement of tRNA from one to the other site on translocation. Rigorous model building, using an interactive computer graphics system, suggests that the tRNA in the P-site could have a different anticodon-loop conformation from that observed in the crystal structure. One of several possible arrangements is shown schematically in figure 9.13. In such a case the movement of tRNA on translocation can be small. (b) The modified base adjacent to the anticodon at the 3'-end is probably playing a role as a base-pair breaker, preventing additional base pairing or slippage beyond the correct, three codon-anticodon base pairs. It may also stabilize the first base pair as well as the kink of the messenger RNA in this model. Such a mechanism will prevent the codes on the messenger RNA from being read out of phase. (c) At the A-site at least, the codon-anticodon complex assumes a short double helical conformation.

All tRNAs participating in the ribosome mediated peptide elongation contain the invariant sequence G-T-Ψ-C in the T-loop. There is evidence from several experiments (for a review see ref. 102) suggesting that procaryotic 5S ribosomal RNAs that contain a common sequence C-C-G-A-A-C recognize this invariant sequence in the T-loop at the A-site. However, in the crystal structure, as shown in figure 9.6, the T-loop is H-bonded not only internally (T54·A58) but also to the D-loop by two base pairs (G18·Ψ55 and G19·C56). Thus all three bases in the sequence T-Ψ-C parti-

Aminoacyl-tRNA

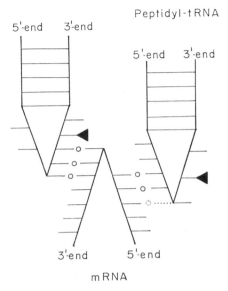

Figure 9.13 Schematic drawing of anticodon loops for a stereochemically possible model for mRNA-tRNA interaction, which requires a relatively small amount of movement when a tRNA translocates from the A-site to the P-site. In this model it is necessary for a tRNA to assume two different anticodon loop conformations, depending on the site it is occupying.

cipate in H-bonding as well as in stacking in the crystal structure and probably are not available for rRNA unless conformational changes are induced by ribosomal components and/or mRNA (103). The recent refinement of yeast tRNA^{Phe} (24) reveals an unusual structural feature involving this sequence, which suggests what may trigger the conformational change of tRNA on 5S rRNA binding. The base pair G19·C56 was found to be not only twisted but also bent by about 30° from the planarity expected for a Watson-Crick pair (see figure 9.14a) and is probably strained and less stable a base pair than a normal G·C pair. One hypothetical mechanism of tRNA-5S RNA recognition could be that the G from the sequence C-C-G-A-A-C in 5S rRNA (or from other parts of rRNA) forms another bent base pair with C56 (dotted figure in figure 9.14b), making a Y-shaped intermediate double base pair, which destabilizes and opens up the C56·G19 base pair, which in turn triggers the unstacking of G57 and zips open at least two tertiary base pairs, Ψ55·G18 and T54·A58, and possibly G53·C61, thus dissociating the T-loop from the D-loop. The now free T-loop assumes a loop conformation similar to the alternate anticodon loop conformation (figure 9.13), and the (G-)T-Ψ-C sequence (figure 9.14c) is then recognized through complementary base pairing by the G-A-A(-C) sequence in 5S rRNA (figure 9.14d). Although such a sequence of events is hard to test experimentally, the implication from the crystal structure is quite reasonable.

Another interesting finding with regard to *E. coli* ribosomes is that two ribosomal proteins, L7 and L12 from the large subunit of the ribosome, have identical amino acid sequences except for the N-terminal residue (104). It is tempting to consider the possibility that these two proteins may be arranged with a twofold symmetry axis between two identical monomers and that they are recognized by the pseudo twofold symmetry of tRNA through the symmetry-recognition process (99) similar to that described in section 9.5.1.

9.5.3 In Processing of tRNA

The transcripts of tRNA genes have been found to be longer than the mature-size tRNA (105) and often contain more than one tRNA (106–108). These precursor tRNAs are either unmodified or undermodified and sometimes lack the sequence C-C-A at the 3′-end, which is an invariant sequence in all mature tRNA. Thus the tRNA processing step provides a convenient additional regulation of tRNA biosynthesis (for a review see ref. 109). Several enzymes are responsible for the processing of the tRNA precursors in *E. coli*. One of the more studied is RNase P, which makes an endonuclease cut to give the correct 5′-end of tRNA both in monomeric and multimeric tRNA precursors. Since RNase P can process the 5′-end of all *E. coli* tRNA precursors, the recognition element cannot be a base sequence within tRNA with the possible exception of the G-T-Ψ-C sequence. Even this sequence is not likely because eucaryotic intiation tRNAs that have the A-U-C sequence instead of T-Ψ-C, are presumably also processed from a precursor by the RNase P equivalent of eucaryotes.

An examination of the known base sequences of tRNA precursors also shows that there is no common sequence at (or near) the cleavage point or at any other region outside the tRNA portion of the precursor. In addition RNA fragments (4 to 14 nucleotides long) containing the RNase P cleavage sites of dimeric tRNA^{Pro}-tRNA^{Ser} precursor are not cleaved by RNase P (110). Although the precise recognition site and the recognition elements of RNase P are not known, it is likely that the enzyme is recognizing the three-dimensional conformation or the invariant bases distributed over the whole structure or part of the structure rather than base sequences of the precur-

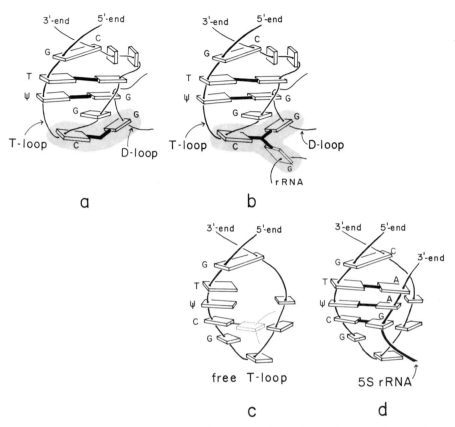

Figure 9.14 A possible sequence of events for 5S rRNA/tRNA interaction. (a) T-loop/D-loop interaction as observed in the three-dimensional structure of yeast tRNA^{Phe}. The bent base pair G19·C56 is shaded. (b) Formation of a Y-intermediate by a base G from 5S rRNA in this particular case. (c) Rearrangement of the intermediate resulting in the dissociation of the D-loop from the T-loop and conformational change of the T-loop. The dotted base is used to suggest that the G·C base pair between the T-loop and rRNA in step (b) may become a planar G·C pair and remain base paired. (d) The G-A-A sequence of 5S rRNA base pairs with the T-Ψ-C sequence of tRNA to form a short, right-handed double helix.

sors or the tRNA portion of the precursor in particular. This is also supported by the indication that the three-dimensional structure of the tRNA portion of the precursor is the same as that of the mature tRNA based on the experimental results that the same bases are modified in *E. coli* tRNA$_1^{Tyr}$ and the tRNA portion of its precursor (111) by methoxamine, which reacts with exposed cytosines, and carbodiimide, which reacts with exposed guanines and uracils.

Another class of data on this subject comes from base-substitution mutants of tRNA precursor for *E. coli* Su$_3^+$ tRNA$_1^{Tyr}$ (112–114; for a review see 109) and of T4-coded tRNAs (McClain, personal communication). These single-base substitution sites are indicated in figure 9.15. One can see that all the single-base substitutions that affect the tRNA biosynthesis of *E. coli* Su$_3^+$ tRNA$_1^{Tyr}$ at the presumed RNase P processing step are either in the double helical stem regions or at bases that form tertiary base pairs such as base 15 and 17 (if one assumes that the tertiary base pairing of this tRNA is similar to that of yeast tRNA^{Phe}). Most of the single-base substitutions that do not affect tRNA biosynthesis (not shown in figure 9.15) are either in the single-stranded regions (bases 35 and 82) or are base changes that allow unusual base pairs with only slight perturbation in the backbone conformation (substitution of U at positions 80 or 81 allows G·U base pairs to be formed in place of G·C pairs). Many of the mutants in which base substitution is in the double helical stems can revert to normal suppressor activity and tRNA biosynthesis by a second mutation on the base across the helical axis to a complementary base, thus recovering normal double helical conformation.

All this evidence, although indirect and limited, suggests that the overall three-dimensional backbone conformation in the stem as well as in tertiary interaction regions, but not in the bases or base sequences, is probably recognized by RNase P. The precise recognition mechanism will have to wait until the RNase P–tRNA complex can be crystallized and studied by X-ray crystallographic methods.

9.5.4 In Modification of tRNA

One of the characteristics of tRNA is that it contains many modified nucleosides, sometimes as much as 20 percent of the total number of nucleosides (for a review see refs. 115, 116). These nucleosides are modified posttranscriptionally (117, 118), and the modifying enzymes are specific for tRNA or tRNA percursors (116). The modification enzymes appear to recognize the conformation as well as the base sequence of the substrate mole-

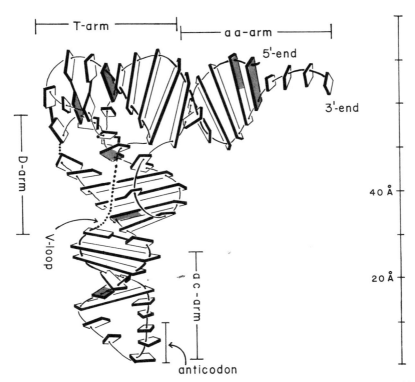

Figure 9.15 Shading indicates single-base substitution mutations that affect tRNA biosynthesis, presumably by blocking RNase P activity on *E. coli* Su$_3^+$ tRNA$_1^{Tyr}$ precursor (112–114) and T4-coded tRNA^{Pro}-tRNA^{Ser} and tRNA^{Glu}-tRNA^{Leu} precursors (McClain, personal communication). It is assumed that the overall three-dimensional structure, excluding the V-loop, of these tRNAs is similar to that of yeast tRNA^{Phe}.

cule (119), and in most cases different enzymes are required for the same modification in a tRNA at two different sites (120). The modification on ribose is always the methylation on O-2′, but the modification on the bases shows very extensive varieties (more than 50 modified bases have been identified so far), and it is evident that the complex modified bases must require more than one modifying enzyme.

Very little is understood about the functional role of these modified nucleosides. The most studied is the modified base located adjacent to the 3′-end of the anticodon. The tRNA lacking modification at this position usually shows some changes in efficiency in protein synthesis. Another interesting finding is that when the equilibrium association constants between

E. coli tRNA^{Glu} (anticodon U-U-C) and tRNA^{Phe} (anticodon G-A-A, complementary to that of tRNA^{Glu}) from various sources are measured, the tRNA^{Phe} with the most extensive modification, yeast tRNA^{Phe} with Y-base, formed the complex 19 times tighter than that containing the simplest modification m^{1}G (121), suggesting that this modification stabilizes codon-anticodon interaction. The stability of a codon-anticodon complex may be achieved by stabilizing the anticodon conformation and/or the codon conformation. It is not obvious, however, how the modified bases at this position accomplish this role.

From the three-dimensional structural point of view, one of the most likely roles for the modified base adjacent to the 3'-end of the anticodon appears to be that of the base-pair breaker; that is, its role may be to ensure that there will be no base pairs between the base adjacent to the 3'-end of an anticodon and that next to the 5'-end of a codon, at the A-site. Such a mechanism can prevent slippage in reading the frame of the codes.

The three-dimensional structure of yeast tRNA shows that most base modifications occur on the surface of the molecule (suggesting that probably no drastic conformational change of tRNA is required for modification to occur) and that base modifications seem to play no significant role in the buildup of the tertiary structure, implying that these are probably for recognition by other proteins.

The methylation on the ribose is always at the O-2' position, and there are three positions in the structure at which O-2' methyl riboses are most commonly found: one in the D-loop and two in the anticodon loop, as shown in figure 9.16a. Since these are three of the most exposed regions in the structure, they may be present to protect tRNA from degradation by ribonucleases, which usually require a O-2' hydroxyl group. The base modifications in all tRNAs in the D$_4$V$_5$ class (class 1) are distributed in all the domains but infrequently in the aa-stem.

9.5.5 In Other tRNA Functions

Transfer RNA plays important roles other than those in ribosome-mediated protein synthesis, such as in cell-wall synthesis, in the regulation of biosynthesis of certain enzymes, in the transport and transfer of amino acid in the absence of ribosomes, in DNA synthesis as a primer for reverse transcriptase, and more. What aspect of tRNA is responsible for such wide varieties of function? One possible answer may be its structural features and the stability of its structure. Transfer RNA is probably one of the most stable small nucleic acids known. It is relatively compact, can easily renature, and

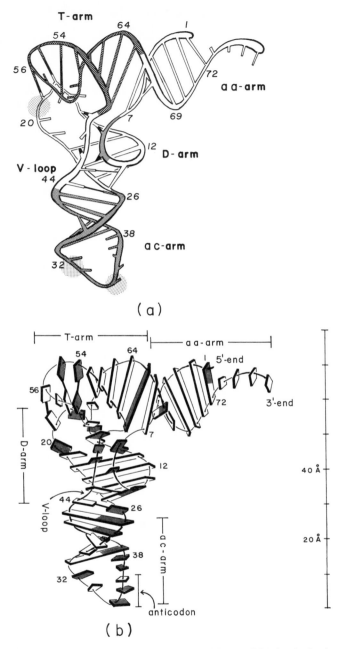

Figure 9.16 The relative locations of nucleotide modification in the three-dimensional structure of tRNA, assuming that all tRNAs have the similar overall structure, excluding the V-loop, as that of yeast tRNA^Phe . (a) The most common locations of 2'-O-methyl ribose found in tRNA are indicated by shaded circles. (b) The locations of modified bases found in tRNAs of class D_4V_5 (class 1) are shaded.

thus probably can be utilized as a landmark or flag that various enzymes will recognize easily. Such may be the case with reverse transcriptase: the enzyme may recognize tRNA-like features formed by the primer tRNA and viral RNA (see figure 9.17)—for example, through symmetry recognition and then base recognition—for specificity similar to that of aminoacyl-tRNA synthetase, and then start reverse transcription. Another possible example is the tRNA sequences located between the 16S and 23S rRNA sequence in *E. coli* rRNA precursor (124, 125). These tRNAs are probably folded into the three-dimensional structures and act as recognition markers for rRNA processing enzymes or even for tRNA processing enzymes (not necessarily RNase P), which in this case also play the role of processing rRNA precursor as a result of processing tRNA precursor present in the spacer region.

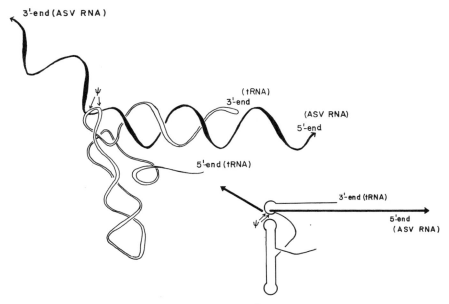

Figure 9.17 A model for the complex between tRNA[Trp] and RNA from avian sarcoma virus (ASV), where part of the tRNA structure was replaced by the viral RNA. The model is based on the assumption that the tRNA[Trp] has the same overall structure as yeast tRNA[Phe], on the observations showing the existence of the complex between the tRNA[Trp] and reverse transcriptase (122), and on the identification of a 16-nucleotide-long fragment, beginning with the penultimate nucleotide at the 3'-end of the tRNA, as the binding sequence to the viral RNA (123). In this model, the unusual Ψ-Ψ sequence in the T-loop of the primer tRNA is located at the very end where the tRNA and the viral RNA make tne final contact. The model suggests that the tRNA-like portion will have approximately the same overall structure as the free tRNA except in the variable loop and portions of the D- and T-loops.

9.6 Conclusion

The three-dimensional structure of yeast tRNA[Phe] reveals an architectural design by which a relatively short nucleic acid can be folded up to form a compact, stable structural entity that has many and diverse functions in living cells. It also explains in a direct and simple way the reason for the presence of most of the conserved and semiconserved bases in all tRNAs and is consistent with the vast amount of other experimental data on tRNA in solution. The structure also has a functional design in that three important functions in ribosomes are associated with three well-separated regions of the structure.

The three-dimensional structure determined by X-ray crystallographic methods is necessarily static, and no knowledge of the dynamic aspect of the molecule can be obtained directly either for a free molecule or for the molecule interacting with other macromolecules. Nevertheless the crystal structure of tRNA provides a solid starting point for studies on tRNA function and sharpens our attention in designing further experiments.

Acknowledgment

The research from the author's laboratory referred to in this review has been supported by grants from the National Institutes of Health, U. S. Department of Health, Education and Welfare (CA-15802), and the National Science Foundation (GB-40814).

References

1. Sigler, P. (1975). Ann. Rev. of Biophysics and Bioengineering *4*, 477.

2. Kim, S.-H. (1976). Prog. Nucl. Acid Res. Mol. Biol. *17*, 181.

3. Rich, A., and RajBhandary, U. L. (1976). Ann. Rev. Biochem. *45*, 805.

4. Cramer, F., van der Haar, F., Saenger, W., and Schlimme, E. (1968). Angew. Chem., Int. Ed. Engl. *7*, 895.

5. Kim, S.-H., and Rich, A. (1968). Science *162*, 1381.

6. Hampel, A., Labanauskas, M., Conners, P. G., Kirkegard, L., RajBhandary, U., Sigler, P., and Bock, R. M. (1968). Science *162*, 1384.

7. Fresco, J. R., Blake, R. D., and Langridge, R. (1968). Nature (London) *220*, 1285.

8. Clark, B. F. C., Doctor, B. P., Holmes, K. C., Klug, A., Marcker, K. A., Morris, S. J., and Paradies, H. H. (1968). Nature (London) *219*, 1222.

9. Hampel, A., and Bock, R. (1970). Biochemistry *9*, 1873.

10. Kim, S.-H., Quigley, G., Suddath, F. L., and Rich, A. (1971). Proc. Nat. Acad. Sci. USA *68*, 841.

11. Ichikawa, T., and Sundaralingam, M. (1972). Nature (London) *236*, 174.

12. Ladner, J. E., Finch, J. T., Klug, A., and Clark, B. F. C. (1972). J. Mol. Biol. *72*, 99.

13. Schevitz, R. W., Navia, M. A., Bantz, D. A., Cornick, G., Rosa, J. J., Rosa, M. D., and Sigler, P. B. (1972). Science *177*, 429.

14. Kim, S.-H., Quigley, G. J., Suddath, F. L., McPherson, A., Sneden, D., Kim, J. J., Weinzierl, J., and Rich, A. (1973). Science *179*, 285.

15. Kim, S.-H., Suddath, F. L., Quigley, G. J., McPherson, A., Sussman, J. L., Wang, A., Seeman, N. C., and Rich, A. (1974). Science *185*, 435.

16. Robertus, J. D., Ladner, J. E., Finch, J. T., Rhodes, D., Brown, R. D., Clark, B. F. C., and Klug, A. (1974). Nature (London) *250*, 546.

17. Stout, C. D., Mizuno, H., Rubin, J., Brennan, T., Rao, S. T., and Sundaralingam, M. (1976). Nucl. Acid Res. *3*, 1111.

18. Holley, R. W., Apgar, J., Everett, G. A., Madison, J. T., Marquisee, M., Merrill, S. H., Penswick, J. R., and Zamir, A. (1965). Science *147*, 1462.

19. RajBhandary, U. L., and Chang, S. H. (1968). J. Biol. Chem. *243*, 598.

20. Kim, S.-H., Sussman, J. L., Suddath, F. L., Quigley, G. J., McPherson, A., Wang, A., Seeman, N. C., and Rich, A. (1974). Proc. Nat. Acad. Sci. USA *71*, 4970.

21. Levitt, M. (1969). Nature (London) *224*, 759.

22. Ladner, J. E., Jack, A., Robertus, J. D., Brown, R. S., Rhodes, D., Clark, B. F. C., and Klug, A. (1975). Proc. Nat. Acad. Sci. USA *72*, 4414.

23. Quigley, G. J., and Rich, A. (1976). Science *194*, 796.

24. Sussman, J. L., Holbrook, S. R., Warrant, R. W., Church, G. M., and Kim, S.-H. Submitted for publication.

24a. Jack, A., Ladner, J. E., and Klug, A. (1976). J. Mol. Biol. *108*, 619.

25. Kim, S.-H., and Sussman, J. L. (1976). Nature *260*, 645.

26. Arnott, S., Hukins, D. W. L., and Dover, S. D. (1972). Bioch. Biophys. Res. Commun. *48*, 1392.

27. Ladner, J., Jack, A., Robertus, J., Brown, R. Rhodes, D., Clark, B. F. C., and Klug, A. (1975). Nucl. Acid Res. *2*, 1629.

28. Quigley, G., Seeman, N. C., Wang, A., Suddath, F. L., and Rich, A. (1975). Nucl. Acid Res. *2*, 2329.

29. Sussman, J. L., and Kim, S.-H. (1976). Bioch. Biophys. Res. Commun. *68*, 89.

30. Sussman, J., Holbrook, S., Church, G., and Kim, S.-H. (1977). Acta Crystallographica, *A33*, 800.

31. Quigley, G. J. Personal communication.

32. Lindahl, T., Adams, A., and Fresco, J. R. (1966). Proc. Nat. Acad. Sci. USA *55*, 941.

33. Stein, A., and Crothers, D. M. (1976). Biochemistry *15*, 157.

34. Holbrook, S., Sussman, J., Warrant, R., Church, G., and Kim, S.-H. (1977). Nucl. Acid Res. *4*, 2811.

35. Suddath, F. L., Quigley, G. J., McPherson, A., Sneden, D., Kim, J. J., Kim, S.-H., and Rich, A. (1974). Nature (London) *248*, 20.

36. Klug, A., Robertus, J. D., Ladner, J. E., Brown, R. S., and Finch, J. T. (1974). Proc. Nat. Acad. Sci. USA *71*, 3711.

37. Lake, J. A., and Beeman, W. W. (1968). J. Mol. Biol. *31*, 115.

38. Krigbaum, W. R., and Godwin, R. W. (1966). Science *154*, 423.

39. Connors, P. G., Labanauskas, M., and Beeman, W. W. (1969). Science *166*, 1528.

40. Ninio, J., Favre, A., and Yaniv, M. (1969). Nature (London) *223*, 1333.

41. Pilz, I., and Kratky, O. (1970). Eur. J. Biol. *15*, 401.

42. Stryer, L. (1968). Science, *162*, 56.

43. Beardsley, K., and Cantor, C. R. (1970). Proc. Nat. Acad. Sci. USA *65*, 39.

44. Yang, C., and Söll, D. (1974). Proc. Nat. Acad. Sci. USA *71*, 2838.

45. Nakanishi, K., Furutachi, N., Funamizu, M., Grunberger, D., and Weinstein, I. B. (1970). J. Am. Chem. Soc. *92*, 7617.

46. Langlois, R., Kim, S.-H., and Cantor, C. R. (1975). Biochemistry *14*, 2554.

47. Uhlenbeck, O. C., Baller, J., and Doty, P. (1970). Nature (London) *225*, 508.

48. Högenauer, G. (1970). Eur. J. Biol. *12*, 527.

49. Uhlenbeck, O. C. (1972). J. Mol. Biol. *65*, 25.

50. Pongs, O., and Reinwald, E. (1973). Eur. J. Biol. *32*, 117.

51. Uhlenbeck, O. C., Chirikjian, J. G., and Fresco, J. R. (1974). J. Mol. Biol. *89*, 495.

52. Freier, S., and Tinoco, I., Jr. (1975). Biochemistry *14*, 3310.

53. Yaniv, M., Favre, A., and Barrell, B. G. (1969). Nature (London) *223*, 1331.

54. Favre, A., Yaniv, A., and Mickelson, A. M. (1969). Biochem. Biophys. Res. Commun. *37*, 266.

55. Bergstrom, D. E., and Leonard, N. J. (1972). Biochemistry *11*, 1.

56. Carré, D. S., Thomas, G., and Favre, A. (1974). Biochimie *56*, 1089.

57. Krauskopf, M., Chen, C.-M., and Ofengand, J. (1972). J. Biol. Chem. *247*, 842.

58. Gamble, R., and Schimmel, P. R. (1974). Proc. Nat. Acad. Sci. USA *71*, 1356.

59. Gamble, R. C., Shoemaker, H. J. P., Jekowsky, E., and Schimmel, P. R. (1976). Biochemistry *15*, 2791.

60. Shoemaker, H. J. P., Gamble, R. C., Budzik, G. P., and Schimmel, P. R. (1976). Biochemistry *15*, 2800.

61. Brown, D. M. (1975). In Basic Principles in Nucleic Acid Chemistry, vol. 2, P. O. P. Ts'o, ed. (New York: Academic Press), p. 2.

62. Cashmore, A. R. (1971). Nature New Biol. *230*, 236.

63. Kucan, Z., Frende, K. A., Kucan, I., and Chambers, R. W. (1971). Nature New Biol. *232*, 177.

64. Singhal, R. P. (1971). J. Biol. Chem. *246*, 5848.

65. Chang, S. E., Cashmore, A. R., and Brown, D. M. (1972). J. Mol. Biol. *68*, 455.

66. Chang, S. E. (1973). J. Mol. Biol. *75*, 533.

67. Chang, S., and Ish-Horowicz, D. (1974). J. Mol. Biol. *84*, 375.

68. Rhodes, D. (1975). J. Mol. Biol. *94*, 449.

69. Piper, P., and Clark, B. F. C. (1974). Nucl. Acid Res. *1*, 45.

70. Goddard, J. P., and Schulman, L. H. (1972). J. Biol. Chem. *247*, 3864.

71. Robertus, J. D., Ladner, J., Finch, J., Rhodes, D., Brown, R. D., Clark, B. F. C., and Klug, A. (1974). Nucl. Acid Res. *1*, 927.

72. Katz, L., and Penman, S. (1966). J. Mol. Biol. *15*, 220.

73. Kearns, D. R., Patel, D. J., and Shulman, R. G. (1971). Nature *229*, 338.

74. Kearns, D. R. (1976). Prog. Nucl. Acid Res. Mol. Biol. *18*, 91.

75. Lightfoot, D. L., Wong, K. L., Kearns, D. R., Reid, B. R., and Shulman, R. G. (1973). J. Mol. Biol. *78*, 71.

76. Shulman, R. G., Hilbers, C. W., Kearns, D. R., Reid, B. R., and Wong, Y. P. (1973). J. Mol. Biol. *78*, 57.

77. Giessner-Prethe, G., and Pullman, B. J. (1970). J. Theoret. Biol. *27*, 87.

78. Wong, Y. P., Kearns, D. R., Reid, B. R., and Shulman, R. G. (1972). J. Mol. Biol. *72*, 725.

79. Bolton, P. H., Jones, C. R., Lerner, D., Wong, L., and Kearns, D. R. (1976). Biochemistry *15*, 4370.

80. Reid, B. R., and Robillard, G. T. (1975). Nature *257*, 287.

81. Robillard, G. T., Tarr, C. E., Vosman, F., and Berendsen, H. J. C. (1976). Nature *262*, 363.

82. Geerdes, H. A. M., and Hilbers, C. W. (1977). Nucl. Acid Res. *4*, 207.

83. Kan, L. S., Ts'o, P. O. P., van der Haar, F., Sprinzl, M., and Cramer, F. (1974). Biochem. Biophys. Res. Commun. *59*, 22.

84. Prinz, H., Maelicke, A., and Cramer, F. (1973). Biochemistry *13*, 1322.

85. Olson, T., Fournier, M. J., Langley, K. H., and Ford, N. C., Jr. (1976). J. Mol. Biol. *102*, 193.

86. Wong, Y. P., Reid, B. R., and Kearns, D. R. (1973). Proc. Nat. Acad. Sci. USA *70*, 2193.

87. Beres, L., and Lucas-Leonard, L. (1973). Biochemistry *12*, 3998.

88. Thomas, G. J., Jr., Chen, M. C., Lord, R. C., Kotsiopoulus, P. S., Tritton, T. R., and Mohr, S. C. (1973). Biochem. Biophys. Res. Commun. *54*, 570.

89. Pongs, O., Werde, P., Erdman, V. A. and Sprinzl, M. (1976). Bioch. Biophys. Res. Commun. *71*, 1025.

90. Kim, S.-H. Adv. Enzym. (in press).

91. Kisselev, L. L., and Favorova, O. O. (1974). Adv. Enzym. *40*, 141.

92. Söll, D., and Schimmel, P. (1974). Enzymes *10*, 489.

93. Roe, B., Sirover, M., and Dudock, B. (1973). Biochemistry *12*, 4146.

94. Giegé, R., Kern, D., Ebel, J.-P., Grosjean, H., DeHenau, S., and Chantrenne, H. (1974). Eur. J. Biol. *45*, 351.

95. Kula, M.-R. (1973). FEBS Letters *35*, 299.

96. Koch, G. L. E., Boulanger, Y., and Hartley, B. S. (1974). Nature (London) *249*, 316.

97. Bruton, C. J., Jakes, R., and Koch, G. L. E. (1974). FEBS Letters *45*, 26.

98. Waterson, R. M., and Konigsberg, W. H. (1974). Proc. Nat. Acad. Sci. USA *71*, 376.

99. Kim, S.-H. (1975). Nature (London) *256*, 679.

100. Rich, A., and Schimmel, P. (1977). Nucl. Acid Res. *4*, 1649.

101. Shoemaker, H. J. P., and Schimmel, P. R. (1976). J. Biol. Chem. *251*, 6823.

102. Erdman, V. (1976). Prog. Nucl. Acid Res. Mol. Biol. *18*, 45.

103. Schwarz, U., Menzel, H. M., and Gassen, H. G. (1976). Biochemistry *15*, 2484.

104. Terhorst, C. P., Möller, W., Laursen, R. and Wittmann-Liebold, B. (1973). Eur. J. Biol. *34*, 138.

105. Altman, S., and Smith, J. D. (1971). Nature New Biol. *233*, 35.

106. McClain, W. H., Guthrie, C., and Barrell, B. G. (1972). Proc. Nat. Acad. Sci. USA *69*, 3703.

107. Schedl, P., and Primakoff, P. (1973). Proc. Nat. Acad. Sci. USA *70*, 2091.

108. Sakano, H., and Shimura, Y. (1975). Proc. Nat. Acad. Sci. USA *72*, 3369.

109. Smith, J. D. (1976). Prog. Nucl. Acid Res. Mol. Biol. *16*, 25.

110. Guthrie, C., Seidman, J. G., Comer, M. M., Bock, R. M., Schmidt, F. J., Barrell, B. G., and McClain, W. H. (1975). Brookhaven Symp. Biol. *26*, 106.

111. Chang, S. E., and Smith, J. D. (1973). Nature New Biol. *246*, 165.

112. Abelson, J. N., Gefter, M. L., Barnett, L., Landy, A., Russel, R. L., and Smith, J. D. (1970). J. Mol. Biol. *47*, 15.

113. Smith, J. D., Barnett, L., Brenner, S., and Russel, R. L. (1971). J. Mol. Biol. *54*, 1.

114. Anderson, K. W., and Smith, J. D. (1972). J. Mol. Biol. *69*, 349.

115. Nishimura, S. (1972). Prog. Nucl. Acid Res. Mol. Biol. *12*, 49.

116. Schäfer, K. P., and Söll, D. (1974). Biochimie *56*, 795.

117. Zeevi, M., and Daniel, V. (1976). Nature *260*, 72.

118. Zubay, G., Cheong, L., and Gefter, M. (1971). Proc. Nat. Acad. Sci. USA *68*, 215.

119. Kuchino, Y., Seno, T., and Nishimura, S. (1971). Biochem. Biophys. Res. Commun. *43*, 476.

120. Cortese, R., Landsberg, S., von der Haar, R. A., Umbarger, H. E., and Ames, B. N. (1974). Proc. Nat. Acad. Sci. USA *71*, 1857.

121. Grosjean, J., Söll, D. G., and Crothers, D. M. (1975). J. Mol. Biol. *103*, 499.

122. Panet, A., Haseltine, W. A., Baltimore, D., Peters, G., Hasada, F., and Dahlberg, J. E. (1975). Proc. Nat. Acad. Sci. USA *72*, 2535.

123. Cordell, B., Stavnezer, E., Friedrich, R., Bishop, J. M., and Goodman, H. M. (1976). J. Virol. *19*, 548.

124. Lund, E., Dahlberg, J. E., Lindahl, L., Jaskunas, S. R., Dennis, P. P., and Nomura, M. (1976). Cell *7*, 165.

125. Yamamoto, M., Lindahl, L., and Nomura, M. (1976). Cell *7*, 179.

10 INTERACTION OF AMINOACYL-tRNA SYNTHETASES AND THEIR SUBSTRATES WITH A VIEW TO SPECIFICITY

Gabor L. Igloi
and
Friedrich Cramer

Abbreviations In schemes and formulas unspecified amino acids are denoted by aa; for example, aa-AMP stands for aminoacyl-adenylate. In the case of specific amino acids the usual three-letter symbols are used: Phe, $tRNA^{Phe}$, and E^{Phe} for phenylalanine, phenylalanine tRNA, and phenylalanyl-tRNA synthetase, respectively. These symbols are replaced by symbols such as X_{xx}, Y_{yy} when unspecified but different amino acids are quoted. The 3'-end of native tRNA is denoted by tRNA-C-C-A, while tRNA-C-C is a tRNA lacking the 3'-terminal AMP. tRNA-C-C-A $(3'NH_2)$ and tRNA-C-C-A$('NH_2)$ are used for tRNAs with 3'-terminal 3'-amino-3'-deoxyadenosine-5'-monophosphate and 2'- or 3'-amino-2'(3')-deoxyadenosine-5'-monophosphate, respectively. Other symbols are used analogously, for example, d for deoxy.

10.1 Aminoacylation, Mechanism, and Specificity

Life depends on specificity of information transfer in replication, transcription, and translation during protein biosynthesis. The aminoacyl-tRNAs are the key intermediates in the translation process; they are responsible for reading the linear information contained in the messenger RNA by virtue of their anticodon and correspondingly arranging the aminoacyl residues for a protein to be synthesized (1).

The aminoacylation of tRNA occurs in two separable steps. First the amino acid is activated with ATP to form the aminoacyl-AMP and pyrophosphate. Then in a second step, carried out by the same enzyme, the aminoacyl residue is tranferred from the aminoacyl-AMP to the 2'- or 3'-OH of the terminal adenosine of tRNA. The position of attachment of the amino acid to the tRNA depends on the enzyme in question and is not identical for all enzymes (2):

$$E + aa + ATP \rightleftharpoons E \cdot aa\text{-}AMP + PP_i, \tag{10.1}$$
$$E \cdot aa\text{-}AMP + tRNA \rightarrow aa\text{-}tRNA + AMP + E. \tag{10.2}$$

The two steps can be followed separately: the first reaction by the pyrophosphate exchange using [^{32}P]pyrophosphate and measuring [^{32}P]ATP. The second part of the reaction is measured by using [^{14}C]amino acid, and the formation of radioactivity labeled aminoacyl-tRNA is monitored. Although the evidence for such a two-step mechanism is considerable (3, 4), some have questioned whether aminoacyl-AMP in an in vivo system is a real intermediate (5).

The aminoacyl-tRNA synthetase must interact with three substrates to bring about the synthesis of aminoacyl-tRNA. Specificity must be expressed toward the amino acid and toward the tRNA, while a certain selectivity must exist for the utilization of ATP rather than any other naturally occurring nucleoside triphosphate. If a noncognate amino acid becomes activated (misactivation),

$$E^{Xxx} + Yyy + tRNA^{Xxx} \rightleftharpoons E^{Xxx} + Yyy\text{-}tRNA^{Xxx}, \tag{10.3}$$

the tRNA corresponding to the enzyme would become esterified with the wrong amino acid, and this would eventually lead to an error in the polypeptide chain sequence. Similarly if a noncognate tRNA becomes aminoacylated (misacylation),

$$E^{Xxx} + Xxx + tRNA^{Yyy} \rightleftharpoons E^{Xxx} + Xxx\text{-}tRNA^{Yyy}, \tag{10.4}$$

the result would again be an altered protein sequence. The problem of misincorporation is especially severe in the synthesis of a sequential biopolymer like protein, since the functionality of the protein usually depends on the correct location of each of the amino acids in the total chain. Any error rate in incorporation of one amino acid therefore becomes amplified by the chain length of the protein and the multiplicity with which each amino acid occurs in a given protein. The error rate of less than 1:3000, which was determined experimentally for the misincorporation of valine into isoleucine sites by rabbit reticulocytes (6), may or may not apply to other possible errors but is considered an upper limit.

This is the task of the synthetases, and thus it is not surprising that they have developed complex processes with which to ensure fidelity in recognizing the correct substrates as well as in correcting misactivated or misacylated substrates.

Bosshard has recently discussed the importance and theoretical aspects of specificity in enzyme catalysis, after considering a number of thermodynamic and kinetic models for specificity, he concluded that no single theory available at present can cover all the types of specificity that are encountered experimentally (7). Thus, although no general theory can cover the specificity of the enzymes dealt with in this survey, the extreme degree of specificity maintained by the aminoacyl-tRNA synthetases has led to increasing interest in this area and to the application of considerable effort in an attempt to clarify this phenomenon.

10.2 Specificity in Activation

10.2.1 Specificity for Amino Acid
Frequently in biochemical reactions the covalent linking of two molecules requires the prior activation of one of the participants in the reaction to a chemically more reactive form. This is also the case for aminoacyl-tRNA formation. First there is an ATP-dependent aminoacyl-tRNA synthetase-catalyzed activation of a specific amino acid:

$$E^{Xxx} + Xxx + ATP \rightleftharpoons E^{Xxx} \cdot Xxx\text{-}AMP. \tag{10.5}$$

The binding of amino acids to the synthetases has been widely studied, and Söll and Schimmel have summarized some of the findings in terms of the number of amino acid binding sites per enzyme (3). However, it has recently become apparent that these figures must be treated with caution. Fersht has established that some of the synthetases exhibit half-site reactivity and that for instance the dimeric enzymes E^{Tyr} (*Bacillus stearothermophilus*) and E^{Tyr} (*Escherichia coli*) can bind a second molecule of tyrosine once the first site is occupied by Tyr-AMP and in the presence of tRNA (8, 9). Perhaps surprisingly the same applies to the monomeric E^{Val}; under functional conditions this enzyme can simultaneously bind Val-AMP, valine, and ATP (8). It has been suggested (8) that this duplication of binding sites on a monomeric enzyme could be due to the observed repeating sequence structure of this enzyme (10). This property of enzymes containing duplicated sequences is, however, not general since for the monomeric E^{Ile} (*E. coli*) in which repeating regions have been found (11), the method of equilibrium partition has been able to detect only one site for isoleucine even under aminoacylating conditions (12). Similarly Hyafil et al. have found only one methionine site per E^{Met} (*E. coli*) subunit (13), although the polypeptide chain contains 200 residues in a repeated sequence. It is therefore of great importance to determine the functional state in which the enzyme exists during the determination of the stoichiometry of binding in order to be able to make definitive statements about the number and function of the binding sites (14).

In contrast to the number of investigations by fast kinetic methods dealing with tRNA binding, the aspect of the binding of small ligands has been largely ignored. Holler and Calvin in a detailed stopped-flow study of E^{Ile} (*E. coli*) concluded that the binding of isoleucine at low temperatures (13°C) could be interpreted as a two-step mechanism, $E + S \rightleftharpoons ES^+ \rightleftharpoons ES$, with a conformational change following the initial complex formation (15). At higher temperatures (25°C) where the first complex was said to be destabilized, the reaction became bimolecular but not diffusion controlled. The relevance of this type of two-step mechanism to the actual process of activation was not apparent until it was included by Holler et al. in their synergistic model (16). For the E^{Tyr} (*B. stearothermophilus*) the binding kinetics of tyrosine fit a simple bimolecular scheme (measured at 25°C) with a rate constant two to three orders of magnitude slower than diffusion controlled (17). A substrate-induced conformation change, although not observed directly, was also suggested for this system.

Observations of the specificity, as measured by exchange of $[^{32}P]$-PP_i into ATP, have confirmed that there is often a high level of specificity in the initial binding of natural amino acids to the enzyme but that some nonnaturally occurring amino acid analogs can often support the exchange reaction and are occasionally even transferred to the tRNA (5, 18).

As far as the initial amino acid binding specificity is concerned, several studies involving the systematic alteration of the amino acid structure have been carried out, and some generalizations can be made about the functional features that are recognized and/or bound.

Amino acid activation is a highly stereospecific process in which only L-amino acids are activated. The frequently quoted exception to this is the activation of D-tyrosine by E^{Tyr} (*E. coli*) (19). Other D-amino acids may or may not be inhibitors of the activation reaction (see, for example, ref. 20 and references therein). To achieve this stereospecificity various model active-site "pockets" and "gloves" have been proposed that bind the a-amino group and the side chain in such a way that only the L-form can be an acceptable substrate (21, 22). The detailed reasoning for these proposals rests on the fact that (a) changes in the a-amino group have generally more drastic effects on binding than modification of the a-carboxyl and (b) from the nature of the reaction the side chain must be involved in recognition and binding. Other conclusions regarding individual enzyme/amino acid pairs have also been possible through the use of substrate analogs. These have shown the importance of the length of the side chain, the nature of a hetero atom in the side chain, and substituents on an aromatic ring (see refs. 20–24). In each case it has been concluded that the optimal structure for maximal binding is that of the natural substrate.

Nevertheless the great similarity between certain amino acids means that the differences in the binding affinity to a given synthetase may not be sufficient to discriminate between them by a simple binding phenomenon. This is emphasized by the now classical observation of Baldwin and Berg that E^{Ile} (E. coli) can catalyze the valine-dependent ATP/PP_i exchange (25). Other examples of the misactivation of natural amino acids have appeared from time to time in the literature (table 10.1), but some caution has to be exercised when dealing with this reaction since it is very sensitive to contamination either in the amino acid or in the enzyme. The homogeneity of the components must therefore be assured before valid deductions can be made. Functional tests exist for ascertaining whether a given misactivation is artifactual or real, and the use of these methods has made it possible to establish that E^{Val} (yeast) activates alanine, serine, cysteine, isoleucine, and threonine and that E^{Phe} (yeast) activates methionine, leucine, and tyrosine, whereas E^{Tyr} (yeast) probably shows absolute specificity in this reaction (25a; see table 10.1).

Table 10.1 Naturally occurring amino acids misactivated by aminoacyl-tRNA synthetases

Enzyme	Amino Acid	Method of Determination	References
Ile (E. coli B)	Val	PP_i exchange,	25
Ile (E. coli K12)	Val, Leu	adenylate	168
Val (E. coli)	Thr, Ala, Ile	formation	23, 169, 170
Val (yeast)	Thr, Ala, Ser, Cys, Ile	PP_i exchange, modified tRNAs	18, 25a
Phe (yeast)	Met, Leu, Tyr		25a
Leu (Aesculus hippocastanum)	Ile	PP_i exchange	171

The possible lack of intrinsic specificity in the amino acid activation makes it important to establish how the overall specificity in aminoacylation is maintained, since none of these misactivated amino acids can be detected as stably esterified to the cognate, native tRNA (25, 25a).

Several correction mechanisms have been suggested recently (18, 26–31). Hopfield considers that a kinetic proofreading occurs where correction of misactivated amino acids takes place prior to transacylation of the amino acid from the aminoacyladenylate to the tRNA, utilizing energy and assuming preferential release of the incorrectly formed aminoacyl-AMP from the enzyme-aminoacyladenylate complex (28). The pathway followed would be

$$E + aa + ATP \overset{a}{\rightleftharpoons} E{\cdot}aa{\cdot}ATP \overset{b}{\underset{PP_i}{\longrightarrow}} E{\cdot}aa\text{-}AMP \overset{d}{\underset{\substack{tRNA \\ c}}{\longrightarrow}} aa\text{-}tRNA + AMP + E$$

$$E + aa + AMP$$

The primary discrimination occurs at the reversible step a; but once bound, the amino acid would be nonspecifically activated by the irreversible step b, which is coupled to ATP hydrolysis. Step c, which is also irreversible, acts as a second or proofreading discriminatory step in which the incorrect aa-AMP is released in preference to the correct adenylate. This process in theory no doubt leads to an increase in specificity (7), but the assumptions on which the process is based do not hold, at least in the case of the aminoacyl-tRNA synthetases. The discrepancy between theory and practice concerns the properties of the E·aa-AMP complex.

First, there is no experimental evidence that the incorrect aa-AMP is less strongly bound than the correct one. Indeed both E·aa-AMP complexes can be isolated as relatively stable entities in the case of the misactivation of valine by E^{Ile} (25). It could, however, be argued that the function of the tRNA, which must be present to achieve correction (25), induces the preferential release of the wrong aa-AMP. This, however, is not clear from Hopfield's scheme.

Second, the important assumption that step c is irreversible, because the released aminoacyladenylate is unstable with respect to hydrolysis to free amino acid and AMP, is evidently not permissible. It is well known that the equilibrium for the binding of aa-AMP by free enzyme very much favors the bound form ($K_D = 10^{-6}$M), although this equilibrium again could be affected by tRNA binding. However, involvement of the spontaneous breakdown of any free aa-AMP in the mechanism during the reaction is unlikely. The half-life of enzyme-free aa-AMP at neutrality varies from 3 min at 25°C for Val-AMP (30) to 5 hrs at 0°C for Ile-AMP (32), whereas addition of tRNA to $E^{Ile} \cdot$Val-AMP leads to a destruction of the adenylate within seconds (30).

Third, the only experimental evidence that Hopfield has for his proposal is the ATP balance for correct and incorrect aminoacylation (33). Since 270 moles of ATP are hydrolyzed for the trapping of one Val-tRNA^{Ile} by EF-tu compared to 1.5 for Ile-tRNA^{Ile}, he argues, the requirement for energy utilization in his scheme is satisfied and only 1 in 270 moles of misactivated valine is transferred to tRNA^{Ile}. This result, however, is exactly what would be predicted by the chemical-proofreading hypothesis. Each correctly activated isoleucine (using 1 ATP) is transferred to tRNA^{Ile}, which dissociates from the enzyme and is complexed by EF-Tu. Similarly each misactivated valine (using 1 ATP) will be transferred to tRNA^{Ile}, but since hydrolytic correction occurs before release of the mischarged tRNA from the synthetase, the EF-Tu can bind only the Val-tRNA^{Ile} molecules that escape hydrolytic correction (which in this case would be 1 in 270).

Finally the proposal of Eldred and Schimmel (26) that a mischarged aminoacyl-tRNA that does not figure in Hopfield's pathway may be a transient intermediate before correction has been confirmed in the case in which E^{Val} misactivates threonine by rapid kinetic measurements (29) and by the use of modified tRNA substrates (18). Previously von der Haar and Cramer had used tRNA modifications to demonstrate the transfer of misactivated valine (by E^{Ile}) to tRNA before correction by chemical proofreading. (31, 34).

Chemical proofreading is based on the fact that tRNA^{Ile}-C-C-A(3'd), lacking the 3'OH of the acceptor terminal ribose, can be aminoacylated with valine by E^{Ile}, that is, that loss of the 3'OH leads to a stable ester (34). Further, a greater than stoichiometric hydrolysis of ATP during aminoacylation of tRNA^{Ile}-C-C-A with valine by E^{Ile}, which was not observed in the presence of tRNA^{Ile}-C-C-A(3'd), means that the native tRNA^{Ile}-C-C-A is continually being esterified and hydrolyzed during the reaction

$$\text{Val-tRNA}^{Ile}\text{-C-C-A} + E^{Ile} \longrightarrow \text{Val} + \text{tRNA}^{Ile}\text{-C-C-A} + E^{Ile}$$

$$PP_i \qquad + \text{AMP} \quad \text{ATP}$$

The absence of this buildup of AMP when tRNA^{Ile}-C-C-A(3'd) is used is also an indication of the formation of a stable ester with this modified tRNA. Thus there exists no barrier against the transfer of misactivated valine to any tRNA^{Ile} as long as the accepting 2'OH in intact. The evidence leads to the conclusion that for correction a water molecule is activated

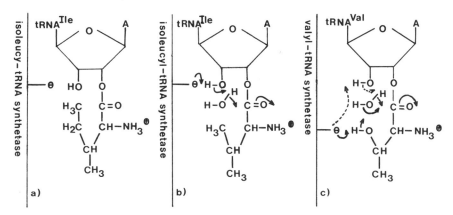

Figure 10.1 Diagrammatic representation of the proposed role of water in the hydrolytic correction of aminoacyl-tRNA by isoleucyl- or valyl-tRNA synthetase. (a) Ile-tRNA^{Ile} and (b) Val-tRNA^{Ile} as proposed by von der Haar and Cramer (31); (c) Thr-tRNA^{Val} showing activation of H_2O via the threonine OH (solid arrows) and via the ribose 3'-OH (dashed arrows) mechanistic pathways (18). The molecular nature of enzyme functional group $-\ominus$ is not defined.

through the enzymatic activation of the essential ribose 3' OH. This water molecule occupies the position of the missing $-CH_3$ group in valine compared to isoleucine and brings about a chemical reaction (hydrolysis) that corrects the initial error in recognition (figure 10.1).

An essentially similar finding was obtained in the E^{Val}/threonine system (18), where an overall comparable chemical process was found that differed from the E^{Ile}/valine case only in its mechanistic detail. Thus the major pathway to activation of the water molecule passes through the threonine β-OH with only a minor contribution from the ribose 3'OH (figure 10.1). A comparison of the theories behind the kinetic and chemical-proofreading schemes has been given by von der Haar (35).

The results of the fast reaction experiments of Fersht (29, 30) tend to favor the chemical-proofreading scheme, with some reservations. Direct comparison of the quench flow data with those of the more chemical approach is in qualitative agreement in the case of E^{Val}/threonine, although the rate of hydrolysis of the incorrectly charged tRNA shows large differences. These discrepancies may arise from the use of macromolecules from different sources (*B. stearothermophilus*: ref. 29; yeast: ref. 18). In the case of E^{Ile}/valine Fersht could not detect the buildup of the transiently mischarged Val-tRNA^{Ile} (30). This could theoretically be due to a slow transfer step from the $E^{Ile} \cdot$ Val-AMP to tRNA followed by a rapid hydrolysis, in contrast to the rapid transfer and slow hydrolysis observed in the E^{Val}/threonine system. However, the measured kinetic constants for the hydrolysis of Val-tRNA^{Ile} by E^{Ile} rule out this possibility, and Fersht proposed two alternative pathways that would account for the apparent lack of a Val-tRNA^{Ile} formation (30). First he suggests that a different, hydrolytically active, enzyme conformation may be required during the transfer of the amino acid from the aa-AMP to the tRNA and that this active conformation has to be acquired in the isolated system of E^{Ile} + Val-tRNA^{Ile} before hydrolysis can occur. This conformational rearrangement would have to be taken into account when considering the overall rate constant of the hydrolysis. Alternatively, there may be a different mechanism by which Val-AMP is hydrolyzed before transfer to the tRNA, and the enzyme-catalyzed deacylation of Val-tRNA^{Ile} may be only a secondary corrective process:

$$E + aa + tRNA + ATP \longrightarrow E\cdot aa\text{-}AMP\cdot tRNA \longrightarrow E\cdot aa\text{-}tRNA + AMP$$
$$PP_i \qquad\qquad\qquad \downarrow$$
$$E + aa + AMP + tRNA$$

This pathway has an important advantage over that of Hopfield: it still exhibits a positive enzyme-catalyzed hydrolysis and not merely a diffusion of the wrong product from the enzyme. However, it does not indicate a function for the tRNA and, in particular, for the all-important ribose cis diol function without which hydrolysis does not occur. Thus the presence of neither tRNA^{Ile} -C-C-A(3'd), nor tRNA^{Ile} -C-C-A(3'NH$_2$) (substrates), nor tRNA^{Ile} -C-C-A(2'd) (a competitive inhibitor for tRNA^{Ile} -C-C-A) brings about a hydrolytic correction of E^{Ile}·Val-AMP (31; von der Haar, personal communication).

The ability of misactivated amino acids to form stable compounds with modified tRNAs, in particular an inert amide link with tRNA-C-C-A(2'NH$_2$) or tRNA-C-C-A(3'NH$_2$) (18, 31, 36), is extremely useful for deciding whether a misactivation is real or a contamination-induced artifact (scheme 10.1) for the reaction

$$E^{Xxx} + Yyy + ATP \rightleftharpoons E^{Xxx}\cdot Yyy\text{-}AMP + PP_i \xrightarrow{\ tRNA^{Xxx}\ }$$
$$E^{Xxx} + AMP + Yyy\text{-}tRNA^{Xxx}.$$

If a true misactivation is occurring, there will be a Yyy-dependent ATP/PP$_i$ exchange, no appearance of [^{14}C]Yyy in tRNA^{Xxx}-C-C-A (since a correction mechanism will prevent this), but a fully stable Yyy-tRNA^{Xxx}-C-C-A('NH$_2$) can be obtained (since the amide bond is nonhydrolyzable). If

Scheme 10.1 Functional tests for the misactivation of natural amino acids

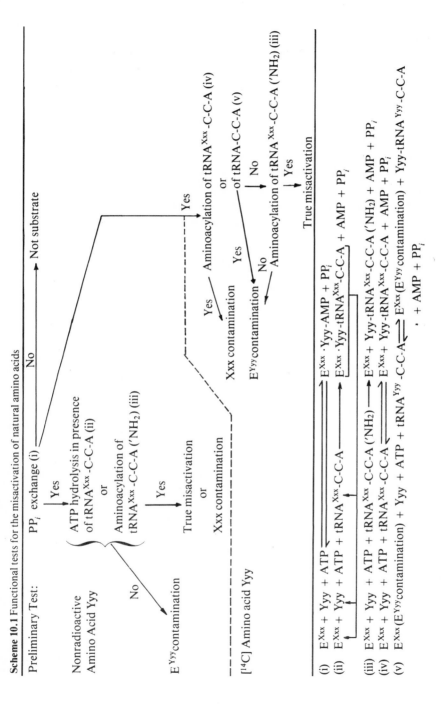

the ATP/PP$_i$ exchange were caused by a contamination E^{Yyy}, none of these amides would be formed, since E^{Yyy} does not generally recognize tRNA^{Xxx}. If the ATP/PP$_i$ exchange is caused by a contaminating Xxx, then both Xxx-tRNA^{Xxx}-C-C-A and Xxx-tRNA^{Xxx}-C-C-A('NH$_2$) will be detected. The purity of the tRNA is easily ascertained by its aminoacylation capacity. With these precautions it has been established that some synthetases can misactivate several natural amino acids and that therefore a correction process may have a general applicability.

As a consequence of the observed transfer of all activated amino acids and analogs to tRNA-C-C-A('d) and/or tRNA-C-C-A('NH$_2$), the chemical-proofreading hypothesis predicts that all activated amino acids (those that participate in ATP/PP$_i$ exchange) can esterify the corresponding tRNA-C-C-A, although the residence time on the native tRNA depends on the amino acid structure and whether correction can occur (18). The distinction usually made between analogs that are activated but not transferred and those that can aminoacylate tRNA is therefore no longer strictly valid. The difference between these amino acids lies in whether it is structurally possible for the synthetase to hydrolyze the aminoacyl-tRNA ester bond and whether a given synthetase has the intrinsic correcting capability.

10.2.2 Specificity for ATP

The activation of amino acids requires the energy-releasing hydrolysis of ATP. Thus specificity also has to be expressed at this level, although in this case the utilization of other naturally occurring energy donors would not influence the fidelity of protein biosynthesis.

The stoichiometry of ATP binding has been determined in several instances, but the comments relating to these types of investigations made in connection with amino acid binding also apply here. That is, although until recently all the results were consistent with one ATP site per monomer (37), the results of Fersht (8) on E^{Tyr} and E^{Val} imply that there may be substrate-dependent changes in the stoichiometry of ATP binding during catalysis in other systems as well. Further, in the case of E^{Met} and E^{Ser} some of the apparent ATP sites have also been implicated in tRNA binding (38, 39).

The nature of the binding sites has been investigated using ATP analogs in a number of instances. Because of the nature of the ATP molecule, a large number of modifications can be carried out and a great variety of nucleoside triphosphates has been tested in the activation and aminoacylation reactions (14, 40–45). It is generally observed by examining the inhibitory properties of the ATP analogs that the 6-NH$_2$ group is essential for binding and/or activity. However, only a few such generalizations can be made from these investigations. Perhaps surprisingly, differences in the structural requirements are sufficiently well-characterized to rule out any artifactual involvement. These differences, which concern mostly the ribose and phosphate regions of the molecule (14, 40–42), are emphasized still further if the effects of a selection of ATP analogs on the synthetase-catalyzed aminoacylation are examined for several enzymes under identical conditions. The detailed studies for several enzymes under identical conditions. The detailed studies of Freist et al. on the enzymes specific for Ser, Phe, Val, Ile, Tyr, Thr, Lys, and Arg (43–45) have made it possible to formulate a hypothesis correlating the presumed solution conformation of the ATP analogs with the subunit structures of the corresponding synthetases. A consideration of these eight synthetases from yeast and approximately twenty ATP analogs led to the proposal that subunit enzymes ($E^{Ser,Phe,Thr,Lys}$) utilize a $Mg^{++} \cdot ATP$ complex (figure 10.2) having an anticonformation, whereas the single-chain synthetases ($E^{Val,Ile,Arg,Tyr}$) accept the syn conformation. This simple relationship is supported by much of the data, but some exceptions do exist, particularly in the interpretation of inhibitor binding. Although the solution structure of the natural substrate $Mg^{++} \cdot ATP$ is still uncertain, it may exist in a syn-anti equilibrium, and then stereoselection by the enzymes may take place (43). Stereoselection of a different kind is certainly observed with some of the synthetases. In this case a new optical active center was introduced into the ATP molecule by substitution of S for O at the a-phosphate (ATPaS) (46). The two diastereoisomers were separated, and it was observed that only one of these forms is utilized by E^{Phe} (yeast) but both are

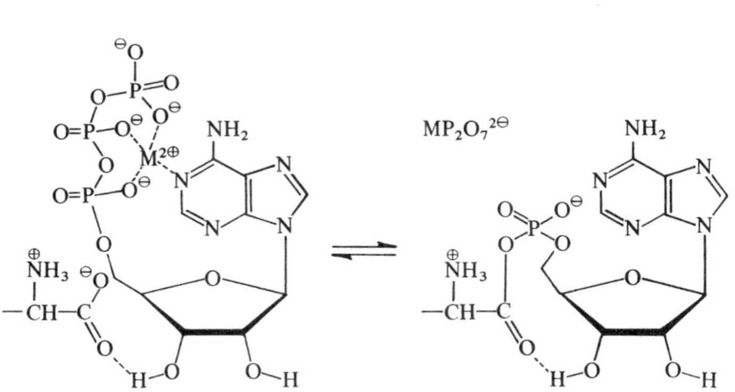

Figure 10.2 Activation of an amino acid starting with a $Mg^{++} \cdot ATP$ complex in syn conformation. From Freist et al. (131).

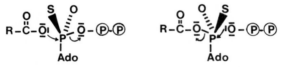

Figure 10.3 Double SN_2 displacement reaction during ATP a S/PP$_i$ exchange. The absolute configuration shown was arbitrarily chosen. From von der Haar et al. (46).

accepted by E^{Phe} (*E. coli*) (47). In the case of the yeast enzyme the analog was used to show that during ATP/PP$_i$ exchange, double inversion of the asymmetric center occurred since there was no overall change in the nature of the optically active center during the exchange reaction (figure 10.3). However, the in vivo function of such a stereoselectivity is not immediately obvious.

It is clear from these studies that not all the synthetases have identical ATP binding sites, even though it has been known for some time that they are all specific for ATP and discriminate against other naturally occurring bases (5). The great sensitivity of the enzymes to small changes in the ATP structure is consistent with the viewpoint that specificity in this case is maintained at a binding level and does not require conformational changes, for example, to enhance the specificity (compare tRNA bindings).

10.2.3 Synergism and the Mutual Effect of Substrate Binding
Binding of the substrates has until recently been studied mostly in isolated systems. A more realistic view of the in vivo process would involve the measurement of the binding affinity of one substrate in the presence of the complete aminoacyl-tRNA synthesizing system. Indeed examination of the effect of one compound on the binding of another, in particular the use of substrate analogs, has demonstrated the importance of treating the complete reaction as a whole.

An interesting aspect revealed by these studies is synergism; that is, the binding of one substrate facilitates the binding of a second substrate. This apparent coupling between binding sites was first characterized by Kosakowsky and Holler (48) for E^{Phe} and later by Fayat and Waller (49) and Blanquet et al. (50) for E^{Met}. The basic observation that led to the idea of synergism was that in each case the binding of the corresponding amino alcohol was enhanced by the presence of ATP, ADP, or AMP, or vice versa. Certain structural requirements have been determined for efficient coupling between sites. Only a bound amino acid can induce synergism in ATP binding so that the factors that govern the binding specificity of amino acids also control the capacity for synergistic coupling. The investigation of the effect of several amino acid analogs and the set of adenosine phosphates and their magnesium salts revealed a relationship in the degree of cooperativity and the balance of the electrostatic charges present in the active site at the time of binding. This led Holler et al. (16) to reintroduce their earlier model of the active separation of charges that may take place when substrates bind to E^{Ile} (24; see figure 10.4). The model applied to E^{Phe} by Holler and extended to E^{Met} by Fayat et al. (51) allows for the separation of an ion pair on the enzyme by neutralization of opposed electrostatic charges on the substrates. The energy required for this charge movement would be detrimental to the binding of the first substrate; but once achieved, this binding would facilitate the binding of the second substrate. This model, which summarizes the observations made for the amino alcohol–dependent binding of adenosine phosphates, must be modified if it is to take into account the properties of the natural substrates (figure 10.4). In this case it is generally observed that there is no synergism between amino acid and ATP binding (13). This has been attributed to the additional negative charge introduced by the a-COOH of the amino acid, which causes unfavorable electrostatic interactions that balance out any cooperative effect. One could then ask whether the observed synergism with substrate analogs has an in vivo significance. The possibility remains that the energetically unfavorable electrostatic crowding, which is predicted for the natural substrates, may be relieved by their reaction to form the enzyme-aminoacyladenylate. The loss of the a-COO^{-}/phosphate repulsion could be the driving force for the activation (13, 51). The model would also give a functional explanation for the two-step binding kinetics of isoleucine observed by stopped flow. The proposed conformational change following the initial binding step could be envisaged as being involved in the ion pair opening step of the synergistic model (16).

The available information concerning synergism leads to the conclusion that this phenomenon does not contribute to the overall specificity in activation. It remains to be seen whether differences can be detected in the cooperative behavior between ATP and a cognate amino alcohol and the same reaction with a noncognate but bound amino alcohol.

10.2.4 Order of Binding

The influence of the presence of one substrate on the enzymatic binding of a second compound is also of mechanistic significance. Thus there has been interest in defining the order in which substrates are bound in the activation

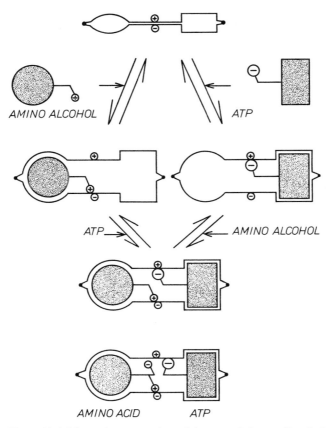

Figure 10.4 Schematic presentation of the synergistic coupling during formation of ternary amino acid:tRNA ligase-ligand complexes. *Top:* Intermolecular coupling between an amino alcohol and ATP or MgATP^{2-}, respectively. In the free enzyme the access to the binding sites specific for amino acid and ATP is impaired by an interacting ion pair. When one of the sites is occupied, the other site is simultaneously opened up together with the ion pair. The second ligand can now enter readily, stabilizing the ternary complex by interactions with fragments of the ion pair and the site adopting the side chain or the adenosine moiety, respectively. *Bottom:* Intramolecular synergism is visualized as a contribution to complex stability via interaction of the amino acid carboxylate with the cationic group of the former ion pair. Addition of ATP (or MgATP^{2-}) to the enzyme/amino acid complex is associated with an accumulation of negative charges and thus repulsion between carboxylate and triphosphate moieties. In this case intermolecular synergism is impaired or completely canceled. From Holler et al. (16).

reaction and in the overall aminoacylation. Kinetic treatment of the amino-acylation reaction (reviewed by Kisselev and Favorova, ref. 4) can give an idea of the sequence of binding events. However, the kinetic laws that govern the reaction are complex, and a simplified interpretation of the rates of the reaction, under a given set of conditions, can lead to ambiguities. Thus no single general binding sequence can cover the activities of all the synthetases, but it is not always clear whether the differences are due to the intrinsic enzymatic properties or whether a mathematical or human artifact can be invoked.

Most kinetic analyses have been carried out by the method of Cleland (52) and have resulted in one of two general schemes. Many of the reactions can be described by a "Bi Uni Uni Bi Ping Pong" mechanism whose details may vary (table 10.2). A common feature of these reaction types is the initial binding of the ATP, followed by the amino acid, release of PP_i, and binding of tRNA. There is more uncertainty about the order of product release, which may or may not be random. Random binding of substrates has in some cases also been established (table 10.2), while more complex mechanisms might be occurring in other instances (53).

It is perhaps significant that E^{Arg}, which requires the presence of tRNA before ATP/PP_i exchange can occur, does not fall into either of the two general classes of mechanism; instead tRNA is bound at an initial stage.

Kinetic analysis, although complicated and indirect, appears to be the only approach now available that can distinguish between binding sequence possibilities. Direct detection of complexes between enzymes and substrates is of little value in this context, since the association in isolated systems does not necessarily reflect a catalytic complex formation. Indeed such binding may result in dead-end complex formation which plays no role in the overall aminoacylation (4).

10.3 Specificity for tRNA in Aminoacylation

10.3.1 The Detection of Enzyme:tRNA Interactions

A wide variety of general methods has been developed to detect the formation of E·tRNA complexes (3). Of the many possibilities only a few have come into regular use, and none is without its disadvantages.

The nitrocellulose membrane filter assay as developed by Yarus and Berg uses the property of these membranes to retain enzyme-bound nucleic acid while permitting uncomplexed material to pass through (54). Although convenient and still widely used (55), retention is rarely quantitative and is suit-

Table 10.2 Order of substrate binding to aminoacyl-tRNA synthetases

Enzyme	Order of Binding	References
Leu (yeast)		172
Leu (*E. coli*)		173, 174
Ser (*E. coli*)		174
Pro (*E. coli*)		175
Thr (Rat liver)		176
Val (*E. coli*)		174

For the group Leu–Val:

ATP   aa   PP$_i$ tRNA   Products

Arg (*Neurospora crassa*) — 177

tRNA Arg ATP → PP$_i$ AMP → Arg-tRNA
                     AMP PP$_i$

Arg (*B. stearothermophilus*) — 178

tRNA ATP ... Arg   Products
ATP tRNA

Phe (*E. coli*)		179
Tyr (*E. coli*)	Random	180
Arg (*E. coli*)		181

able only for detecting strongly bound tRNA. For this reason filtration is often carried out at low pH—where the complexes are more stable but where dangers of denaturation of the protein exist (56)—and at low temperatures to further optimize the binding interactions. Equilibrium gel filtration (as exemplified by ref. 9) is also insensitive; it does not detect weak binding and requires large amounts of material but is useful as a rapid semi-quantitative analysis. Fluorimetric methods are not perhaps the most widely used of the available techniques. They are highly sensitive, require only small quanitities, and are readily automated and computer-linked; perhaps most important is that the binding interaction can be determined under a wide range of conditions and in the presence of other substrates. The use of the intrinsic protein fluorescence in determining binding parameters between tRNAs and synthetases was first introduced by Hélène et al. (57) and has been developed and refined to make it usable in stopped-flow and temperature-jump kinetics. The disadvantages of fluorimetry are twofold. First, inherent in its great sensitivity is the sensitivity to external influences and the need for pure compounds. Second, not all E:tRNA interactions change the protein tryptophan fluorescence (58), although in some cases this lack of intrinsic fluorescence change can be overcome by monitoring the fluorescence of a covalently or noncovalently bound label (59–61). Ultracentrifugal analysis has also been used for several years to study these complexes (62). Although it requires more material than fluorimetry, it can detect both the presence and the absence of complex formation, and corrections can be directly applied for inactive material or impurities (56).

No single method is adequate for the complete characterization of an E·tRNA complex, but complementary use of these techniques can lead to a detailed description of the binding phenomena.

10.3.2 Examples of Aminoacyl-tRNA Synthetase:tRNA Interactions

Serine system The binding of tRNA^{Ser} (yeast) to E^{Ser} (yeast) has been investigated by several methods but was initiated by the fluorimetric studies of Rigler et al. (63). Since then analysis of the binding by ultracentrifugation (56, 59) and protection against nuclease digestion (64) have supplemented the various fluorimetric measurements (39, 60, 65, 66). It was generally established that given the varying conditions of pH and ionic strength, the K_D lay in the range considered typical for this type of interaction (10^{-6}M–10^{-7}M), but there was initially some disagreement about the number of binding sites taking part in this association. It has now become apparent that many of the differences can be accounted for by considering nonspecific binding at low pH (58), the concentration range observed (56, 59), or refinements in the knowledge of physical constant such as molecular weight and the extinction coefficient (56). Thus there now seems to be sufficient evidence to support the idea that E^{Ser} (yeast) bears two sites for tRNA^{Ser} (yeast). These sites, which were considered equivalent and independent (39, 65, 67), although there were indications of cooperativity (58, 60), have recently been shown to be nonequivalent (56, 66). The nonequivalence gave rise to two observations: binding constants differ by a factor of 100 (56), and one site is sensitive to temperature, salt, and ATP (66). An increase in these parameters caused the stoichiometry to drop from 2:1 to 1:1. Rigler et al. (66) interpreted the two types of binding site as possibly representing "binding to an unspecific site mediated mainly through hydrophilic interactions" as suggested by the effects of pH, temperature, and ionic strength. These two types of sites would presumably be different from sites occupied by heterologous tRNA^{Ser} (rat liver), for example, where in the absence of ATP up to nine tRNAs are bound (39). Binding to the second (recognition) site would be mediated by a hydrophilic as well as a hydrophobic interaction. Krauss et al. (56), who also consider the possibility of intrinsically different binding sites, argue against it in terms of the subunit structure of the enzyme which would require an asymmetric association between the subunits for the appearance of intrinsically nonequivalent sites. Krauss et al. suggest the alternate view that the binding of one tRNA to a priori identical sites disfavors the binding of the second one. Such an effect would also explain the variability of the number of sites observed with different E:tRNA ratios but does not exclude the possibility that the variation is due to the enzyme monomer-dimer equilibrium as suggested by Rigler et al. (66) although not observed by Pingoud et al. (59).

This different interpretation of the nature of the two binding sites is also reflected in the design of and conclusions reached by rapid kinetic techniques. Early experiments in this direction (59, 63) by stopped-flow methods were in agreement with a single bimolecular step $A + B \rightleftharpoons AB$ with a near-diffusion-controlled association. A slow rearrangement of the initial complex was not observed, but a fast conformational change could not be excluded (59). Temperature-jump kinetics (with a much better time resolution than stopped-flow) could resolve the simple association reaction into (at least) two steps (66, 68); the simplest step that fits the data is $A + B \rightleftharpoons AB^+ \rightleftharpoons AB$. However, because of the existence of two sites on seryl-tRNA synthetase, the conditions used for carrying out the rapid kinetics must be adjusted so that in any one experiment the binding of tRNA to one site is considered while the contribution from the other site can be ignored. The choice of these conditions was dominated by the different interpretation of the binding phenomena and was thus governed by different criteria.

It seems that the relaxation measured by Rigler et al. (66) was made up of two components and hence that the numerical values obtained for the individual rate constants cannot be directly compared with those of Riesner et al. (68) where the components of the binding reaction were separately evaluated. Nevertheless saturation behavior of the concentration dependence of $1/\tau$ was observed in both investigations and interpreted as indicative of two (or more) consecutive steps during enzyme:tRNA association. There is also general agreement that the initial binding, whether to the first or the second site, is near-diffusion-controlled, but for the second step (presumably a conformational change) the reported rate constants differ. Riesner et al. (68) (scheme 10.2) find that the equilibrium constant for this conformational change after binding of the first tRNA is 2.3, increasing the overall binding constant by a factor of 3.5 (since the overall K_{Bind} is related to the equilibrium constants by $K_{Bind} = K_1 (1 + K_2)$) and making this equilibrium essential for tight binding. For the binding of the second tRNA, however, the equilibrium for the conformational change lies in favor of the initially formed complex ($K = 0.1$), and the second step will only occur at high substrate concentrations. Rigler et al. (66) found that the second equilibrium contributes a tenfold increase in the K_{Bind} for the binding of tRNA^{Ser} under the conditions used, converting the binding constant from a value corresponding to a weak nonspecific binding to a strong specific binding, a conclusion similar to that of Riesner et al. (68), who found that the second equilibrium brings about a tighter binding that would occur when a cognate tRNA is bound.

Clearly there is still considerable disagreement in the kinetic details of the reaction, but a multistep binding process has certainly been established.

Tyrosine system   Jakes and Fersht (9) attempted to resolve the controversy between the studies of Krajewska-Grynkiewicz et al. (69) and Chousterman and Chapeville (70). Using the same method, nitrocellulose filtration, these investigators determined the number of binding sites for tRNA^{Tyr} on the $E.$ $coli$ tyrosyl-tRNA synthetase as two and one, respectively. Jakes and Fersht (9), using fluorimetry and equilibrium gel filtration, also concluded that the stoichiometry for both $E.$ $coli$ and $B.$ $stearothermophilus$ enzyme was 1:1. It soon became apparent, however, that there existed a second, weaker tRNA site (71), which was not within the accessible range of their measurements. The fluorimetric determination of two nonequivalent tRNA^{Tyr} sites on E^{Tyr} was also confirmed for the $E.$ $coli$ enzyme by ultracentrifugal analysis (56) with $K_{D1} = 10^{-6}$ M and $K_{D2} = 4 \times 10^{-6}$ M at pH 7.2 with the usual pH dependence. The presence of two nonequivalent sites can be interpreted in several ways; in this case anticooperativity was not only predicted on the basis of the intrinsic symmetric nature of the E^{Tyr} ($B.$ $stearothermophilus$) crystal structure (72), but also demonstrated directly (73). It was shown by stopped-flow methods using fluoresceinisothiocyanate derivative of tRNA^{Tyr} (74) that the rate of dissociation of the tRNA bound first, after occupation of the second tRNA site, was identical with the measured rate of dissociation from the weak site. This was interpreted as an effect caused by

Scheme 10.2 Kinetic constants of the two-step binding mechanism in the recognition of tRNA by the synthetase

Binding of first tRNA

$$E + tRNA \underset{k_{21}^{1} = 220\ \text{sec}^{-1}}{\overset{k_{12}^{1} = 2.7 \times 10^{8}\,\text{M}^{-1}\text{sec}^{-1}}{\rightleftharpoons}} (E \cdot tRNA)^{+} \underset{k_{32}^{1} = 330\ \text{sec}^{-1}}{\overset{k_{23}^{1} = 760\ \text{sec}^{-1}}{\rightleftharpoons}} E \cdot tRNA$$

Recombination $K_{1}^{1} = 1.2 \times 10^{6}\,\text{M}^{-1}$ Conformational change $K_{2}^{1} = 2.3$

Binding of second tRNA

$$E \cdot tRNA + tRNA \underset{k_{21}^{11} = 270\ \text{sec}^{-1}}{\overset{k_{12}^{11} = 0.9 \times 10^{8}\,\text{M}^{-1}\text{sec}^{-1}}{\rightleftharpoons}} (E \cdot tRNA \cdot tRNA)^{+} \underset{k_{32}^{11} = 1250\ \text{sec}^{-1}}{\overset{k_{23}^{11} = 120\ \text{sec}^{-1}}{\rightleftharpoons}} E \cdot tRNA \cdot tRNA$$

Recombination $K_{1}^{11} = 3.3 \times 10^{5}\,\text{M}^{-1}$ Conformational change $K_{2}^{11} = 0.1$

Source: Riesner et al. (68).

anticooperativity rather than by unequal, independent binding. In the latter case the rate of dissociation of tRNA from the first site (strong binding) would not have been influenced by the occupation of the second (weak) site. In an anticooperative mode, however, either saturation of the enzyme by two molecules of tRNA would induce symmetry with respect to the binding sites or the interchange between weak and strong sites would become fast compared with the rate of dissociation, making the tRNA molecules indistinguishable, and the observed dissociation of any one of these tRNAs would then occur by the fast dissociation process associated with weak binding. (A similar consideration established anticooperativity between sites on the dimeric E^{Met} (*E. coli*) using initiator $tRNA^{Met}$s (*E. coli* and rabbit liver) that give rise to different fluorescence amplitudes on binding to the enzyme (75).)

The binding relaxation curve fitted to a single exponential by Pingoud et al. (73) was resolved into two processes by the temperature-jump technique of Krauss et al. (76) in a manner similar to that used in the E^{Ser} case. The fast conformational transition that has been associated with the step following the initial near-diffusion-controlled association could play a role in the interplay between the weak and strong sites.

Phenylalanine system Early studies on the binding of $tRNA^{Phe}$ (yeast) to E^{Phe} (yeast) indicated a 1:1 stoichiometry (references in ref. 3) and an affinity that was little affected by ATP or phenylalanine (58, 77). Fasiolo et al. (78) on the other hand found a tRNA:enzyme ratio of 2:1, and this was confirmed by Krauss et al. (56). The binding to the two sites was found to be equivalent with $K_D = 3 \times 10^{-8}$ M at pH 7.2. It was postulated that this identity of binding sites might be correlated with the physically larger size of E^{Phe} (MW 260,000), where two tRNAs may be simultaneously accommodated (E^{Tyr}, MW 85,000; E^{Ser}, MW 120,000). This equivalence of binding sites was also observed by Fasiolo et al. (55) using nitrocellulose filtration; but under catalytic conditions in the presence of ATP, PP_i, and phenylalanine, negative cooperativity was said to be triggered and the stoichiometry fell to 1:1 with a decreased tRNA affinity (1.5×10^{-7} M at pH 5.8). The presence of phenylalanine or ATP alone was apparently not sufficient to induce the triggering (78) or to alter the affinity (77). The E^{Phe} (*E. coli*) on the other hand apparently has two tRNA sites that show a difference in their affinity for tRNA even in the absence of other substrates (79).

Stopped-flow and temperature-jump kinetics monitoring the fluorescence of the Y-base in yeast $tRNA^{Phe}$ (76) reveal, as for E^{Tyr} and E^{Ser}, a substrate-dependent saturation of $1/\tau$ and therefore a multi (two) step binding process, namely a near-diffusion-controlled association followed by a fast conformation rearrangement. Differences between the enzymes discussed so far occur only at the level of the absolute values of the rate and equilibrium constants of the individual steps.

The binding of noncognate tRNAs   Interactions of synthetases with noncognate tRNAs have been known for some time (62, 63, 80, 81). Although many descriptions of such interactions are available, detailed studies of these complexes have only recently been carried out. Pachman et al. (58) described the cross interaction between E^{Ser} (yeast):tRNA^{Phe} (yeast) and E^{Phe} (yeast):tRNA^{Ser} (yeast), while Lam and Schimmel (82) compared the binding equilibria between E^{Ile} (*E. coli*) and E^{Val} (yeast) and several cognate and noncognate tRNAs by fluorimetry. In the case of the isoleucine enzyme the affinity of the enzyme for noncognate tRNAs varied from a fivefold weakening in the binding for tRNA^{Val} (yeast) to a 10^4-fold weakening for tRNA^{Glu} (*E. coli*), and a similar range was seen for the valine enzyme. The much reduced binding of tRNA^{Phe} to E^{Ile} had also been found by Yarus (83) and reflects in that case a reduction in the rate of association from 1.4×10^8 $1 \cdot M^{-1}$ for tRNA^{Ile} to 7×10^3 $1 \cdot M^{-1}$ for tRNA^{Phe}. The E^{Val} (*B. stearothermophilus*) and E^{Val} (*E. coli*) on the other hand show remarkable insensitivity to the nature of the noncognate tRNA and will bind with approximately equal affinity (as measured by the K_m in misacylation) to several tRNAs from *E. coli* and from yeast (84, 85). An intermediate effect is shown by E^{Val} (yeast) (82, 85), for which the affinity of noncognates is only up to a factor of 100 weaker than that of the cognate tRNA. Some enzymes are intrinsically more specific in their binding of tRNAs than others. E^{Phe} (yeast) will bind a number of noncognate tRNAs (76, 86, 87), but E^{Ser} (yeast) and E^{Tyr} (*E. coli*) discriminate much more efficiently between cognates and noncognates (56, 58, 65).

On the E^{Ser} (yeast) both tRNA sites are available for noncognate binding (58), but for E^{Phe} (yeast) Krauss et al. (76) found that only one tRNA^{Tyr} can be bound per enzyme molecule and that binding of tRNA^{Phe} prevents E^{Phe}:tRNA^{Tyr} interaction, suggesting an overlap of binding sites. Rapid kinetic studies on the latter system also show that (a) the process is always two to three times faster than the fastest relaxation observed for the cognate reaction and (b) the behavior of $1/\tau$ versus concentration justified the assumption of a single-step binding reaction without the subsequent fast conformational changes described for the specific interaction. In the absence of

Mg^{++}, where it is known that $tRNA^{Tyr}$ (*E. coli*) does not bind to E^{Tyr} (*E. coli*), the binding relaxation for the E^{Phe} (yeast):$tRNA^{Tyr}$ (*E. coli*) interaction was similar to that in the presence of Mg^{++}. In addition, however, a much slower process was observed and was attributed to a binding-related unfolding of the tRNA tertiary structure that resulted in the rearrangement of the tRNA to optimize the efficiency of the nonspecific interaction. The binding in the heterologous but cognate system of E^{Phe} (*E. coli*):$tRNA^{Phe}$ (yeast) appears to behave similarly to the cognate homologous couple.

10.3.3 Factors Affecting the Binding of tRNA to the Synthetase

The same external influences known to perturb binding equilibria during specific interactions also cause changes in the noncognate binding but not always along the same lines.

It is a generally observed phenomenon that the affinity of an aminoacyl-tRNA synthetase for tRNA increases with decreasing pH and salt concentration and is usually almost independent of the temperature (5, 56, 58, 59, 66, 76, 82, 88). The effect of salts on complex formation was investigated by Bonnet et al. (89), who applied a quantitative analysis to the binding of $tRNA^{Phe}$ (yeast) to E^{Phe} (yeast) and of $tRNA^{Val}$ (yeast) to E^{Val} (yeast) at varying salt concentrations under conditions that minimized tRNA conformational changes. The effect of salts on the enzyme conformation, which has been observed at low salt concentrations (90), was not considered. It was suggested that 40 to 50 percent of the total binding energy is contributed by ionic interactions and that in the case of E^{Phe} six such ionic interactions take part. The effect of salt on noncognate associations was found to be similar to that observed for the specific binding (91).

The effect of organic solvents, which are known to strengthen the binding (92), could also be explained in terms of enhancement of ionic interactions without postulating a conformational change (89). The considerable change in the binding interaction between pH 5 and 7, where the conformation of the tRNA is unaffected (76), has been attributed to an enzyme histidine ionization process (82, 88), which gives rise to, for example, a tryptophan fluorescence intensity pH-dependence in E^{Ile}. Magnesium, which depresses the pK values of some of the abnormal bases in $tRNA^{Ile}$ (93), did not alter the pH-dependence of the association reaction, and a pH effect on the tRNA was therefore considered unlikely. The increase in binding when the pH and ionic strength are lowered is a generally observed phenomenon and may also be accompanied by an increased nonspecific binding (91), an increased stoichiometry (58), or even by binding of non-tRNA polynucleotides such as poly d(A-T) (76) or 5S rRNA (87), although the binding of the noncognate $tRNA^{Val}$ (yeast) to E^{Lys} (yeast) was found to be independent of pH (94).

In addition to the general trend seen in the salt- and pH-dependence of enzyme:tRNA association, the limited information available suggests differences in the effect of temperature on the different complexes (60, 66, 76, 77, 82).

The effect of magnesium, which exerts its influence at concentrations far below those of neutral salts, varies with the system. In some cases its presence is essential for strong complex formation (59, 66, 81), whereas in others complex formation occurs in the absence of Mg^{++} (58, 71, 76–78, 88, 91). At the other end of the scale Mg^{++} has been found to destabilize the E^{Lys}:$tRNA^{Val}$ (yeast) complexes (94). In the same way the association constant for noncognate tRNAs binding to E^{Val} (yeast) and to E^{Phe} (yeast) was decreased by Mg^{++} (76, 91). Because of its known conformational effects (95), the behavior of Mg^{++} in synthetase:tRNA association has often been attributed to conformational stabilization. For instance $tRNA^{Tyr}$ (*E. coli*), which does not bind to its cognate synthetase in the absence of Mg^{++}, is known to have a stable tertiary structure at room temperature only in the presence of Mg^{++} (56), whereas $tRNA^{Phe}$ maintains its tertiary structure under binding conditions in the absence of Mg^{++}-induced conformational changes. However, Mg^{++} coordinated to tRNA may also be required for direct interaction with the enzyme (96, 97), so that the possibility of a more complex function for Mg^{++} exists. In particular the fact that Mg^{++} often decreases the binding efficiency of noncognates while it increases the rate of misacylation suggests that Mg^{++} may also play a part in the kinetics of the reaction.

The binding of tRNA to synthetases clearly does not occur as an isolated phenomenon but is part of the process of aminoacylation. It is therefore to be expected that the presence of the other substrates in this reaction may affect the macromolecular interaction (see the observation of Fasiolo et al. (55) on the E^{Phe}).

10.3.4 Regions of the Aminoacyl-tRNA Synthetase Involved in tRNA Binding

Some details of the tRNA regions involved in synthetase:tRNA interaction are now being elucidated, but very little is known about the sequences or areas in the synthetase that take part in tRNA binding. This is clearly correlated with our current lack of knowledge of the primary, let alone the tertiary, structures of these enzymes. The solution to this problem still awaits the cocrystallization and X-ray analysis of enzyme:tRNA complexes. This goal, which has often been mentioned in reviews (37), apparently remains as elusive as ever. Other less absolute methods of determining the regions of the synthetases that take part in tRNA binding have not been widely used.

Affinity labeling has found only limited application in the case of enzyme:tRNA interactions. When such attempts have been made to learn more about the nature of the functional groups on the enzyme involved in such complexes, there has been only mixed success. The first and best characterization of covalently bound E·tRNA complex was made by Bruton and Hartley in the E^{Met} system (98). In that case the methionyl-tRNA^{Met} was reacted with p-nitrophenylchloroformate to give a reactive p-nitrophenylcarbamyl-methionyl-tRNA^{Met}, which was shown to be bound through a urea linkage to a single lysine residue presumably at or near the catalytic site of the enzyme. All other reported affinity-labeling attempts (99–103) merely established that the chemically reactive tRNA was covalently bound to its cognate synthetase and that some protection against this binding could be achieved by the presence of substrates. In only one study was the affinity label not linked to the 3'-terminal aminoacyl residue of aminoacyl-tRNA (103), and in this case a photoreagent was introduced by alkylation of s^4U_8 of tRNA^{Phe}. The modified tRNA was a substrate in aminoacylation, with an unchanged K_D but an increased K_m, and could be photolinked to E^{Phe} (*E. coli*).

The scope of photolabeling has been extended by the work of Schimmel, who used the intrinsic photolability of the nucleotide bases in tRNA to achieve direct cross-linking with the synthetases (104–107). The resulting complexes have so far given information only about proposed contact sites on the tRNA part of the E·tRNA complex.

10.3.5 The Implication of the Binding Mechanism for Specificity

For the enzyme:tRNA pairs that have been investigated in detail two characteristics of the binding process have been resolved. These are negative cooperativity between two binding sites and a two-step association sequence. Past experience has shown that it is not always possible to generalize a property of a limited number of synthetases to the whole group of enzymes, so our discussion is limited to the three enzymes E^{Phe} (yeast), E^{Ser} (yeast), E^{Tyr} (*E. coli*) with which these two characteristics have been associated. Can these properties contribute to the maintenance of specificity in aminoacylation?

Negative cooperativity A number of models have suggested pathways permitting negative cooperativity. At a structural level it has been shown that the binding sites of the synthetases specific for Tyr, Ser, and Met are intrinsically identical and have nonequivalence induced on binding of the first tRNA. This may be a result of steric incompatibility of two tRNAs on one enzyme, or it may arise from a tRNA-induced conformational change that lowers the binding affinity of the second tRNA.

At a kinetic level Fersht (8) has described three mechanisms that relieve the unfavorable interactions at the second site by product formation; the relief of strain increases the reaction rate by lowering the activation energy. Similarly Bosshard (7) suggests that in the case of nonequivalent sites the extra binding energy from the second site could be used to advance the formation of the transition state at the first site and contribute to the catalytic efficiency. Fasiolo et al. (55) favor a flip-flop mechanism in which a chemical step at one site is linked with a different chemical step at the other, the sites alternating in activity but maintaining their asymmetry. They admit, however, that for the overall aminoacylation there are 32 possible flip-flop models and that a definite mechanism cannot yet by proposed. The flip-flop sequence does not appear to hold for E^{Tyr} (*B. stearothermophilus*) where a symmetrical $E \cdot (Tyr\text{-}AMP)_2$ can be formed (8). The enhanced catalytic activity brought about by the anticooperativity is observed in the cases of E^{Tyr} (yeast) (73) and E^{Met} (*E. coli*) (75), where it leads to large increases in the rate of exchange between free and bound tRNA. However, there is general agreement that negative cooperativity cannot contribute to specificity (7, 8).

Multistep binding process There is considerable evidence that the two-step binding process, deduced from the behavior of the association relaxation times, can significantly affect the specificity of synthetase:tRNA interaction.

The observation of Pachmann et al. (58) that nonspecific and specific interactions occurred at the same sites, together with the fact that ionic forces contribute significantly to the tRNA binding, suggested to these workers that nonspecific interactions generally precede the specific recognition step. This was also the conclusion reached by von der Haar (90) on the basis of nonspecific affinity elution chromatography and by Knorre (108) on theoretical grounds. These findings on the isolated enzyme:tRNA systems emphasized the work on misacylation in which the occurrence of nonspecific interactions had already been functionally investigated. The existence of the postulated two steps—nonspecific recombination followed by a specific recognition step—was established by rapid kinetics (66, 68, 76). The work of Krauss et al. (76) confirmed the proposal of Rigler et al. (66) who described the overall reaction as

$$E + tRNA \xrightarrow[\text{recombination}]{\text{scanning}} (E{\cdot}tRNA)^+ \xrightarrow[\text{conformational change}]{\text{identification}} E{\cdot}tRNA$$

in which a noncognate tRNA that cannot induce the conformational change leaves the recombination site before it can be aminoacylated. This effect would be amplified by the increased rate of dissociation compared with the cognate system (76). The high turnover of the binding step assures that all tRNAs can be channeled to the recombination site where noncognates are rejected because of their inability to trigger the conformational change. Such a triggering activity was first demonstrated by von der Haar and Gaertner (109) in steady-state kinetic studies on the inhibitory properties of various 3'-end adenosine modified $tRNA^{Phe}$ in ATP/PP_i exchange and aminoacylation. An intact terminal adenosine was required for the initiation of the conformational activation of the enzyme. That this triggering is related to the change seen by rapid kinetics was recently demonstrated by Krauss et al. (110), who found that $tRNA^{Phe}$-C-C associated with E^{Phe} in a single near-diffusion-controlled step without a subsequent fast structural rearrangement (compre the case of the noncognate tRNA). It would therefore seem that unless the terminal adenosine is in exactly the correct enzymatic environment after the initial nonspecific binding, triggering of aminoacylation will not occur and the wrong tRNA will be released.

The universality of this mechanism is, however, not assured. The enzymes E^{Tyr} (E. coli) and E^{Ser} (yeast) do not appear to bind noncognate tRNAs, and therefore some kind of recognition must already occur at the recombination step in these systems.

10.3.6 Misacylation and Endogenous Hydrolysis

Misacylations are considered to be those reactions in which the specificity for the recognition of the cognate tRNA has been wholly or partially lost. Three such pathways can be envisaged.

Type I: $E^{Xxx}(a) + Xxx + tRNA^{Yyy}$ (b) $\longrightarrow$ Xxx-$tRNA^{Yyy}$ (b),

where a and b represent the type of organism from which the enzyme and tRNA are isolated. One can subdivide type I into (i) species a = b, giving a homologous misacylation; (ii) a $\neq$ b and Xxx = Yyy, resulting in a heterologous acylation; and (iii) a $\neq$ B and Xxx $\neq$ Yyy, forming a heterologous misacylated tRNA.

Type II: $E^{Xxx} + Yyy + tRNA^{Yyy} \longrightarrow Yyy\text{-}tRNA^{Yyy}$.

Type III: $E^{Xxx} + Yyy + tRNA^{Zzz} \longrightarrow Yyy\text{-}tRNA^{Zzz}$.

Of these, type I misacylations have been studied and used in investigations of enzyme:tRNA recognition processes and thus will be the subject of the following discussion. However, types II and III, which also involve misactivation, are feasible with enzymes that activate a second naturally occurring amino acid such as E^{Ile} or E^{Val}. The fourth alternative,

$$E^{Xxx} + Yyy + tRNA^{Xxx} \longrightarrow Yyy\text{-}tRNA^{Xxx},$$

is prevented by natural correction processes. A heterologous version of these latter types can also be proposed.

As Jacobson's review indicates (111), it was through investigations of tRNA isoacceptors that the occurrence of type I(ii) heterologous acylations was discovered. The inability of crude enzyme preparations from one organism to aminoacylate a certain tRNA in a tRNA extract from another organism to the same extent as in the homologous system indicated the existence of multiple forms of tRNA that accepted the same amino acid (isoacceptors) but were not all recognized by the heterologous synthetase. This finding was the basis for the investigations concerned with the species specificity of the synthetases. The access to the primary structure of the tRNAs and the availability of the components of the reaction in a pure form in the 1960s meant that a less empirical approach to investigations of synthetase specificity could be taken. From this point on, systematic studies on recognition and misacylation were reported.

If one aminoacyl-tRNA synthetase catalyzes the aminoacylation of several tRNAs, then it would appear that comparing the tRNA sequences should reveal regions common to all these tRNAs, and such regions would be the much-sought recognition region on the polynucleotide. In practice the success of this approach has been limited. However, by careful examination of mischarging, several important observations concerning the specificity of aminoacylation have been made. This discussion focuses mainly on the results from three types of investigations.

The work of Dudock and co-workers utilized misacylation as an analytical tool to determine which tRNAs are recognized by one particular synthetase. This work has been followed by a detailed analysis and chemical or enzymatic modifications of the proposed recognition regions. The groups of Ebel and of Yarus have concentrated on the kinetics of the misacylation process and have in this way helped to elucidate the more dynamic aspects of the recognition process.

The special reaction conditions required to induce acylation of a tRNA by a noncognate synthetase, first introduced by Jacobson (111) and later developed by Giegé et al. (112, 113), were used in a preparative way by Eldred and Schimmel (26) and by Yarus (83) for the synthesis of tRNAs misacylated using E^{Ile} (*E. coli*). It was found that high Mg^{++}/ATP ratios as well as the presence of organic solvents (DMSO or MeOH) stimulated misacylation. The effects of different buffers, salt concentrations, and pHs were also considered and in some instances used to learn more about the enzyme:tRNA interactions. The occurrence of such misacylations in individual purified systems and the occurrence of isoleucylation to a low but significant extent under the usual aminoacylation conditions (83) raised the questions of whether these reactions were important in vivo, how widespread misacylations were, and by what mechanism the cell ensures that these mistakes are not transmitted during protein biosynthesis.

The misacylation by an isolated enzyme of a mixture of tRNAs was investigated by Kern et al. (114), Giege et al. (84, 113), and Yarus and Mertes (115). The former studies included the misacylation of bulk *E. coli* or yeast tRNA by E^{Phe} (yeast) or E^{Val} (yeast). In both cases the number of tRNAs heterologously misacylated (E (yeast) + tRNA (*E. coli*)) was greater than the homologous mistakes where, in fact, misacylation by E^{Val} could not be detected. The fact that the extent of misacylation in bulk homologous tRNA was less than the corresponding value for certain isolated tRNAs was interpreted to mean that the presence of the cognate tRNA in the tRNA mixture gave rise to competition effects; through these effects the cognate tRNA was recognized better by the enzyme and the aminoacylation of noncognate tRNAs was prevented. This competition would take place to a lesser extent in heterologous systems. The same conclusion was reached by Yarus (116), who interpreted his findings in terms of a model that correlated the chemical properties of an amino acid with the ease with which its corresponding tRNA could be isoleucylated. This led him to ascribe tRNAs to families based on the hydrophobicities of their cognate amino acids and the ease with which these tRNAs could be misisoleucylated. Similarly the theoretical approach by Crothers et al. (117) showed that it is possible to correlate the fourth nucleotide from the 3'-end with the subdivision of the amino acids into hydrophobic and hydrophilic groups. The nucleotide A corresponded to the former type, while G could be associated with the latter group. These families were said to be separated from one another by inter-cognate barriers that could be lowered by organic solvents (115). The concept of families was reintroduced by Giegé et al. (84), who found that one set of tRNAs comprised the ones most easily aminoacylated by several different synthetases. The idea of interacting families of tRNAs and amino-acyl-tRNA synthetases was also proposed.

Yarus (83) suggested that the observed number of misacylations implied that the interaction of a particular aminoacyl-tRNA synthetase with a set of noncognate tRNAs must arise from the recognition and binding of structural features common to these tRNAs. This idea was extended by Kern et al. (114), who proposed that the high specificity in aminoacylation was maintained more at the level of the aminoacylation reaction rather than by a highly specific binding. The difficulty in carrying out misacylations under normal reaction conditions would then arise not only from a weaker enzyme:tRNA binding but perhaps also from a reduction in the rate of the transacylation step due to an incorrect enzyme:tRNA contact. (This concept parallels the multistep binding process found by physical methods.) The use of special conditions, such as organic solvents and high Mg^{++} concentrations, would then cause conformational changes favoring transacylation.

The investigation of the kinetics of misacylation have thrown more light on these proposals. Giegé et al. (84, 113), Bonnet and Ebel (91), and Ebel et al. (85) find that in several systems the K_m for misacylation is only slightly greater than for the cognate charging, whereas the V_{max} is up to 10^3 slower for misacylation. The contribution to specificity is therefore from the rate of aminoacylation rather than from the difference in binding affinities. Further, by defining specificity as the ratio of the aminoacylation velocities V_{corr}/V_{incorr} Ebel et al. have calculated a specificity of the order of 10^4 for the isolated homologous E^{Val}:tRNA^{Asp} (yeast) systems under normal aminoacyl-ating conditions (table 10.3) almost entirely on the basis of the differences in V_{max}. For heterologous misacylations the specificity factor is usually smaller than in the homologous case. It seems that in the case of correct aminoacylation, where the partners in the reaction are in the optimal relative conformation, solvents cause alterations that reduce the reaction rate (table 10.3), whereas in the case of noncognate pairs the same conformational changes may optimize the enzyme:tRNA interaction (84).

Yarus (83), on the other hand, found that for the isoleucylation of tRNA^{Phe} (E. coli) by E^{Ile} (E. coli) the K_m for misacylation was 10^5 times higher than that of the correct reaction; he also found a decrease in V_{max} similar to that found by Ebel et al. (85) and Giegé et al. (84, 113), suggesting

Table 10.3 Increase in specificity due to the relative rates of correct and incorrect aminoacylations

| Enzyme | tRNA | Specificity Function (Sp)[a] Conditions | | References |
		Normal	+DMSO[b]	
Homologous				
Val (yeast)	Ala (yeast)	3.3×10^4	n.d.[c]	85
Arg (yeast)	Asp (yeast)	2.7×10^4	n.d.	85
Heterologous				
Val (*E. coli*)	Phe (yeast)	2.5×10^4	1.0×10^3	113
Val (*E. coli*)	Ala (yeast)	no reaction	3.2×10^3	113
Val (*B. ste.*)[d]	Ile (*E. coli*)	3.3×10^2	2.3	84
Val (*B. ste.*)	Met$_f$ (*E. coli*)	2.5×10^3	1.7×10^2	84
Val (*B. ste.*)	Met$_f$ (yeast)	1.6×10^2	3.4	84
Val (*B. ste.*)	Phe (yeast)	88	5.5	84

[a] Calculated according to Ebel et al. (85).
[b] DMSO = Dimethylsulfoxide.
[c] Not determined.
[d] *B. stearothermophilus*.

that in this case it is the initial higher affinity of the enzyme for the correct tRNA that contributes to the specificity. Thus the effect of solvents is probably not as simple as was first supposed.

Calculations of the competitive effects of cognate tRNAs on the velocity of misacylation (compare those on the isolated systems) based solely on the differences in binding efficiencies (116) or on a combination of V_{max} differences and binding competition (85) show that the binding effects contribute only a ten- to twentyfold increase in specificity compared with the 10^5-fold effect of the different aminoacylation rates. The importance of V_{max} in specificity is emphasized by the accumulation of evidence that noncognate enzyme:tRNA interactions can be similar in strength to those of the cognate pair. Nevertheless it is to be anticipated that in the cell, where the concentrations of a tRNA and of a synthetase are said to be equal at approximately 10^{-6} M for *E. coli* (118), the majority of unacylated tRNAs will exist as cognate enzyme:tRNA complexes. Even though the chance of misacylation under these conditions is minimal and since the velocity effects should be an additional barrier against the incorporation of the wrong amino acid, it has been suggested that the intrinsic hydrolytic capacity of the synthetases could act as another correction mechanism (27).

Many, but not all (31), aminoacyl-tRNA synthetases have the endogenous property of hydrolyzing cognate aminoacyl-tRNAs in the absence of AMP and PP_i (119), and not in the reverse of aminoacylation. Some have the capability of removing the wrong amino acid from the cognate tRNA:

$$E^{Zzz} + Xxx\text{-}tRNA^{Zzz} \longrightarrow E^{Zzz} + tRNA^{Zzz} + Xxx. \qquad (10.6)$$

The fact that this reaction in certain cases proceeds at a much greater rate than the corresponding cognate reactions is the reason this hydrolysis was suggested as a corrective mechanism (26). We should immediately distinguish between two reactions that may superficially seem identical: namely chemical proofreading and the endogenous hydrolytic capacity of the synthetases. The scheme just described appears to be part of the reaction described previously as chemical proofreading. Whereas in chemical proofreading the same enzyme corrects its own misactivation and transfer in a concerted process (without release of the aminoacylated tRNA), in the reaction in equation 10.6 $tRNA^{Zzz}$ is misacylated with Xxx by E^{Xxx} (not E^{Zzz} as occurs in the proofreading system). The misacylated tRNA is released from E^{Xxx} and is hydrolyzed by E^{Zzz} (scheme 10.3). It is tempting to attribute such a function to the endogenous hydrolytic capacity of the synthetases, but considerable existing evidence suggests that the hydrolysis of aminoacyl-tRNA by the cognate synthetases does not fulfill a corrective function. Although E^{Ile} (*E. coli*) will specifically deacylate Val-tRNA^{Ile} without attacking Phe-tRNA^{Ile} (26), Bonnet et al. (120), Bonnet and Ebel (121), and Sourgoutchov et al. (122) showed that there was a general lack of specificity in this reaction. Yarus (123) observed the hydrolysis of 15 out of 18 correctly aminoacylated tRNAs by E^{Ile} (*E. coli*). A common but not exclusive feature of the hydrolytic reaction is the inability to inhibit it by free amino acids or aminoacyladenylates. In other words if the aminoacyl part of the aminoacyl-tRNA occupied the amino acid binding site on the enzyme, then addition of free amino acid would be expected to inhibit hydrolysis. In general it is clear that it is the tRNA moiety that brings about the interaction between noncognate aminoacyl-tRNAs and synthetases (91) and that many more such noncognate pairs can be detected by monitoring the hydrolytic reaction. (123). The fact that the aminoacyl moiety does not seem to reside in the amino acid binding site is perhaps not surprising in view of the large number of bound different aminoacyl-tRNAs and because of the well-known high specificity for the amino acid in the aminoacylation reaction.

Scheme 10.3 Distinction between chemical proofreading and endogenous hydrolysis

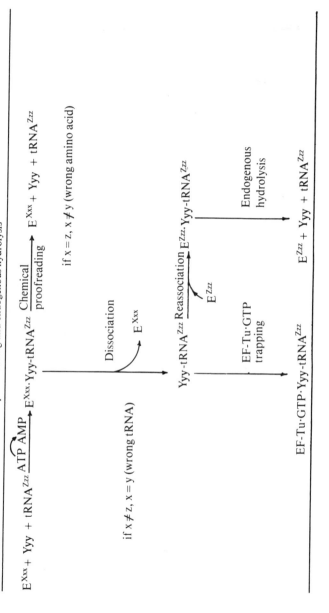

The insensitivity of the hydrolysis to free amino acid may also be related to the fact that hydrolysis is said to occur from the nonaccepting OH of the terminal ribose (31). The possibility therefore exists that the positioning of a free amino acid in the binding site in the vicinity of the accepting ribose OH may still be accomplished with little or no effect on the tRNA binding.

The apparent lack of specificity in the endogenous hydrolytic reaction argues against the early proposal of Yarus (27) that it could be involved in maintaining specificity within the cell. There is now further quantitative and qualitative evidence that rules out the possibility of such a correcting process. Bonnet and Ebel (121) analyzed the contribution to specificity (defined in this case as the ratio between the concentration of correctly and incorrectly aminoacylated tRNA) afforded by a hydrolytic pathway. They concluded that deacylation could contribute only a 0.1 percent increase in specificity beyond that already inherent in the system. However, they assume that in vivo there are equilibria between aminoacylation and deacylation (as is observed in vitro; see ref. 124) and incorporation of the amino acid into the protein. This is clearly not an in vivo situation. The presence of large amounts of elongation factor in cells (125), with its high affinity for aminoacyl-tRNA—$K_D \approx 10^{-7}M$ (126)— means that once an aminoacylated tRNA is released from the synthetase it will immediately be complexed by the elongation factor and protein synthesis will proceed. Thus a wrongly aminoacylated tRNA has only a minimal chance of reassociation with a hydrolytically competent synthetase. This abolishment of deacylation by EF-Tu has been observed (33, 119). It should again be emphasized that a hydrolytic correction that takes place before release of the aminoacyl-tRNA from the synthetase, for example chemical proofreading, would not be affected by the presence of EF-Tu (scheme 10.3).

The source of specificity must therefore lie at a stage prior to release of aminoacyl-tRNA from the synthetase and most probably resides in the rate and binding differences. However, this leaves open the question of what function the endogenous hydrolytic property of the synthetases has in vivo.

10.3.7 Recognition

Although the goal in this field for many years has been to identify the recognition region of tRNA, we are still in a state of considerable uncertainty. The evidence showing that substantial noncognate tRNA binding can occur means that identification of the specifying regions (the parts of the tRNA molecule involved in the highly specific interaction with the enzyme) is particularly difficult. Nevertheless alternative hypotheses have implicated almost every region of the tRNA in the recognition process (4, 127). What is becoming apparent is that the different enzymes probably express their recognition by subtle differences in the structural regions of tRNA with which they interact.

The recognition problem has frequently been discussed (3, 4, 128, 129), and several theoretical approaches toward its understanding have been made (130, 131). The methods potentially available for the solution of the recognition problem have been described (3). Evidence for the possible involvement of certain regions of the tRNA in recognition has also been reviewed (4, 127). Rather than merely listing recent experimental results, we shall restrict ourselves to a discussion of some examples that demonstrate the use of some of the techniques.

The recognition of tRNA$_f^{Met}$ Chemical modification of tRNA is a frequently and sometimes indiscriminately applied technique for probing the functional regions of tRNA. In the case of tRNA$_f^{Met}$ it has been refined to such a degree that interpretation of the observations leads to implication of not merely certain regions of this tRNA in recognition but even certain atoms within the nucleosides. Schulman and Pelka (132) have used results from over 30 published modification reactions on tRNA$_f^{Met}$ which have been carried out under nondenaturing conditions (thus eliminating the problem of renaturation) and which have brought about alterations at 34 of 77 nucleotides. The interpretation correlates these results with the known properties of the E^{Met}:tRNA$_f^{Met}$ system and with the probable three-dimensional structure of the tRNA. The structural modifications considered can be classified according to their effects on the function of the tRNA (figure 10.5).

Individual modification of most bases in the dihydrouridine loop and stem (as long as this does not cause the introduction of bulky groups) leads to little or no loss of methionine acceptor activity. Enzymatic excision of oligonucleotide sequences from this region can lead to inactivation, but this has been attributed more to the overall effect on the conformation of the whole molecule than to a specific attack on a recognition site.

The anticodon loop (figure 10.5) on the other hand yields much more positive information. Conversion of the anticodon C$_{35}$ to U$_{35}$ using bisulphite does not cause a conformational change but results in complete inactivation of the tRNA both in aminoacylation and binding. By selective chemical modifications it has been possible to trace the cause of this inactivation to the alteration at the N3 position of cytidine C$_{35}$. It is proposed that interaction with E^{Met} occurs at this position. Modification of the adjacent nu-

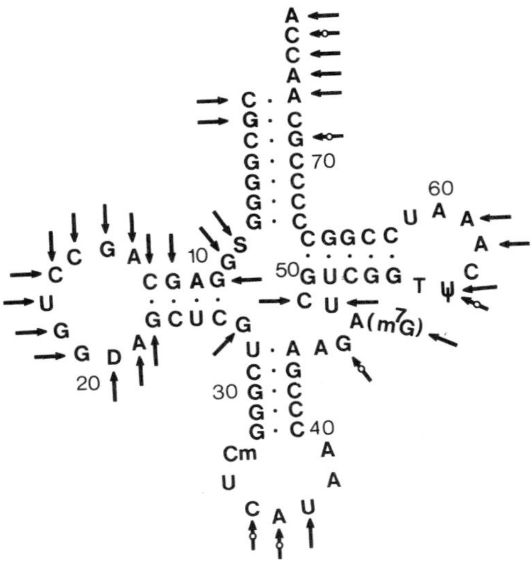

Figure 10.5 Modified nucleotides in tRNA$_f^{Met}$. Solid arrows indicate residues that have been modified without loss of methionine acceptor activity. Arrows with open circles indicate sites where modification has resulted in loss of methionine acceptance. Data from Schulman and Pelka (132).

cleotide A$_{36}$ also leads to inactivation, but whether this is due to a direct effect as in C$_{35}$ or whether it merely prevents the enzyme:C$_{35}$ interaction is not clear. The third base of the anticodon, U$_{37}$, can be altered with no loss of activity, although with some reagents a change in the kinetic behavior was observed.

Four of the five nucleotides in the variable loop have been modified. Of these, three can be changed with no apparent effect on the function of the tRNA. Photooxidation of G$_{46}$, however, leads to an inactivation which still has no unique explanation. Possibly it may interact directly with E^{Met}, or the loss of the purine ring may cause changes in the local conformation of the tRNA to which the enzyme is known to be sensitive.

Several bases in the acceptor stem region have been modified; the most clear-cut effect is observed on photooxidation of G$_{71}$, which results in complete inactivation and loss of the ability to bind to the enzyme. This inactivation is probably not due to the photochemical destruction of the base pairing, since disruption of the adjacent G$_2$-C$_{72}$ link by photooxidation of C$_{72}$ does not affect the viability of the tRNA. The results point to a direct G$_{71}$-enzyme link, although an interference with binding of the enzyme to a neighboring base is a possibility that cannot be ruled out.

Modification of the -C-C-A end does not have as dramatic an effect as would have been anticipated from its known importance in aminoacylation (2). Treatment of the terminal A_{77} with chloracetaldehyde, which blocks the N1 and 6-NH_2 groups, results only in a lower V_{max} for aminoacylation. C_{76} on the other hand is very sensitive to the same reagent; and although the ability to bind to the enzyme is not lost, aminoacylation is prevented. In this case the affected residue is not essential for recognition (binding) but is important in the catalytic mechanism. The tRNA is insensitive to modification at C_{75}.

To summarize, on the basis of chemical modifications, three regions of $tRNA_f^{Met}$ have been postulated as recognition regions (figure 10.5). These are the anticodon, the variable loop, and the acceptor stem. Of these the site of enzyme:tRNA interaction in the anticodon has been pinpointed as the N3 position of C_{35}, whereas the other sites have been localized to one or two nucleotides. In addition a site at the -C-C-A terminus, though not essential for binding, is important for catalysis. The identification of the recognition sites in ref. 132 takes into account most of the results obtained on the same system by various chemical methods as well as the crystallographic information from $tRNA^{Phe}$ which is predicted to form the general basis of the three-dimensional structure of all other tRNAs (127). The conclusions agree with those of Dube (133); using nuclease digestion to determine the regions of the tRNA that are protected by the synthetase, Dube found in his three-point attachment model that E^{Met} protects the anticodon, part of the 3'-stem, and the extra loop from ribonuclease attack. Bruton and Clark (134) on the other hand found that the binding of a complementary nucleotide triplet A-U-G to the anticodon had no effect on the rate of aminoacylation, suggesting the noninvolvement of the anticodon. The fact that modification of any one of the three sites on $tRNA_f^{Met}$ abolishes the amino acid acceptor ability means that all these sites must be simultaneously available for recognition. Other examples of tRNAs in which modification of a single site causes total loss of activity exist (see references in ref. 132), and in the case of $tRNA_f^{Met}$ it has been suggested that this total inactivation by a single base change is unlikely to be due to a binding difference. It is more probable that the correct binding of the three regions plays an active part in the bringing about a conformational change to an active E·tRNA complex. This would be in agreement with the kinetic and physical data on enzyme:tRNA interactions which have demonstrated the existence of such a triggering process.

The recognition region of tRNA ^{Phe} (yeast)   The ability of one enzyme to aminoacylate a set of noncognate tRNAs must mean that these tRNAs contain certain sequence homologies, parts of which contain regions recognized by the synthetase. Sequence analysis of such a set of tRNAs should therefore reveal the identity of the recognition region. There are a number of theoretical and practical problems associated with this approach. As stressed by Yarus (83), the relevance of comparing a collection of primary structures which have been designated cloverleaf forms by theoretical methods to the real three-dimensional system is questionable. Further, a recognition site composed of, for example, an electrostatic charge distribution would not be identified by this method. On the other hand such comparisons have been done extensively in the case of tRNA^{Phe} and have been relatively consistent with the data available. This may be not so surprising if, as has been suggested (127), all tRNAs are found to have similar tertiary structures and if the primary sites of recognition are individual bases, as is the case for tRNA$_f^{Met}$. In practice not all noncognate aminoacylations are usable for such analysis. It is well known that E^{Phe} (yeast) will, in the presence of organic solvents and high Mg^{++} and low salt concentrations aminoacylate practically all tRNAs from yeast. Obviously all yeast tRNAs cannot contain the recognition region for E^{Phe} (yeast). For the purposes of sequence comparison, therefore, the tRNAs in question should be aminoacylatable under as nearly normal assay conditions as possible without recourse to extremes where little is known about the changes that occur in the tRNA structure and even less about the effects on the enzyme. This aspect of the method has frequently been criticized in view of the relatively arbitrary distinction between normal and special aminoacylation conditions (4). By definition homologous systems are inappropriate since mischarging here occurs only at very slow rates and to low extents, but heterologous enzyme:tRNA pairs are suitable. The heterologous aminoacylation by E^{Phe} (yeast) of several tRNAs has been investigated most thoroughly in this way.

Phenylalanyl-tRNA synthetase from yeast at pH 5.8 and at high concentrations will phenylalanylate tRNA^{Val} (*E. coli*), tRNA^{Phe} (wheat), tRNA^{Phe} (*E. coli*), tRNA$_{2a}^{Val}$ (*E. coli*), tRNA$_{2b}^{Val}$ (*E. coli*), tRNA^{Ile} (*E. coli*), tRNA^{Met} (*E. coli*), tRNA$_{1a}^{Ala}$ (*E. coli*), tRNA$_2^{Ala}$ (*E. coli*), and tRNA^{Lys} (*E. coli*) (135–139) to at least 80 percent (86). Competitive inhibition of the homologous acylation by tRNA$_2^{Ala}$ showed that the same site was involved in all the aminoacylations. Comparison of the primary structures of the first eight of these tRNAs with that of tRNA^{Phe} (yeast) showed relatively few areas common to all eight (86). On the basis of the homologies it was postulated that a region of the dihydrouridine stem and an adenosine at the fourth position from the 3'-end were required for recognition by E^{Phe} (yeast) (figure 10.6). This is in agreement with the observation that tRNAs lacking one or the other of these features are, under the same conditions, not aminoacylated by this enzyme. For instance $tRNA_3^{Gly}$ (*E. coli*), which has the correct dihydrouridine stem sequence but has U at the fourth position from the 3'-end, is neither aminoacylated nor bound by E^{Phe} (yeast). Conversely many tRNAs that are not aminoacylated by E^{Phe} have the required A at the fourth position from the 3'-end but have the inappropriate dihydrouridine sequence. Further analysis of the kinetics of the heterologous misacylations allowed Roe et al. (86) to classify these tRNAs into three classes according to their V_{max} in aminoacylation by E^{Phe} at pH 8.2 (the K_m for all these tRNAs was between $0.8 \times 10^{-7}M$ and $6 \times 10^{-7}M$). This classification could be correlated with other structural features, namely the size of the dihydrouridine loop and the presence or absence of N2 methylation at G_{10} (table 10.4, as of 1973). From the table it is evident that Roe et al. considered the loop size to have a major influence on the kinetics of aminoacylation and that this influ-

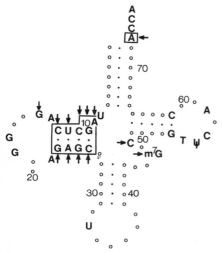

Figure 10.6 Composite tRNA for yeast E^{Phe}. Nucleotides that are not the same in the tRNAs whose structure was known in table 10.5 are shown in the composite by a dot. Nucleotides found in the same position in all tRNAs are denoted by a letter. Nucleotides that are common to the tRNAs in table 10.5 are marked by arrows. The extra dot near the dihydrouridine loop denotes the fact that some of these tRNAs have one extra nucleotide in this region. The boxes enclose the proposed recognition regions. From Roe et al. (86).

Table 10.4 Correlation between tRNA structure and kinetics of aminoacylation with phenylalanine by phenylalanyl-tRNA synthetase

tRNA		Rate of Aminoacylation	Size of hU Loop	Presence of m^2G_{10}
Phe	(yeast)	fast	8	yes
Phe	(wheat)	fast	8	yes
Phe	(*E. coli*)	intermediate	8	no
Val_1	(*E. coli*)	intermediate	8	no
Ala_1	(*E. coli*)	intermediate	8	no
Lys	(*E. coli*)	intermediate	not known	not known
Ala_2	(*E. coli*)	slow	not known	
$Val_{2A, 2B}$	(*E. coli*)	slow	9	
Ile	(*E. coli*)	slow	9	
Met	(*E. coli*)	slow	9	

Source: Roe et al. (86).

ence was further amplified by methylation of the G_{10} residue. The evidence for the involvement of this modified base in catalysis is convincing (140). Specific enzymatic methylation at this location of tRNA^{Phe} (*E. coli*) (intermediate class) with N2-methylguanine methylase increased the V_{max} of the heterologous reaction with E^{Phe} (yeast) by a factor of ten with a minor decrease in K_m, and the same effect was seen with tRNA$_1^{Ala}$. The methylated tRNA^{Phe} (*E. coli*) was still a substrate for its homologous E^{Phe} (*E. coli*) with an unaltered K_m but a tenfold decrease in V_{max}. It would appear therefore that methylation of G_{10} is a factor that enhances the catalysis and therefore contributes to the fidelity of the reaction. These interpretations allow one to predict certain structural features for the tRNAs in table 10.4 whose primary structures were not known at that time. Transfer RNA^{Lys} (*E. coli*), for instance, should have the sequence

-C-U-C-G-A
‥‥‥‥‥
-G-A-G-C

as part of the dihydrouridine stem, an A at the fourth position from the 3′-end, and, from the kinetics, an eight-membered dihydrouridine loop and an unmethylated G_{10}. We now know that this is exactly the case (141). Whether tRNA$_2^{Ala}$ behaves in the predicted way remains to be seen.

Are other regions of the tRNA also involved in recognition, and how do other lines of evidence agree with this model? Confirmation of the nonessential role of the anticodon in recognition in this case has been obtained. The anticodon has been excised enzymatically (142) with only partial loss of activity. However, binding of the native tRNA^{Phe} to the E^{Phe} protects both dihydrouridine and anticodon loops from nuclease digestion (64) and affects the fluorescence of the Y-base near the anticodon (77). Further, excision of the Y-base decreases the binding by greater than tenfold (143) without affecting the amino acid acceptor ability. The anticodon in this case is apparently not a recognition region (compare E^{Met}) but contributes to the binding efficiency.

Removal of the 3'-terminal seven nucleotides, including the essential A at the fourth position, does not prevent binding to the synthetase as measured by protection against nuclease digestion (64) or by ultracentrifugal complex detection (144). The A at the fourth position may therefore have the same function as has been attributed to the C$_{76}$ of tRNA$_f^{Met}$, namely taking an active part in the catalytic process. Alternatively, since tRNA^{Gly} (E. coli) does not bind to E^{Phe}, it is possible that the fourth position has a discriminator function permitting only A to bind (compare ref. 117). However, Yaneva and Beltchev (145) prepared a tRNA^{Lys} (yeast) lacking the 3'-terminal quartet of nucleotides including the G at the fourth position. Although the dihydrouridine region of this modified tRNA corresponds to the sequence required by the Dudock hypothesis for recognition by E^{Phe} and despite the binding experiments on similarly shortened tRNA^{Phe} (64, 144), no binding of this tRNA^{Lys} was detected. Binding in this case was measured in terms of the inhibition by the modified tRNA of the aminoacylation of tRNA^{Phe}, but it is not clear whether lack of inhibition in this system corresponds to lack of binding.

Three other positions on the composite tRNA structure are common to all phenylalanylated tRNAs (figure 10.6). These are the nucleotides G$_{15}$, m^{7}G$_{46}$, and C$_{48}$. Of these the sites at 15 and 48 are common to all tRNA sequences as purines and pyrimidines, respectively; in the case of tRNA^{Phe} (yeast) they are involved in tertiary hydrogen bonding with each other (127). A tertiary interaction is also observed between m^{7}G$_{46}$ and G$_{22}$. These nucleotides which are present in the composite picture may therefore be important in structure rather than in recognition. This is emphasized by enzymatic modification of tRNA^{Phe} (yeast) where the nucleotides Y$_{37}$ to m^{7}G$_{46}$ were removed (146). The remaining molecule that would presumably have an altered conformation through the loss of two tertiary interactions A$_{44}$·m$_2^2$G$_{26}$ and m^{7}G$_{46}$·G$_{22}$ was still recognized by the enzyme in that 15 percent aminoacylation was observed.

Other chemical studies on this system are in general agreement with this hypothesis (86 and references therein). In particular photo cross-linking of E^{Phe} with $tRNA^{Phe}$ (106), which tends to give information regarding regions of the tRNA in the vicinity of the contact areas, shows no evidence of close contact between the enzyme and the anticodon region. However, the recent sequence determination of $tRNA^{Met}$ (yeast) (147), which has all the required recognition bases, means that some modification of the hypothesis may be required to account for the discrimination by E^{Phe} (yeast) between $tRNA^{Phe}$ (yeast) and $tRNA^{Met}$ (yeast).

Although it is possible to argue that the method of sequence comparison provides information concerning the binding regions of tRNA rather than the specific recognition sites (4), it is now well known that E^{Phe} will bind strongly noncognate tRNAs without being able to aminoacylate them (76). These noncognate tRNAs do not contain the proposed recognition regions identical to $tRNA^{Phe}$, and it therefore seems that these regions must play a part in the enzyme:tRNA interaction in addition to binding.

To summarize, on the basis of the evidence available, a relatively self-consistent picture has emerged regarding the recognition of $tRNA^{Phe}$ (yeast) by E^{Phe} (yeast). So far two regions of the tRNA have been found to be important (figure 10.6). As was the case for $tRNA^{Met}$, both regions must be recognized for aminoacylation to take place. Binding occurs in the absence of A at the fourth position from the 3'-end so that this position may either act as a catalytic promoter (144) or as a discriminating site for A-containing tRNAs (86). The sequence

-C-U-C-G-A
 . . . .
-G-A-G-C

of the dihydrouridine stem is important for recognition, although it is still not clear whether only certain parts of this sequence are recognized (86) or whether the local conformation about this region is the critical factor (148).

The recognition regions of some other tRNAs Sequence comparisons of tRNAs misacylated by valyl-tRNA synthetase from yeast (114), from *E. coli* (84), and from *B. stearothermophilus* (84) have implicated regions of the dihydrouridine stem, the terminal part of the amino acid acceptor stem, and the extra loop for all three enzymes. In addition for the *B. stearothermophilus* enzyme the recognition of the anticodon was also suggested. Such an interaction, although not predicted by primary structure comparison, has also been proposed for the yeast enzyme by a number of chemical modification reactions. These have included loss of activity on removal of A-C from the anticodon I-A-C, and a similar effect was seen on modification of the cytidine residues with chemical reagents (4). Loss of amino acid acceptance after conversion of cytidine to uridine with bisulphite (149) was also traced to modification of the anticodon cytidine together with C_{72} or C_{75}. If the role of the acceptor region proves essential, it may again be catalytic since removal of the last seven nucleotides from the 3'-end of yeast tRNA^{Val} results in only a threefold decrease in the binding strength (144).

Transfer RNA^{Glu} (*E. coli*) is another tRNA for which the amino acid acceptor activity is sensitive to modification at the anticodon region. Benzoylation of 2-thio-5(methylaminomethyl)uridine$_{35}$ leads to 75 percent loss of activity (150). Conversion of this s^{2}mam^{5}U$_{35}$ to uridine by treatment with CNBr followed by loss of cyanate causes a reduction in activity (151), although a mutant tRNA^{Glu} lacking sulphur is still active (152). Bisulphite conversion of cytidine to uridine in tRNA^{Glu} results in inactivation which is correlated with modification of C_{33} and the anticodon C_{37} and is also related to a 3'-terminal cytidine alteration (151).

Transfer RNA^{Gly} (*E. coli*) can also be inactivated by anticodon manipulation (153), while tRNA^{Ala} from yeast is another tRNA that is relatively insensitive to modification in that are (154).

Recently the nucleotide sequences of six tRNA^{Gly} that are aminoacylated by E^{Gly} (*E. coli*) have been compared. The only common features were in the first two base pairs of the acceptor stem and two cytidines in the anticodon. It was suggested that these may be recognition sites (155). This would be in agreement with Sabban et al. (156), who found a reduced aminoacylation activity after bisulphite induced cytidine-to-uridine changes at the anticodon of tRNA^{Gly} (*E. coli*).

In the case of E^{Ile} (*E. coli*) the photo cross-linking method of Schimmel (107) indicates an interaction between the synthetase and tRNA^{Ile} at the acceptor and dihydrouridine stem regions. The results from this somewhat drastic method (enzyme activity is rapidly lost during the course of the experiment) were confirmed by the enzyme-induced perturbation in the pattern of tritium exchange from solution into the tRNA purine bases (157). The areas of the anticodon as well as the acceptor TΨC-stem have also been implicated in recognition by this method. The reason that photo cross-linking studies did not suggest the anticodon interaction is not clear. However, such a recognition site was also postulated by Yarus (83) on the basis of sequence comparisons and through the oligonucleotide binding studies of Schimmel et al. (104).

The introduction of *E. coli* mutants for the investigation of recognition regions (158, 159) has facilitated the observation of the effect of specific alterations of the tRNA structure on the in vivo specificity of aminoacylation. Isolation of mutant su_3 suppressor $tRNA^{Tyr}$ having mutations G_1, A_2, U_{81}, or G_{82} were shown to be aminoacylatable with glutamine by E^{Gln} and, except for G_{82}, could also accept tyrosine (160, 161). On the basis of a sequence comparison between $tRNA_{AI}^{Tyr}$ and $tRNA^{Gln}$ it was suggested that recognition of the tRNA by the two synthetases was probably based on a regional (acceptor stem) basis rather than on a single nucleotide, although the fourth nucleotide from the 3'-end was thought to be important. The fact that a second alteration had occurred in the same tRNA, namely G_{35} to C_{35} in the anticodon, to convert it to the amber suppressor meant that this area could not be ruled out as another possible recognition region. Indeed Berg (162) and Yaniv et al. (163) showed that the su_7^+ $tRNA^{Trp}$ with a single base change in the anticodon compared with the wild-type $tRNA^{Trp}$ was also aminoacylated with glutamine. This mutation was said to cause loss of tryptophan acceptance, but Celis et al. (164) have found that small amounts of tryptophan are inserted into peptides by this tRNA in addition to glutamine, suggesting a closely related recognition in $tRNA^{Gln}$ and $tRNA^{Trp}$.

10.4 Recognition Regions and Correlation with the Tertiary Structure of tRNA: General Conclusions

On the basis of the examples discussed in preceding sections, certain conclusions can be reached regarding the putative recognition sites of tRNAs. Certain areas of the cloverleaf structure have so far not been implicated in recognition. These include the dihydrouridine loop, which may have a catalytic function in some instances, and the T Ψ C-loop as well as perhaps nucleotides common to all tRNAs. Similarly there is no evidence that modified or rare bases contribute to recognition although they may affect the catalytic process. The remaining four features of tRNA, namely the acceptor stem (in particular the fourth nucleotide from the 3'-end), the dihydrouridine stem, the anticodon, and the variable loop have all been suggested as possible recognition sites. It is also apparent that it is not sufficient for the enzyme to recognize one region; instead several, often widely separated, recognition and catalytic sites must interact with the synthetase to bring about aminoacylation through the triggering of the enzyme:tRNA complex into a catalytically competent form.

It is not yet clear whether one can generalize the definition of the recognition region. In some cases it has been established that individual nucleotides are involved, whereas in others a local conformation of several nucleotides may make up the recognition region. The importance of a catalytic or discriminatory site, usually associated with a nucleotide near the 3'-end, further complicates the system but may provide the basis for the enhanced velocity of aminoacylation between cognate pairs.

How do these proposed areas of recognition correlate with the three-dimensional structure of tRNA? We have to date only the crystal structure of tRNA^{Phe} (yeast) on which to base comparisons. However, there is evidence that (a) the crystal structure reflects the structure in solution (165) and (b) the structure of other tRNAs will have the same general features (127). From this evidence it has been suggested that interaction of tRNA with a synthetase occurs along one side of the L-shaped molecule (figure 10.7) where parts of most of the regions mentioned may lie (127). The synthetases are thought to bear a groove that in some cases may extend from the amino acid acceptor region to the anticodon and in others stop short of the anticodon, in this way bringing about the observed differences between synthetase:tRNA pairs. A further flexibility in the interaction has been suggested as arising from the angle at which the tRNA molecule is inserted into the groove, making interactions with the otherwise inaccessible regions such as the variable loop more likely. Nonspecific tRNA binding would then be due to, for example, phosphate/enzyme interaction (as also indicated by the ionic nature of such complexes), whereas specificity would arise from the correct positioning of the recognition site in the enzyme groove and the associated catalytic events. This model could then also include the symmetry hypothesis of Kim (166) which proposes that the pseudo twofold symmetry of the tRNA would be recognized by the synthetases which often themselves have a symmetry element either in the subunit structure or in a duplicated primary sequence (figure 10.8). A topographical study of an enzyme:tRNA complex that has met with some success is the low-angle X-ray scattering work of Österberg et al. on the lysine system from yeast (94). The results of this investigation are, however, in apparent disagreement with the model of Rich and Schimmel (127) in that two dimeric enzyme molecules seem to be complexed to one tRNA in an overall A_4B structure (figure 10.9) and that the protein/nucleic acid interactions occur along the opposite side of the tRNA to that proposed by Rich and Schimmel (figure 10.7). It is not yet clear whether this behavior is unique to the lysine sys-

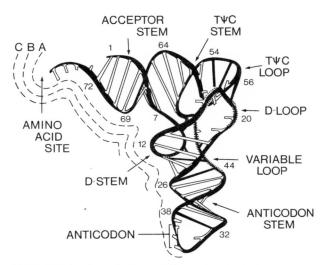

Figure 10.7 A schematic diagram illustrating the folding of the polynucleotide chain in yeast phenylalanine tRNA and its proposed interaction with the synthetase surface. The ribose-phosphate backbone is drawn as a continuous tube and cross bars indicate the bases. Nucleotide numbers are given. Secondary base-base hydrogen bonding interactions have unshaded cross bars, while tertiary ones are black. The regions with variable numbers of nucleotides in different tRNA molecules are shown in dotted outline. Three different synthetase surfaces are shown by the dashed lines (A–C). These represent different sizes of the recognition region. Interactions are not intended to be viewed as simply occurring on an edge of the flattened tRNA molecule; they also involve parts of the tRNA on either side of the indicated lines. The plane of the tRNA is not necessarily perpendicular to the surface of the synthetase. From Rich and Schimmel (127).

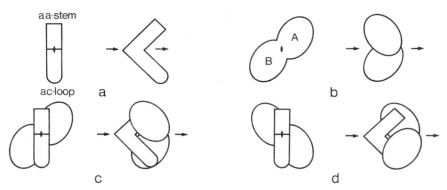

Figure 10.8 Two orthogonal views of (a) tRNA, (b) two symmetry-related unit domains (hypothetical) in an aminoacyl-tRNA synthetase. Parts (c) and (d) are two alternative hypothetical models for the tRNA·enzyme complex formed by coinciding the pseudo twofold axes of both molecules. The pseudo twofold axes are indicated by arrows when on the plane of the page and by ◖ when perpendicular to it. From Kim (162, 182).

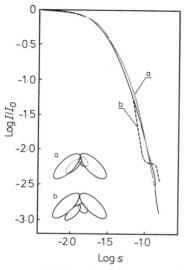

Figure 10.9 Experimentally observed scattering curve from the complex of lysine:tRNA ligase + tRNA[Lys] from yeast (solid curve) compared with theoretical scattering curves calculated for models involving two major ellipsoids representing the ligase molecules and two minor ellipsoids (tRNA). The a-axes of the ligase ellipsoids form a right angle, and the tRNA ellipsoids situated within this angle interact with the enzyme ellipsoids. The a-axes of all four ellipsoids are in the same plane with the b-axes parallel. (a) About 50 percent of the tRNA ellipsoid volume is within the enzyme ellipsoid volume; (b) the tRNA ellipsoids barely touch the enzyme ellipsoids. Both models give a radius of gyration of 57.5 Å; $s = (4\pi/\lambda) \sin \vartheta$, where λ is the wavelength used, $\lambda = 1.54$ Å, and ϑ the half-scattering angle. From Österberg et al. (94).

tem or whether it is a general phenomenon, but recent neutron small-angle scattering investigations have suggested a similar behavior for E^{Val} (yeast) at low tRNA concentrations (167).

Acknowledgments

We are grateful to Drs. H. Faulhammer, W. Freist, D. Gauss, and F. von der Haar for critical reading of this article and to Dr. D. Gauss for painstaking editorial work on the final manuscript.

References

1. Ofengand, J. (1977). In Molecular Mechanisms of Protein Biosynthesis, H. Weissbach and P. Pestka eds. (New York: Academic Press), pp. 7–79.

2. Sprinzl, M., and Cramer, F. (1978). Prog. Nucl. Acid Res. Mol. Biol. (in press).

3. Söll, D., and Schimmel, P. R. (1974). In The Enzymes, vol. 10, P. D. Boyer, ed. (New York: Academic Press), pp. 489–538.

4. Kisselev, L. L., and Favorova, O. O. (1974). Adv. Enzymol. *40*, 141–238.

5. Loftfield, R. B. (1972). Prog. Nucl. Acid Res. Mol. Biol. *12*, 87–128.

6. Loftfield, R. B., and Vanderjagt, D. (1972). Biochem. J. *128*, 1353–1356.

7. Bosshard, H. R. (1976). Experientia *32*, 949–1090.

8. Fersht, A. R. (1975). Biochemistry *14*, 5–12.

9. Jakes, R., and Fersht, A. R. (1975). Biochemistry *14*, 3344–3350.

10. Koch, G. L. E., Boulanger, Y., and Hartley, B. S. (1974). Nature *249*, 316–320.

11. Kula, M.-R. (1973). FEBS Letters *35*, 299–302.

12. Hustedt, H., Flossdorf, J., and Kula, M.-R. (1977). Eur. J. Biochem. *74*, 199–202.

13. Hyafil, F., Jacques, Y., Fayat, G., Fromant, M., Dessen, P., and Blanquet, S. (1976). Biochemistry *15*, 3678–3685.

14. Di Natale, P., Schechter, A. N., Leporo, G. C., and De Lorenzo, F. (1976). Eur. J. Biochem. *62*, 293–298.

15. Holler, E., and Calvin, M. (1972). Biochemistry *11*, 3741–3752.

16. Holler, E., Hammer-Raber, B., Hanke, T., and Bartmann, P. (1975). Biochemistry *14*, 2496–2503.

17. Fersht, A. R., Mulvey, R. S., and Koch, G. L. E. (1975). Biochemistry *14*, 13–18.

18. Igloi, G. L., von der Haar, F., and Cramer, F. (1977). Biochemistry *16*, 1696–1702.

19. Calendar, R., and Berg, P. (1967). J. Mol. Biol. *26*, 39–54.

20. Leporo, G. C., Di Natale, P., Guarini, L., and De Lorenzo, F. (1975). Eur. J. Biochem. *56*, 369–374.

21. Santi, D. V., and Danenberg, P. V. (1971). Biochemistry *10*, 4813–4820.

22. Flossdorf, J., Pratorius, H. J., and Kula, M.-R. (1976). Eur. J. Biochem. *66*, 147–155.

23. Owens, S. L., and Bell, F. E. (1970). J. Biol. Chem. *245*, 5515–5523.

24. Holler, E., Rainey, P., Orme, A., Bennett, E. L., and Calvin, M. (1973). Biochemistry *12*, 1150–1159.

25. Baldwin, A. N., and Berg, P. (1966). J. Biol. Chem. *241*, 839–845.

25a. Igloi, G. L., von der Haar, F. and Cramer, F. (1978). Biochemistry (submitted).

26. Eldred, E. W., and Schimmel, P. R. (1972). J. Biol. Chem. *247*, 2961–2964.

27. Yarus, M. (1972). Proc. Nat. Acad. Sci. USA *69*, 1915–1919.

28. Hopfield, J. J. (1974). Proc. Nat. Acad. Sci. USA *71*, 4135–4139.

29. Fersht, A. R., and Kaethner, M. M. (1976). Biochemistry *15*, 3342–3346.

30. Fersht, A. R. (1977). Biochemistry *16*, 1025–1030.

31. von der Haar, F., and Cramer, F. (1976). Biochemistry *15*, 4131–4138.

32. Igloi, G. L. (1973). D. Phil. Thesis, University of York, England.

33. Hopfield, J. J., Yamane, T., Yue, V., and Coutts, S. M. (1976). Proc. Nat. Acad. Sci. USA *73*, 1164–1168.

34. von der Haar, F., and Cramer, F. (1975). FEBS Letters *56*, 215–217.

35. von der Haar, F. (1977). FEBS Letters *79*, 225–228.

36. Fraser, T. H., and Rich, A. (1975). Proc. Nat. Acad. Sci. USA *72*, 3044–3048.

37. Kalousek, F., and Konigsberg, W. (1975). In MTP International Reviews of Science, Biochemistry Series One, H. R. V. Arnstein, ed. (London: Butterworths), vol. 7, pp. 57–88.

38. Lawrence, F., Blanquet, S., Poiret, M., Robert-Gero, M., and Waller, J.-P. (1973). Eur. J. Biochem. *36*, 234–243.

39. Maelicke, A., Engel, G., Cramer, F., and Staehelin, M. (1974). Eur. J. Biochem. *42*, 311–314.

40. Santi, D. V., Danenberg, P. V., and Montgomery, K. A. (1971). Biochemistry *10*, 4821–4824.

41. Lawrence, F., Shire, D. J., and Waller, J.-P. (1974). Eur. J. Biochem. *41*, 73–81.

42. Marutzky, R., Flossdorf, J., and Kula, M.-R. (1976). Nucl. Acids Res. *3*, 2067–2077.

43. Freist, W., von der Haar, F., Sprinzl, M., and Cramer, F. (1976). Eur. J. Biochem. *64*, 389–393.

44. Freist, W., von der Haar, F., Faulhammer, H., and Cramer, F. (1976). Eur. J. Biochem. *66*, 493–497.

45. Freist, W., Sternbach, H., von der Haar, F., and Cramer, F. (1977). Eur. J. Biochem. (in press).

46. von der Haar, F., Cramer, F., Eckstein, F., and Stahl, K.-W. (1977). Eur. J. Biochem. *76*, 263–267.

47. Pimmer, J., Holler, E., and Eckstein, F. (1976). Eur. J. Biochem. *67*, 171–176.

48. Kosakowski, H. M., and Holler, E. (1973). Eur. J. Biochem. *38*, 274–282.

49. Fayat, G., and Waller, J. P. (1974). Eur. J. Biochem. *44*, 335–342.

50. Blanquet, S., Fayat, G., and Waller, J.-P. (1975). J. Mol. Biol. *94*, 1–15.

51. Fayat, G., Fromant, M., and Blanquet, S. (1977). Biochemistry *16*, 2570–2579.

52. Cleland, W. W. (1963). Biochim. Biophys. Acta *67*, 104–137; 173–187; 188–196.

53. Knorre, D. G., Malygin, E. G., Slinko, M. G., Timoshenko, V. I., Zinoviev, V. V., Kisselev, L. L., Kochkina, L. K., and Favorova, O. O. (1974). Biochimie *56*, 845–855.

54. Yarus, M., and Berg, P. (1967). J. Mol. Biol. *28*, 479–490.

55. Fasiolo, F., Ebel, J.-P., and Lazdunski, M. (1977). Eur. J. Biochem. *73*, 7–15.

56. Krauss, G., Pingoud, A., Boehme, D., Riesner, D., Peters, F., and Maass, G. (1975). Eur. J. Biochem. *55*, 517–529.

57. Hélène, C., Brun, F., and Yaniv, M. (1969). Biochem. Biophys. Res. Commun. *37*, 393–398.

58. Pachmann, U., Cronvall, E., Rigler, R., Hirsch, R., Wintermeyer, W., and Zachau, H. G. (1973). Eur. J. Biochem. *39*, 265–273.

59. Pingoud, A., Riesner, D., Boehme, D., and Maass, G. (1973). FEBS Letters *30*, 1–5.

60. Rigler, R., Cronvall, E., Ehrenberg, M., Pachmann, U., Hirsch, R., and Zachau, H. G. (1971). FEBS Letters *18*, 193–198.

61. Bartmann, P., Hanke, T., Hammer-Raber, B., and Holler, E. (1974). Biochemistry *13*, 4171–4175.

62. Lagerkvist, U., and Rymo, L. (1969). J. Biol. Chem. *244*, 2476–2483.

63. Rigler, R., Cronvall, E., Hirsch, R., Pachmann, U., and Zachau, H. G. (1970). FEBS Letters *11*, 320–323.

64. Hörz, W., and Zachau, H. G. (1973). Eur. J. Biochem. *32*, 1–14.

65. Engel, G., Heider, H., Maelicke, A., von der Haar, F., and Cramer, F. (1972). Eur. J. Biochem. *29*, 257–262.

66. Rigler, R., Pachmann, U., Hirsch, R., and Zachau, H. G. (1976). Eur. J. Biochem. *65*, 307–315.

67. Maelicke, A., and Cramer, F. (1975). Eur. J. Biochem. *52*, 171–178.

68. Riesner, D., Pingoud, A., Boehme, D., Peters, F., and Maass, G. (1976). Eur. J. Biochem. *68*, 71–80.

69. Krajewska-Grynkiewicz, K., Buoncore, V., and Schlesinger, S. (1973). Biochim. Biophys. Acta *312*, 518–527.

70. Chousterman, S., and Chapeville, F. (1973). Eur. J. Biochem. *35*, 51–56.

71. Bosshard, H. R., Koch, G. L. E., and Hartley, B. S. (1975). Eur. J. Biochem. *53*, 493–498.

72. Irwin, M. J., Nyborg, J., Reid, B. R., and Blow, D. M. (1976). J. Mol. Biol. *105*, 577–586.

73. Pingoud, A., Boehme, D., Riesner, D., Kownatzki, R., and Maass, G. (1975). Eur. J. Biochem. *56*, 617–622.

74. Pingoud, A., Kownatzki, R., and Maass, G. (1977). Nucl. Acids Res. *4*, 327–338.

75. Blanquet, S., Dessen, P., and Iwatsubo, M. (1976). J. Mol. Biol. *103*, 765–784.

76. Krauss, G., Riesner, D., and Maass, G. (1976). Eur. J. Biochem. *68*, 81–93.

77. Krauss, G., Römer, R., Riesner, D., and Maass, G. (1973). FEBS Letters *30*, 6–10.

78. Fasiolo, F., Remy, P., Pouyet, J., and Ebel, J.-P. (1974). Eur. J. Biochem. *50*, 227–236.

79. Bartmann, P., Hanke, T., and Holler, E. (1975). Biochemistry *14*, 4777–4786.

80. Lapointe, J., and Söll, D. (1972). J. Biol. Chem. *247*, 4975–4981.

81. Blanquet, S., Petrissant, G., and Waller, J.-P. (1973). Eur. J. Biochem. *36*, 227–233.

82. Lam, S. S. M., and Schimmel, P. R. (1975). Biochemistry *14*, 2775–2780.

83. Yarus, M. (1972). Biochemistry *11*, 2352–2361.

84. Giegé, R., Kern, D., Ebel, J.-P., Grosjean, H., de Henau, S., and Chantrenne, H. (1974). Eur. J. Biochem. *45*, 351–362.

85. Ebel, J.-P., Giegé, R., Bonnet, J., Kern, D., Befort, N., Bollack, C., Fasiolo, F., Gangloff, J., and Dirheimer, G. (1973). Biochimie *55*, 547–557.

86. Roe, B., Sirover, M., and Dudock, B. (1973). Biochemistry *12*, 4146–4154.

87. von der Haar, F. (1976). Hoppe-Seyler's Z. Physiol. Chem. *357*, 819–823.

88. Hélène, C., Brun, F., and Yaniv, M. (1971). J. Mol. Biol. *58*, 349–365.

89. Bonnet, J., Renaud, M., Raffin, J. P., and Remy, P. (1975). FEBS Letters *53*, 154–158.

90. von der Haar, F. (1976). Eur. J. Biochem. *64*, 395–398.

91. Bonnet, J., and Ebel, J.-P. (1975). Eur. J. Biochem. *58*, 193–201.

92. Yarus, M. (1972). Biochemistry *11*, 2050–2060.

93. Lynch, D. C., and Schimmel, P. R. (1974). Biochemistry *13*, 1852–1861.

94. Österberg, R., Sjöberg, B., Rymo, L., and Lagerkvist, U. (1975). J. Mol. Biol. *99*, 383–400.

95. Lynch, D. C., and Schimmel, P. R. (1974). Biochemistry *13*, 1841–1852.

96. Weiner, L. M., Backer, J. M., and Rezvukhin, A. I. (1975). Biochim. Biophys. Acta *383*, 316–324.

97. Backer, J. M., Vocel, S. V., Weiner, L. M., Oshevskii, S. I., and Lavrik, I. O. (1975). Biochem. Biophys. Res. Commun. *63*, 1019–1026.

98. Bruton, C. J., and Hartley, B. S. (1970). J. Mol. Biol. *52*, 165–178.

99. Santi, D. V., Marchant, W., and Yarus, M. (1973). Biochem. Biophys. Res. Commun. *51*, 370–375.

100. Bartmann, P., Hanke, T., Hammer-Raber, B., and Holler, E. (1974). Biochem. Biophys. Res. Commun. *60*, 743–747.

101. Lavrik, I. O., and Khutoryanskaya, L. Z. (1974). FEBS Letters *39*, 287–290.

102. Gorshkova, I. I., Lavrik, O. I., Nevinskii, G. A., and Khutoryanskaya, L. Z. (1975). Mol. Biol. (Moscow) *9*, 409–414.

103. Gorshkova, I. I., Knorre, D. G., Lavrik, O. I., and Nevinsky, G. A. (1976). Nucl. Acids Res. *3*, 1577–1589.

104. Schimmel, P. R., Uhlenbeck, O. C., Lewis, J. B., Dickson, L. A., Eldred, E. W., and Schreier, A. A. (1972). Biochemistry *11*, 642–646.

105. Schoemaker, H. J. P., and Schimmel, P. R. (1974). J. Mol. Biol. *84*, 503–513.

106. Schoemaker, H. J. P., Budzik, G. P., Giegé, R., and Schimmel, P. R. (1975). J. Biol. Chem. *250*, 4440–4444.

107. Budzik, G. P., Lam, S. S. M., Schoemaker, H. J. P., and Schimmel, P. R. (1975). J. Biol. Chem. *250*, 4430–4439.

108. Knorre, D. G. (1975). FEBS Letters *58*, 50–52.

109. von der Haar, F., and Gaertner, E. (1975). Proc. Nat. Acad. Sci. USA *72*, 1378–1382.

110. Krauss, G., Riesner, D., and Maass, G. (1977). Nucl. Acids Res. *4*, 2253–2262.

111. Jacobson, K. B. (1971). Prog. Nucl. Acid Res. Mol. Biol. *11*, 461–488.

112. Giegé, R., Kern, D., Ebel, J.-P., and Taglang, R. (1971). FEBS Letters *15*, 281–285.

113. Giegé, R., Kern, D., and Ebel, J.-P. (1972). Biochimie *54*, 1245–1255.

114. Kern, D., Giegé, R., and Ebel, J.-P. (1972). Eur. J. Biochem. *31*, 148–155.

115. Yarus, M., and Mertes, M. (1973). J. Biol. Chem. *248*, 6744–6749.

116. Yarus, M. (1972). Nature New Biol. *239*, 106–108.

117. Crothers, D. M., Seno, T., and Söll, D. G. (1972). Proc. Nat. Acad. Sci. USA *69*, 3063–3067.

118. Joseph, D. R., and Muench, K. H. (1971). J. Biol. Chem. *246*, 7602–7609.

119. Schreier, A. A., and Schimmel, P. R. (1972). Biochem. *11*, 1582–1589.

120. Bonnet, J., Giegé, R., and Ebel, J.-P. (1972). FEBS Letters *27*, 139–144.

121. Bonnet, J., and Ebel, J.-P. (1974). FEBS Letters *39*, 259–262.

122. Sourgoutchov, A., Blanquet, S., Fayat, G., and Waller, J.-P. (1974). Eur. J. Biochem. *46*, 431–438.

123. Yarus, M. (1973). J. Biol. Chem. *248*, 6755–6758.

124. Dietrich, A., Kern, D., Bonnet, J., Giegé, R., and Ebel, J.-P. (1976). Eur. J. Biochem. *70*, 147–158.

125. Furano, A. V. (1976). Eur. J. Biochem. *64*, 597–606.

126. Pingoud, A., Urbanke, C., Krauss, G., Peters, F., and Maass, G. (1977). Eur. J. Biochem. *78*, 403–409.

127. Rich, A., and Schimmel, P. R. (1977). Nucl. Acids Res. *4*, 1649–1665.

128. Cramer, F. (1971). Prog. Nucl. Acid Res. Mol. Biol. *11*, 391–421.

129. Chambers, R. W. (1971). Prog. Nucl. Acid Res. Mol. Biol. *11*, 489–525.

130. Ivanov, V. I. (1975). FEBS Letters *59*, 282–286.

131. Freist, W., and Cramer, F. (1976). J. Theor. Biol. *58*, 401–416.

132. Schulman, L. H., and Pelka, H. (1977). Biochemistry *16*, 4256–4265.

133. Dube, S. K. (1973). Nature New Biol. *243*, 103–105.

134. Bruton, C. J., and Clark, B. F. C. (1974). Nucl. Acids Res. *1*, 217–221.

135. Dudock, B. S., DiPeri, C., and Michael, M. S. (1970). J. Biol. Chem. *245*, 2465–2468.

136. Dudock, B., DiPeri, C., Scileppi, K., and Reszelbach, R. (1971). Proc. Nat. Acad. Sci. USA *68*, 681–684.

137. Taglang, R., Waller, J.-P., Befort, N., and Fasiolo, F. (1970). Eur. J. Biochem. *12*, 550–557.

138. Roe, B., Sirover, M., Williams, R., and Dudock, B. (1971). Arch. Biochem. Biophys. *147*, 176–177.

139. Roe, B., and Dudock, B. (1972). Biochem. Biophys. Res. Commun. *49*, 399–406.

140. Roe, B., Michael, M., and Dudock, B. (1973). Nature New Biol. *246*, 135–138.

141. Chakraburtty, K., Steinschneider, A., Case, R. V., and Mehler, A. H. (1975). Nucl. Acids Res. *2*, 2069–2075.

142. Hörz, W., Meyer, D., and Zachau, H. G. (1975). Eur. J. Biochem. *53*, 533–539.

143. Krauss, G., Peters, F., and Maass, G. (1976). Nucl. Acids Res. *3*, 631–639.

144. Bonnet, J., Befort, N., Bollack, C., Fasiolo, F., and Ebel, J.-P. (1975). Nucl. Acids Res. *2*, 211–221.

145. Yaneva, M., and Beltchev, B. (1976). Biochimie *58*, 889–890.

146. Wintermeyer, W., and Zachau, H. G. (1970). FEBS Letters *11*, 160–164.

147. Koiwai, O., and Miyazaki, M. (1976). J. Biochem. (Tokyo) *80*, 937–950.

148. Kumar, S. A., Krauskopf, M., and Ofengand, J. (1973). J. Biochem. (Tokyo) *74*, 341–353.

149. Chambers, R. W., Aoyagi, S., Furukawa, Y., Zawadzka, H., and Bhanot, O. S. (1973). J. Biol. Chem. *248*, 5549–5551.

150. Cedegren, R. J., Beauchemin, N., and Toupin, J. (1973). Biochemistry *12*, 4566–4570.

151. Singhal, R. P. (1974). Biochemistry *13*, 2924–2932.

152. Seno, T., Agris, P. F., and Söll, D. (1974). Biochim. Biophys. Acta *349*, 328–338.

153. Chakraburtty, K. (1975). Nucl. Acids Res. *2*, 1793–1804.

154. Bhanot, O. S., and Chambers, R. W. (1977). J. Biol. Chem. *252*, 2551–2559.

155. Marcu, K. B., Mignery, R. E., and Dudock, B. S. (1977). Biochemistry *16*, 797–806.

156. Sabban, E., Bhanot, O. S., and Chambers, R. W. (1977). Fed. Proc. *36*, 706.

157. Schoemaker, H. J. P., and Schimmel, P. R. (1976). J. Biol. Chem. *251*, 6823–6830.

158. Hooper, M. L., Russell, R. L., and Smith, J. D. (1972). FEBS Letters *22*, 149–155.

159. Shimura, Y., Aono, H., Ozeki, H., Sarabhai, A., Lamfrom, H., and Abelson, J. (1972). FEBS Letters *22*, 144–148.

160. Smith, J. D., and Celis, J. E. (1973). Nature New Biol. *243*, 66–71.

161. Celis, J. E., Hooper, M. L., and Smith, J. D. (1973). Nature New Biol. *244*, 261–264.

162. Berg, P. (1973). Harvey Lectures *67*, 247–272.

163. Yaniv, M., Folk, W. R., Berg, P., and Söll, L. (1974). J. Mol. Biol. *86*, 245–260.

164. Celis, J. E., Coulondre, C., and Miller, J. H. (1976). J. Mol. Biol. *104*, 729–734.

165. Robertus, J. D., Ladner, J. E., Finch, J. T., Rhodes, D., Brown, R. S., Clark, B. F. C., and Klug, A. (1974). Nucl. Acids Res. *1*, 927–932.

166. Kim, S.-H. (1976). Prog. Nucl. Acid Res. Mol. Biol. *17*, 182–216.

167. Giegé, R., Jacrot, B., Moras, D., Thierry, J.-C., and Zaccai, G. (1977). Nucl. Acids Res. *4*, 2421–2427.

168. Kondo, M., and Woese, C. R. (1969). Biochemistry *8*, 4177–4182.

169. Bergmann, F. H., Berg, P., and Dieckmann, M. (1961). J. Biol. Chem. *236*, 1735–1740.

170. Loftfield, R. B., and Eigner, E. A. (1966). Biochim. Biophys. Acta *130*, 426–448.

171. Anderson, J. W., and Fowden, L. (1970). Biochem. J. *119*, 691–697.

172. Lin, C.-S., Irwin, R., and Chirikjian, J. G. (1975). J. Biol. Chem. *250*, 9299–9303.

173. Rouget, P., and Chapeville, F. (1968). Eur. J. Biochem. *4*, 305–309.

174. Myers, G., Blank, H. U., and Söll, D. (1971). J. Biol. Chem. *246*, 4955–4964; correction *247*, 6011.

175. Papas, T. S., and Mehler, A. H. (1971). J. Biol. Chem. *246*, 5924–5928.

176. Allende, C. C., Chaimovich, H., Gatica, M., and Allende, J. E. (1970). J. Biol. Chem. *245*, 93–101.

177. Nazario, M., and Evans, J. A. (1974). J. Biol. Chem. *249*, 4934–4942.

178. Parfait, R., and Grosjean, H. (1972). Eur. J. Biochem. *30*, 242–249.

179. Santi, D. V., Danenberg, P. V., and Satterly, P. (1971). Biochemistry *10*, 4804–4812.

180. Santi, D. V., and Pena, V. A. (1971). FEBS Letters *13*, 157–160.

181. Papas, T. S., and Peterkofsky, A. (1972). Biochemistry *11*, 4602–4608.

182. Kim, S.-H. (1975). Nature *256*, 679–681.

INDEX